Lecture Notes i

T0280420

The Lecture Notes in Physics

The series Lecture Notes in Physics (LNP), founded in 1969, reports new developments in physics research and teaching – quickly and informally, but with a high quality and the explicit aim to summarize and communicate current knowledge in an accessible way. Books published in this series are conceived as bridging material between advanced graduate textbooks and the forefront of research and to serve three purposes:

- to be a compact and modern up-to-date source of reference on a well-defined topic

- to serve as an accessible introduction to the field to postgraduate students and nonspecialist researchers from related areas

- to be a source of advanced teaching material for specialized seminars, courses and schools

Both monographs and multi-author volumes will be considered for publication. Edited volumes should, however, consist of a very limited number of contributions only. Proceedings will not be considered for LNP.

Volumes published in LNP are disseminated both in print and in electronic formats, the electronic archive being available at springerlink.com. The series content is indexed, abstracted and referenced by many abstracting and information services, bibliographic networks, subscription agencies, library networks, and consortia.

Proposals should be sent to a member of the Editorial Board, or directly to the managing editor at Springer:

Christian Caron
Springer Heidelberg
Physics Editorial Department I
Tiergartenstrasse 17
69121 Heidelberg / Germany
christian.caron@springer.com

J. B. Weiss
A. Provenzale (Eds.)

Transport and Mixing
in Geophysical Flows

 Springer

Jeffrey B. Weiss
Department of Atmospheric
and Oceanic Sciences
University of Colorado
Boulder, CO 80309-0311, USA
Jeffrey.Weiss@colorado.edu

Antonello Provenzale
CNR - Istituto di Scienze
dell'Atmosfera e del
Clima Corso Fiume 4
10133 Torino, Italy
A.Provenzale@isac.cnr.it

J. B. Weiss and A. Provenzale (Eds.), *Transport and Mixing in Geophysical Flows*, Lect. Notes Phys. 744 (Springer, Berlin Heidelberg 2008), DOI 10.1007/ 978-3-540-75215-8

ISBN: 978-3-642-09448-4   e-ISBN: 978-3-540-75215-8

Lecture Notes in Physics ISSN: 0075-8450

Cover design: eStudio Calamar S.L., F. Steinen-Broo, Pau/Girona, Spain

Printed on acid-free paper

9 8 7 6 5 4 3 2 1

springer.com

Preface

Geophysical flows are characterized by their ability to transport heat, momentum and material quantities large distances. The entire climate system can be seen as a heat engine, and the solar energy absorbed at the tropics is transported, by the ocean currents and the global atmospheric circulation, toward the poles. As a result, the surface temperature of the earth is more uniform than it would be in the absence of a fluid envelope. Stratospheric currents induce global transport, and on smaller scales, regional circulations stir and mix nutrients and pollutants in ocean basins and in the troposphere. Transport and mixing also affect the functioning of marine ecosystems, where they are responsible for nutrient and plankton advection, and can determine the primary productivity of entire oceanic regions.

In this volume, we have collected the lecture notes of some of the contributions presented during the 2004 Summer Course on "Transport in Geophysical Flows" of the French–Italian School on "Fundamental Processes in Geophysical and Environmental Flows," directed by A. Provenzale and J. B. Weiss. Some time has elapsed since the school, and the notes of the lectures have been revised by the authors to include more recent material and new perspectives. Although necessarily incomplete, this collection of notes provides an updated view of some of the currently active research areas in the study of geophysical transport processes.

Transport in fluids can be approached from two complementary perspectives. In the Eulerian view of mixing, the focus is on the the concentration field. Advection stretches and folds the concentration field, and sharpens gradients, while diffusion smoothes the field. In the Lagrangian view, fluid parcels are followed around as they move with the flow, experiencing chaotic or stochastic motion. The lectures consider both passive particles, which they are carried freely by the flow, and reactive particles, where chemical or biologically reactions change the character of the particles.

The first part of this volume describes theoretical and idealized problems in transport. The local theory of mixing is discussed in the notes by Jean-Luc Thiffeault. This approach focuses on linear velocity fields and provides insight into the fundamental process of transport. The lecture notes of Guido Boffetta, Andrea Mazzino and Angelo Vulpiani, and of Yves Pomeau describe several aspects of the transport of passive and reactive particles. The transport of radiation in the atmosphere is an essential part of the climate system, yet it is often ignored by fluid dynamicists. Ed Spiegel's lecture notes remedy this by providing an introduction to radiative transfer. In geophysical flows, coherent vortices are ubiquitous and their impact on transport is discussed in the notes of Provenzale, Babiano, Bracco, Pasquero and Weiss.

The second part of the volume contains lectures on more realistic experimental and observational aspects of transports. The notes of Wells, Clercx and van Heijst describe transport in a rotating tank which achieves quasi-two-dimensional flow. Experiments are conducted using both passive and chemically reacting tracers. Many geophysical flows are dominated by structures such as vortices, jets and fronts, which generate transport barriers and inhomogeneous mixing. The lecture notes of Noboru Nakamura describe a technique to quantify transport in structured flows by using the passive tracer field itself as a coordinate. The statistics of Lagrangian data, such as those obtained from atmospheric balloons and freely drifting ocean floats, are described by Joe LaCasce. The final chapter by Marina Lévy describes the interaction between biological reactions and ocean mesoscale turbulence.

Boulder, CO, USA and Torino, Italy,   *Jeffrey B. Weiss*
May, 2007   *Antonello Provenzale*

Contents

Coherent Vortices and Tracer Transport

A. Provenzale, A. Babiano, A. Bracco, C. Pasquero, and J. B. Weiss...101

Part II Experiments and Observations

Dispersion and Mixing in Quasi-two-dimensional Rotating Flows

M. G. Wells, H. J. H. Clercx and G. J. F. van Heijst 119

Quantifying Inhomogeneous, Instantaneous, Irreversible Transport Using Passive Tracer Field as a Coordinate

N. Nakamura...137

Lagrangian Statistics from Oceanic and Atmospheric Observations

J. H. LaCasce 165

Part I

Theory

Scalar Decay in Chaotic Mixing

J.-L. Thiffeault

Department of Mathematics, University of Wisconsin, Madison, WI, USA
jeanluc@mailaps.org

Abstract. I review the *local theory* of mixing, which focuses on infinitesimal blobs of scalar being advected and stretched by a random velocity field. An advantage of this theory is that it provides elegant analytical results. A disadvantage is that it is highly idealised. Nevertheless, it provides insight into the mechanism of chaotic mixing and the effect of random fluctuations on the rate of decay of the concentration field of a passive scalar.

1 Introduction

The equation that is in the spotlight is the *advection–diffusion equation*

$$\partial_t \theta + \boldsymbol{v} \cdot \nabla \theta = \kappa \nabla^2 \theta \qquad (1)$$

for the time-evolution of a distribution of concentration $\theta(\boldsymbol{x}, t)$, being *advected* by a velocity field $\boldsymbol{v}(\boldsymbol{x}, t)$ and *diffused* with diffusivity κ. The concentration θ is called a *scalar* (as opposed to a vector). We will restrict our attention to incompressible velocity fields, for which $\nabla \cdot \boldsymbol{v} = 0$. For our purposes, we shall leave the exact nature of θ nebulous: it could be a temperature or the concentration of salt, dye, chemicals, isotopes, or even plankton. The only assumption for now is that this scalar is *passive*, which means that its value does not affect the velocity field $\boldsymbol{v}$. Clearly, this is not strictly true of some scalars like temperature, because a varying buoyancy influences the flow, but is often a good approximation nonetheless.

The advection–diffusion equation is linear, but contrary to popular belief that does not mean it is simple! Because the velocity (which is regarded here as a given vector field) is a function of space and time, the advection term (the second term in (1)) can cause complicated behaviour in θ. Broadly speaking, the advection term tends to create sharp gradients of θ, whilst the diffusion term (the term on the right-hand side of (1)) tends to wipe out gradients. The evolution of the concentration field is thus given by a delicate balance of advection and diffusion.

J.-L. Thiffeault: *Scalar Decay in Chaotic Mixing*, Lect. Notes Phys. **744**, 3–35 (2008)
DOI 10.1007/978-3-540-75215-8_1   © Springer-Verlag Berlin Heidelberg 2008

The advection term in (1) is also known as the *stirring* term, and the interplay of advection and diffusion is often called *stirring and mixing*. As we shall see, the two terms have very different roles, but both are needed to achieve an efficient mixing.

To elicit some broad features of mixing, we will start by deriving some properties of the advection–diffusion equation. First, it conserves the total quantity of θ. If we use angular brackets to denote the average of θ over the fixed domain of interest V, i.e.

$$\langle \theta \rangle := \frac{1}{V} \int_V \theta \, \mathrm{d}V,$$

then we find directly from (1) that

$$\partial_t \langle \theta \rangle + \langle \boldsymbol{v} \cdot \nabla \theta \rangle = \kappa \langle \nabla^2 \theta \rangle. \tag{2}$$

Because the velocity field is incompressible, we have

$$\boldsymbol{v} \cdot \nabla \theta = \nabla \cdot (\theta \, \boldsymbol{v}),$$

and also $\nabla^2 \theta = \nabla \cdot (\nabla \theta)$. Thus, we can use the divergence theorem to write (2) as

$$\partial_t \langle \theta \rangle = -\frac{1}{V} \int_S \theta \boldsymbol{v} \cdot \hat{\boldsymbol{n}} \, \mathrm{d}S + \kappa \frac{1}{V} \int_S \nabla \theta \cdot \hat{\boldsymbol{n}} \, \mathrm{d}S, \tag{3}$$

where S is the surface bounding V, $\mathrm{d}S$ the element of area, and $\hat{\boldsymbol{n}}$ the outward-pointing normal to the surface. For a closed flow, two possibilities are now open to us: (i) the domain V is periodic or (ii) $\boldsymbol{v}$ and $\nabla \theta$ are both tangents to the surface S. In the first case, the terms on the right-hand side of (3) vanish because boundary terms always vanish with periodic boundary conditions (a bit tautological, but true!). In the second case, both $\boldsymbol{v} \cdot \hat{\boldsymbol{n}}$ and $\nabla \theta \cdot \hat{\boldsymbol{n}}$ vanish. Either way,

$$\partial_t \langle \theta \rangle = 0 \tag{4}$$

so that the mean value of θ is constant. Since V is constant, this also implies that the total amount of θ is conserved. The second set of boundary conditions we used implies that there is no fluid flow or flux of θ through the boundary of the volume. It is thus natural that the total θ is conserved! For periodic boundary conditions, whatever leaves the volume re-enters on the other side, so it also makes sense that θ is conserved. Because of (4) and because we can always add a constant to θ without changing its evolution (only derivatives of θ appear in (1)), we will always choose

$$\langle \theta \rangle = 0 \tag{5}$$

without loss of generality. In words: the mean of our scalar vanishes initially, so by (4) it must vanish for all times.

Now let's look at another average of θ: rather than averaging θ itself, which has yielded an important but boring result, we average its square. The *variance* is defined by

$$\mathrm{Var} := \langle \theta^2 \rangle - \langle \theta \rangle^2, \qquad (6)$$

where the second term on the right vanishes by (5). To obtain an equation for the time-evolution of the variance, we multiply (1) by θ and integrate:

$$\langle \theta\, \partial_t \theta \rangle + \langle \theta\, \boldsymbol{v} \cdot \nabla \theta \rangle = \kappa \langle \theta\, \nabla^2 \theta \rangle.$$

We rearrange on the left and integrate by parts on the right, to find

$$\left\langle (\partial_t + \boldsymbol{v} \cdot \nabla)\, \tfrac{1}{2}\, \theta^2 \right\rangle = \kappa \left\langle \nabla \cdot (\theta\, \nabla \theta) - |\nabla \theta|^2 \right\rangle.$$

Now there are some boundary terms that vanish under the same assumptions as before, and we get

$$\partial_t \mathrm{Var} = -2\kappa \left\langle |\nabla \theta|^2 \right\rangle. \qquad (7)$$

Notice that, once again, the velocity field has dropped out of this averaged equation. However, now the effect of diffusion remains. Moreover, it is clear that the term on the right-hand side of (7) is *negative-definite* (or zero): this means that the variance always decreases (or is constant). The only way it can stop decreasing is if $\nabla \theta$ vanishes everywhere, that is, θ is constant in space. But because we have assumed $\langle \theta \rangle = 0$, this means that $\theta = 0$ everywhere. In that case, we have no choice but to declare the system to be *perfectly mixed*: there are no variations in θ at all anymore. Equation (7) tells us that variance tends to zero, which means that the system inexorably tends to the perfectly mixed state, without necessarily ever reaching it. Variance is thus a useful measure of mixing: the smaller the variance, the better the mixing.

There is a problem with all this: (7) no longer involves the velocity field. But if variance is to give us a measure of mixing, shouldn't its time-evolution involve the velocity field? Is this telling us that stirring has no effect on mixing? Of course not, as any coffee-drinker will testify, whether he/she likes it with milk or sugar: stirring has a huge impact on mixing! So what's the catch?

The catch is that (7) is not a closed equation for the variance: the right-hand side involves $|\nabla \theta|^2$, which is not the same as θ^2. The extra gradient makes all the difference. As we will see, under the right circumstances the stirring velocity field creates very large gradients in the concentration field, which makes variance decrease much faster than it would if diffusivity were acting alone. In fact, when κ is very small, in the best stirring flows the gradients of θ scale as $\kappa^{-1/2}$, so that the right-hand side of (7) becomes independent of the diffusivity. This, in a nutshell, is the essence of *enhanced mixing*.

Several important questions can now be raised:

- How fast is the approach to the perfectly mixed state?
- How does this depend on κ?
- What does the concentration field look like for long times? What is its spectrum?

- How does the probability distribution of θ evolve?
- Which stirring fields give efficient mixing?

The answers to these questions are quite complicated, and not fully known. In the following sections we will attempt to give some hints of the answers and provide some references to the literature.

This is not meant to be a comprehensive review article, so entire swaths of the literature are missing. We focus mainly on *local* or *Lagrangian theories*, which involve deterministic and stochastic approaches for quantifying stretching using a local idealisation of the flow. The essential feature here is that the advection–diffusion equation is solved along fluid trajectories. These theories trace their origins to Batchelor [1], who treated constant matrices with slow time dependence, and Kraichnan [2, 3], who introduced fast (delta-correlated) time dependence. Zeldovich et al. [4] approached the problem from the random-matrix theory angle in the magnetic dynamo context. More recently, techniques from large-deviation theory [5, 6, 7, 8] and path integration [9, 10, 11, 12, 13] have allowed an essentially complete solution of the problem. It is this work that will be reviewed here, as it applies to the decay of the passive scalar (and not the PDF of concentration or its power spectrum). We will favour expediency over mathematical rigour and try to give a flavour of what these local theories are about without describing them in detail.

The story will proceed from here as follows: in Sect. 2 advection of a blob by a linear velocity field is considered, with diffusion included. This problem has an exact solution, but it can be made simpler in the limit of small diffusivity. Solutions are examined for a straining flow in two and three dimensions (Sects. 2.2 and 2.4), as well as a shear flow in two dimensions (Sect. 2.3). Randomness is added in Sect. 3: the strain associated with the velocity field is assumed to vary, and the consequences of this for a single blob (Sect. 3.1) and a large number of blobs (Sect. 3.2) are explored. Practical implementation is discussed in Sect. 4, and a simple model for a micromixer is analysed in Sect. 4.1. Finally, the limitations of the theory are presented in Sect. 4.2.

2 Advection and Diffusion in a Linear Velocity Field

We will start by considering what happens to a passive scalar advected by a linear velocity field. The overriding advantage of this configuration is that it can be solved analytically, but that is not its only pleasant feature. Like most good toy models, it serves as a nice prototype for what happens in more complicated flows. It also serves as a building block for what may be called the *local theory* of mixing (Sect. 3).

The perfect setting to consider a linear flow is in the limit of large Schmidt number. The Schmidt number is a dimensionless quantity defined as

$$Sc := \nu/\kappa$$

where ν is the kinematic viscosity of the fluid and κ the diffusivity of the scalar. The Schmidt number may be thought of as the ratio of the diffusion time for the scalar to that for momentum in the fluid. Alternatively, it can be regarded as the ratio of the (squared) length of the smallest feature in the velocity field to that in the scalar field. This last interpretation is due to the fact that if θ varies in space more quickly than $\sqrt{\kappa}$, then its gradient is large and diffusion wipes out the variation. The same applies to variations in the velocity field with respect to ν. Hence, for large Schmidt number the scalar field has much faster variations than the velocity field. This means that it is possible to focus on a region of the domain large enough for the scalar concentration to vary appreciably, but small enough that the velocity field appears linear. Because there are many cases for which Sc is quite large, this motivates the use of a linear velocity field. In fact, large Sc is the natural setting for chaotic advection. It is also the regime that was studied by Batchelor and leads to the celebrated Batchelor spectrum [1]. The limit of small Sc is the domain of *homogenisation theory* and of turbulent diffusivity models. We shall not discuss such things here.

2.1 Solution of the Problem

We choose a linear velocity field of the form

$$v = x \cdot \sigma(t), \qquad \mathrm{Tr}\,\sigma = 0,$$

where σ is a traceless matrix because $\nabla \cdot v$ must vanish. Inserting this into (1), we want to solve the initial value problem:

$$\partial_t \theta + x \cdot \sigma(t) \cdot \nabla\theta = \kappa \nabla^2 \theta, \qquad \theta(x,0) = \theta_0(x). \qquad (8)$$

Here the coordinate x is really a deviation from a reference fluid trajectory. (In Appendix 1 we derive (8) from (1) by transforming to a comoving frame and assuming the velocity field is smooth.) We will follow closely the solution of Zeldovich et al. [4], who solved this by the method of "partial solutions". Consider a solution of the form

$$\theta(x,t) = \hat{\theta}(k_0,t) \exp(\mathrm{i}k(t) \cdot x), \qquad k(0) = k_0, \quad \hat{\theta}(k_0,0) = \hat{\theta}_0(k_0), \qquad (9)$$

where k_0 is some initial wavevector. We will see if we can make this into a solution by a judicious choice of $\hat{\theta}(k_0,t)$ and $k(t)$. The time derivative of (9) is

$$\partial_t \theta = (\partial_t \hat{\theta} + \mathrm{i}\,\partial_t k \cdot x\,\hat{\theta}) \exp(\mathrm{i}k(t) \cdot x)$$

and we have

$$v \cdot \nabla\theta = \mathrm{i}\,(x \cdot \sigma \cdot k)\,\hat{\theta} \exp(\mathrm{i}k(t) \cdot x).$$

Putting these together into (8) and cancelling out the exponential gives

$$\partial_t \hat{\theta} + \mathrm{i}\,x \cdot (\partial_t k + \sigma \cdot k)\,\hat{\theta} = -\kappa\,k^2 \hat{\theta}.$$

This must hold for all x, and neither $\hat{\theta}$ nor k depends on x, so we equate powers of x. This gives the two evolution equations

$$\partial_t k = -\sigma \cdot k, \tag{10a}$$

$$\partial_t \hat{\theta} = -\kappa\, k^2 \hat{\theta}. \tag{10b}$$

We can write the solution to (10a) in terms of the *fundamental solution* $\mathcal{T}(t,0)$ as

$$k(t) = \mathcal{T}(t,0) \cdot k_0,$$

where

$$\partial_t \mathcal{T}(t,0) = -\sigma(t) \cdot \mathcal{T}(t,0), \qquad \mathcal{T}(0,0) = \mathrm{Id},$$

and Id is the identity matrix. The advantage of doing this is that we can use the same fundamental solution for all initial conditions. We will usually write $\mathcal{T}_t$ to mean $\mathcal{T}(t,0)$. Note that because $\mathrm{Tr}\,\sigma = 0$, we have

$$\det \mathcal{T}_t = 1.$$

If σ is not a function of time, then the fundamental solution is simply a matrix exponential,

$$\mathcal{T}_t = \exp(-\sigma\, t),$$

but in general the form of $\mathcal{T}_t$ is more complicated.

Now that we know the time dependence of k, we can express the solution to (10) as

$$k(t) = \mathcal{T}_t \cdot k_0, \tag{11a}$$

$$\hat{\theta}(k_0, t) = \hat{\theta}_0(k_0) \exp\left\{ -\kappa \int_0^t \left(\mathcal{T}_s \cdot k_0 \right)^2 \mathrm{d}s \right\}. \tag{11b}$$

We can think of $\mathcal{T}_t$ as transforming a Lagrangian wavevector k_0 to its Eulerian counterpart k. Thus (11b) expresses the fact that $\hat{\theta}$ decays diffusively at a rate determined by the cumulative norm of the wavenumber k experienced during its evolution.

The full solution to (8) is now given by superposition of the partial solutions:

$$\theta(x, t) = \int \hat{\theta}(k_0, t) \exp(\mathrm{i}k(t) \cdot x)\, \mathrm{d}^3 k_0$$

$$= \int \hat{\theta}_0(k_0) \exp\left\{ \mathrm{i}x \cdot \mathcal{T}_t \cdot k_0 - \kappa \int_0^t \left(\mathcal{T}_s \cdot k_0 \right)^2 \mathrm{d}s \right\} \mathrm{d}^3 k_0, \tag{12}$$

where $\hat{\theta}_0(k_0)$ is the Fourier transform of the initial condition $\theta_0(x)$.[1] Assuming the spatial mean of θ vanishes, the variance (6) is

---

[1] We are using the convention

$$\text{Var} = \int \theta^2(\boldsymbol{x}, t)\, \mathrm{d}^3 x = \int |\hat{\theta}(\boldsymbol{k}_0, t)|^2\, \mathrm{d}^3 k_0 ,$$

which from (11b) becomes

$$\text{Var} = \int |\hat{\theta}_0(\boldsymbol{k}_0)|^2 \exp\left\{ -2\kappa \int_0^t (\mathcal{T}_s \cdot \boldsymbol{k}_0)^2\, \mathrm{d}s \right\} \mathrm{d}^3 k_0 . \tag{13}$$

We thus have a full solution of the advection–diffusion equation for the case of a linear velocity field and found the time-evolution of the variance. But what can be gleaned from it? We shall look at some special cases in the following section.

2.2 Straining Flow in 2D

We now take an even more idealised approach: consider the case where the velocity gradient matrix σ is constant. Furthermore, let us restrict ourselves to two-dimensional (2D) flows. After a coordinate change, the traceless matrix σ can only take two possible forms:

$$\sigma^{(2a)} = \begin{pmatrix} \lambda & 0 \\ 0 & -\lambda \end{pmatrix} \quad \text{and} \quad \sigma^{(2b)} = \begin{pmatrix} 0 & 0 \\ U' & 0 \end{pmatrix} . \tag{14}$$

Case (2a) is a purely straining flow that stretches exponentially in one direction and contracts in the other. Case (2b) is a linear shear flow in the x_1 direction. We assume without loss of generality that $\lambda > 0$ and $U' > 0$. The form $\sigma^{(2b)}$ is known as the Jordan canonical form and can only occur for degenerate eigenvalues. Since by incompressibility the sum of these identical eigenvalues must vanish, they must both vanish. The corresponding fundamental matrices $\mathcal{T}_t = \exp(-\sigma t)$ are

$$\mathcal{T}_t^{(2a)} = \begin{pmatrix} \mathrm{e}^{-\lambda t} & 0 \\ 0 & \mathrm{e}^{\lambda t} \end{pmatrix} \quad \text{and} \quad \mathcal{T}_t^{(2b)} = \begin{pmatrix} 1 & 0 \\ -U't & 1 \end{pmatrix} . \tag{15}$$

These are easy to compute: in the first instance one merely exponentiates the diagonal elements, in the second the exponential power series terminates after two terms, because the square of $\sigma^{(2b)}$ is zero.

Let us consider case (2a), a flow with constant stretching (the case considered by Batchelor [1]). The action of the fundamental matrix on $\boldsymbol{k}_0$ for case (2a) is

$$\hat{\theta}(\boldsymbol{k}) = \frac{1}{(2\pi)^d} \int \theta(\boldsymbol{x})\, \mathrm{e}^{-i\boldsymbol{k}\cdot\boldsymbol{x}}\, \mathrm{d}^d x ,$$

$$\theta(\boldsymbol{x}) = \int \hat{\theta}(\boldsymbol{k})\, \mathrm{e}^{i\boldsymbol{k}\cdot\boldsymbol{x}}\, \mathrm{d}^d k ,$$

for the Fourier transform in d dimensions.

$$\mathcal{T}_t^{(2a)} \cdot k_0 = \left(e^{-\lambda t} k_{01} , \, e^{\lambda t} k_{02} \right), \tag{16}$$

with norm

$$\left(\mathcal{T}_t^{(2a)} \cdot k_0 \right)^2 = e^{-2\lambda t} k_{01}^2 + e^{2\lambda t} k_{02}^2 . \tag{17}$$

The wavevector $k(t) = \mathcal{T}_t^{(2a)} \cdot k_0$ grows exponentially in time, which means that the length scale is becoming very small. This only occurs in the direction x_2, which is sensible because that direction corresponds to a contracting flow. Picture a curtain being closed: the bunching up of the fabric into tight folds is analogous to the contraction. (Of course, it is difficult to close a curtain exponentially quickly forever!) The component of the wavevector in the x_1 direction decreases in magnitude, which corresponds to the opening of a curtain.

Let's see what happens to one Fourier mode. By inserting (17) in (11b), we have

$$\hat{\theta}(k_0, t) = \hat{\theta}_0(k_0) \exp \left\{ -\kappa \int_0^t \left(e^{-2\lambda s} k_{01}^2 + e^{2\lambda s} k_{02}^2 \right) ds \right\} .$$

The time-integral can be done explicitly, and we find

$$\hat{\theta}(k_0, t) = \hat{\theta}_0(k_0) \exp \left\{ -\frac{\kappa}{2\lambda} \left(\left(e^{2\lambda t} - 1 \right) k_{02}^2 - \left(e^{-2\lambda t} - 1 \right) k_{01}^2 \right) \right\} .$$

For moderately long times ($t \gtrsim \lambda^{-1}$), we can surely neglect $e^{-2\lambda t}$ compared to 1, and 1 compared to $e^{2\lambda t}$,

$$\hat{\theta}(k_0, t) \simeq \hat{\theta}_0(k_0) \exp \left\{ -\frac{\kappa}{2\lambda} \left(e^{2\lambda t} k_{02}^2 + k_{01}^2 \right) \right\} . \tag{18}$$

Actually, this assumption of moderately long time is easily justified physically. If $\kappa k^2 / \lambda \ll 1$, where k is the largest initial wavenumber (that is, the smallest initial scale), then the argument of the exponential in (18) is small, unless

$$e^{2\lambda t} \gtrsim Pe \tag{19}$$

where the *Péclet number* is

$$Pe = \frac{\lambda}{\kappa \, k^2} . \tag{20}$$

Thus the assumption that $e^{2\lambda t}$ is large is a consequence of Pe being large, since otherwise the exponential in (18) is near unity and can be ignored—variance is approximately constant. We can turn (19) into a requirement on the time:

$$\lambda t \gtrsim \log Pe^{1/2} . \tag{21}$$

It is clear from (21) that λ^{-1} sets the time scale for the argument of the exponential in (18) to become important. The Péclet number influences this time scale only weakly (logarithmically). This is probably the most important physical fact about chaotic mixing: *small diffusivity has only a logarithmic effect.*

Thus vigorous stirring always has a chance to overcome a small diffusivity, no matter how small: we need just stir a bit longer.

Note that the variance is given by

$$\text{Var} = \int |\hat{\theta}_0(\boldsymbol{k}_0)|^2 \exp\left\{-\frac{\kappa}{\lambda}\left(e^{2\lambda t} k_{02}^{\,2} + k_{01}^{\,2}\right)\right\} d^2 k_0 \,,$$

which is approximately constant for $t \ll \lambda^{-1} \log Pe^{1/2}$. This does *not* mean that the concentration field

$$\theta(\boldsymbol{x}, t) = \int \hat{\theta}(\boldsymbol{k}_0, t)\, e^{i\boldsymbol{k}(t)\cdot\boldsymbol{x}}\, d^2 k_0 \tag{22}$$

is constant, even if $\hat{\theta}(\boldsymbol{k}_0, t)$ is, because $\boldsymbol{k}(t) = \mathcal{T}_t^{(2a)} \cdot \boldsymbol{k}_0$ is a function of time from (16). This time dependence becomes important for $t \gtrsim \lambda^{-1}$.

The Péclet number may be thought of as the ratio of the advection time of the flow to the diffusion time for the scalar. It is usually written as

$$Pe := UL/\kappa \,, \tag{23}$$

where U is a typical velocity and L a typical length scale. Our velocity estimate in (20) is λ/k, and our length scale is k, which are both natural for the problem at hand. Just like large Sc, large Pe is the natural setting for chaotic advection. In fact, if Pe is small then diffusion is faster than advection, and stirring is not really required! Large Pe means that diffusion by itself is not very effective, so that stirring is required. We shall always assume that Pe is large.

We return to (18): the striking thing about that equation is its prediction for the rate of decay of the concentration field. Roughly speaking, (18) predicts

$$\theta(\boldsymbol{x}, t) \sim \exp\left\{-Pe^{-1}\, e^{2\lambda t}\right\} \tag{24}$$

for $\lambda t \gg 1$. Equation (24) is the exponential of an exponential—a *superexponential* decay. This is *extremely* fast decay. In fact, unnaturally so: it is hard to imagine a physically sensible system that could mix this quickly. Something more has to be at work here.

If we examine (18) closely, we see that the culprit is the term

$$e^{2\lambda t}\, k_{02}^{\,2} \,, \tag{25}$$

which grows exponentially fast. This term has its origin in the Laplacian in the advection–diffusion equation (1): the contracting direction of the flow (the x_2 direction) leads to an exponential increase in the wavenumber via the curtain-closing mechanism. This is exactly the mechanism for enhanced mixing we advertised on p. 5: very large gradients of concentration are being created, exponentially fast. This mechanism is just acting too quickly for our taste!

So what's the problem? We are doing the wrong thing to obtain our estimate (24). This estimate tells us how fast a typical wavevector decays, and it says that this occurs very quickly. What we really want to know is what modes survive superexponential decay the longest, and at what rate *they* decay. Clearly the concentration in most wavenumbers gets annihilated almost instantly, once the condition (21) is satisfied. But a small number remains: those are the modes with wavevector closely aligned to the x_1 (stretching) direction or equivalently that have a very small projection on the x_2 (contracting) direction. To overcome the exponential growth in (25), we require

$$k_{02} \sim e^{-\lambda t}, \tag{26}$$

that is at any given time we need consider only wavenumbers satisfying (26), since the concentration in all the others has long since been wiped out by diffusion. The consequence is that the k_{02} integral in (22) is dominated by these surviving modes. To see this, we blow up the k_{02} integration by making the coordinate change $\widetilde{k}_{02} = k_{02} e^{\lambda t}$ in (22):

$$\theta(\boldsymbol{x}, t) = e^{-\lambda t} \int_{-\infty}^{\infty} dk_{01} \int_{-\infty}^{\infty} d\widetilde{k}_{02} \, \hat{\theta}_0(k_{01}, \widetilde{k}_{02} e^{-\lambda t})$$

$$\times \, e^{i\boldsymbol{k}(t) \cdot \boldsymbol{x}} \exp\left\{ -\frac{\kappa}{2\lambda} \left(\widetilde{k}_{02}^{\,2} + k_{01}^2 \right) \right\}. \tag{27}$$

The decay factor $e^{-\lambda t}$ has appeared in front. For small diffusivity, we can neglect the k_{01}^2 term in the exponential (it just smooths out the initial concentration field a little).[2] We can then take the inverse Fourier transform of $\hat{\theta}_0(k_{01}, \widetilde{k}_{02} e^{-\lambda t})$,

$$\hat{\theta}_0(k_{01}, \widetilde{k}_{02} e^{-\lambda t}) = \frac{1}{(2\pi)^2} \int \theta_0(\widetilde{\boldsymbol{x}}) \exp\left(-ik_{01}\widetilde{x}_1 - i\widetilde{k}_{02} e^{-\lambda t} \widetilde{x}_2 \right) d\widetilde{x}_1 \, d\widetilde{x}_2,$$

and insert this into (27). We then interchange the order of integration: the k_{01} integral gives a δ-function, and the $\widetilde{k}_{02}$ integral gives a Gaussian. The final result is

$$\theta(\boldsymbol{x}, t) = e^{-\lambda t} \int_{-\infty}^{\infty} \theta_0(e^{-\lambda t} x_1, \widetilde{x}_2) \, G\left(x_2 - e^{-\lambda t} \, \widetilde{x}_2 \,;\, \ell \right) d\widetilde{x}_2, \tag{28}$$

where

$$G(x; \ell) := \frac{1}{\sqrt{2\pi\ell^2}} \, e^{-x^2/2\ell^2} \tag{29}$$

is a normalised Gaussian distribution with standard deviation ℓ, and we defined the length

$$\ell := \sqrt{\kappa/\lambda}.$$

---

[2] We require the initial condition to be smooth at small scales. Here's why: for small κ, k_{01} needs to be large to matter in the argument of the exponential. But a smooth θ decays exponentially with k_{01}, so there is no variance in these modes anyway.

If the initial concentration decays for large $|x_2|$ (as when we have a single blob of dye), then (28) can be simplified to

$$\theta(\boldsymbol{x},t) = \mathrm{e}^{-\lambda t}\, G\left(x_2;\, \ell\right) \int_{-\infty}^{\infty} \theta_0(\mathrm{e}^{-\lambda t} x_1, \tilde{x}_2)\, \mathrm{d}\tilde{x}_2 \,. \tag{30}$$

So the x_1 dependence in (30) is given by the "stretched" initial distribution, averaged over x_2. The important thing to notice is that

$$\theta(\boldsymbol{x},t) \sim \mathrm{e}^{-\lambda t} \,. \tag{31}$$

This is a much more reasonable estimate for the decay of concentration than (24)! The concentration thus decays exponentially at a rate given by the rate of strain (or stretching rate) in our flow. The exponential decay is entirely due to the narrowing of the domain for eligible (i.e. nondecayed) modes. This "domain of eligibility" is also known as the *cone* or the *cone of safety* [4, 14]. (In two dimensions it is more properly called a wedge.) The concentration associated with wavevectors that fit within this cone is temporarily shielded from being diffusively wiped out, but as the aperture of the cone is shrinking exponentially more and more modes leave the safety of the cone as time progresses.

Notice that (31) is independent of κ. This brings us to the second most important physical fact about chaotic mixing: *the asymptotic decay rate of the concentration field tends to be independent of diffusivity.* But note that a nonzero diffusivity is crucial in forcing the alignment (26). The only effect of the diffusivity is to lengthen the wait before exponential decay sets in, as given by the estimate (21). But this effect is only logarithmic in the diffusivity.

We can also try to think of (30) in real rather than Fourier space. Consider an initial distribution of concentration. Our straining flow will stretch this distribution in the x_1 direction and contract it in the x_2 direction. Gradients in x_2 will thus become very large, so that eventually diffusion will limit further contraction in the x_2 direction and the distribution will stabilise with width $\sqrt{\kappa/\lambda}$ (see Fig. 1). This is what the Gaussian prefactor in (30) is telling us: the asymptotic distribution has "forgotten" its initial shape in x_2. We say that the contracting direction has been *stabilised.*

2.3 Shear Dispersion in 2D

So far we have only considered case (2a) in (14). For case (2b), we have from (15)

$$\mathcal{T}_t^{(2b)} \cdot \boldsymbol{k}_0 = (k_{01}, k_{02} - U't\, k_{01})$$

with norm

$$\left(\mathcal{T}_t^{(2b)} \cdot \boldsymbol{k}_0\right)^2 = k_{01}^2 + \left(k_{02} - U't\, k_{01}\right)^2 \,. \tag{32}$$

Inserting (32) in (12), we have

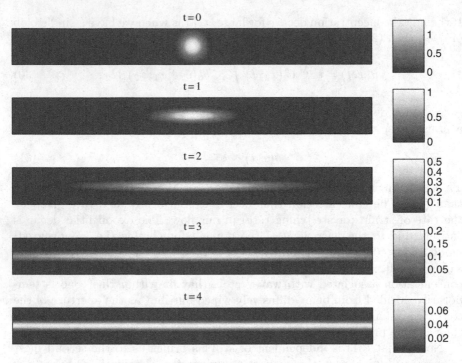

Fig. 1. A patch of dye in a uniform straining flow. The amplitude of the concentration field decreases exponentially with time. The length of the filament increases exponentially, whilst its width is stabilised at $\ell = \sqrt{\kappa/\lambda}$

$$\theta(\boldsymbol{x}, t) = \int \hat{\theta}_0(\boldsymbol{k}_0)\, e^{i\boldsymbol{k}(t)\cdot\boldsymbol{x}} \exp\left\{-\kappa \int_0^t \left(k_{01}^2 + (k_{02} - U's\,k_{01})^2\right) ds\right\} d^2 k_0 .$$

We can then explicitly do the time-integral in the exponential:

$$\theta(\boldsymbol{x}, t) = \int \hat{\theta}_0(\boldsymbol{k}_0)\, e^{i\boldsymbol{k}(t)\cdot\boldsymbol{x}}$$

$$\times \exp\left\{-\kappa\, k_{01}^2\, t - \frac{\kappa}{3U'k_{01}}\left((U't\,k_{01} - k_{02})^3 + k_{02}^3\right)\right\} d^2 k_0 . \quad (33)$$

The enhancement to diffusion in this case is reflected in the cubic power of time in the exponential. This is not as strong as the exponential enhancement of case (2a), but is nevertheless very significant. This phenomenon is known as *shear dispersion* or *Taylor dispersion*. The mechanism is often called the *venetian blind* effect. Assuming the initial distribution θ_0 depends only on k_{01}, the lines of constant concentration which are initially vertical are tilted by the shear flow, in a manner reminiscent of venetian blinds. The distance between the lines of constant concentration decreases with time as $(U't)^{-1}$, which gives an effective enhancement to diffusion. The time required to overcome a weak diffusivity is thus

$$U't \gtrsim (k_{01}^2 \kappa/U')^{-1/3}. \tag{34}$$

If we use k_{01}^{-1} as a length scale and U' as a time scale, we can define a Péclet number $Pe := U'/(k_{01}{}^2 \kappa)$ and rewrite (34) as

$$U't \gtrsim Pe^{1/3} \tag{35}$$

which should be compared to (21), the corresponding expression for the case (2a). Here there is a power law dependence on the Péclet number, rather than logarithmic, so we may have to wait a long time for diffusion to become important. This makes the linear velocity field approximation more likely to break down.

Let us consider the time-asymptotic limit $U't \gg 1$: we might then be tempted to neglect everything but the $U't\,k_{01}$ term in the argument of the exponential in (33). However, this would be a mistake. To see more clearly what happens, define the dimensionless time $\tau := U't$ and the length $\chi^2 = \kappa/U'$. Equation (33) then becomes

$$\theta(\boldsymbol{x}, t) = \int \hat{\theta}_0(\boldsymbol{k}_0)\, e^{i\boldsymbol{k}(t)\cdot\boldsymbol{x}}$$
$$\times \exp\left\{-\chi^2 \left(k_{01}^2\tau + k_{02}^2\tau + \tfrac{1}{3}k_{01}^2\tau^3 - k_{01}k_{02}\tau^2\right)\right\} \mathrm{d}^2k_0. \tag{36}$$

The first two terms in the exponential are just what is expected of regular diffusion in the absence of flow. The next term is the enhancement to diffusion along the x_1 direction: it will force the modes $k_{01} \sim \tau^{-3/2}$ to be dominant, since everything else will be damped away. Similarly, the last term forces $k_{02} \sim \tau^{-1/2}$. Assuming these scalings, the only term that can be neglected for $\tau \gg 1$ is the very first one, $k_{01}^2\tau$.

We make the change of variable $\widetilde{k}_{01} = \tau^{3/2}k_{01}$, $\widetilde{k}_{02} = \tau^{1/2}k_{02}$ in (36):

$$\theta(\boldsymbol{x}, t) = \tau^{-2} \int \hat{\theta}_0(\widetilde{k}_{01}\,\tau^{-3/2},\, \widetilde{k}_{02}\,\tau^{-1/2})\, e^{i(\tau^{-3/2}x_1 - \tau^{-1/2}x_2)\widetilde{k}_{01} + i\tau^{-1/2}x_2\widetilde{k}_{02}}$$
$$\times \exp\left\{-\chi^2 \left(\widetilde{k}_{02}^{\,2} + \tfrac{1}{3}\widetilde{k}_{01}^{\,2} - \widetilde{k}_{01}\widetilde{k}_{02}\right)\right\} \mathrm{d}\widetilde{k}_{01}\, \mathrm{d}\widetilde{k}_{02}. \tag{37}$$

If we approximate $\hat{\theta}_0(\widetilde{k}_{01}\,\tau^{-3/2},\, \widetilde{k}_{02}\,\tau^{-1/2}) \simeq \hat{\theta}_0(0,0)$, we can do the integrals in (37) and find

$$\theta(\boldsymbol{x}, t) \simeq 2\sqrt{3}\pi\, \chi^{-2}\,\tau^{-2}\, \hat{\theta}_0(0,0)\, \exp\left\{-\frac{3x_1^2 - 3x_1 x_2\tau + x_2^2\tau^2}{\chi^2\tau^3}\right\}. \tag{38}$$

For moderate values of x_1 ($x_1 \ll \chi\tau$), we have

$$\theta(\boldsymbol{x}, t) \simeq 2\sqrt{3}\pi\, \chi^{-2}\,\tau^{-2}\, e^{-x_2^2/\chi^2\tau}\, \hat{\theta}_0(0,0). \tag{39}$$

The width in the x_2 direction of an initial distribution thus increases as $\chi\tau^{1/2} = \sqrt{\kappa t}$. This is independent of U' and is exactly the same as expected from pure diffusion. The width in the x_1 direction in (38) increases as $\chi\tau^{3/2} = U't\sqrt{\kappa t}$ (see Fig. 2).

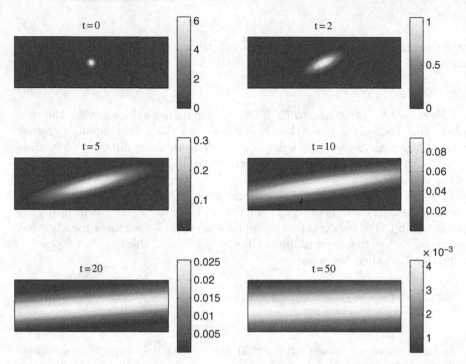

Fig. 2. A patch of dye in a uniform shearing flow. The amplitude of the concentration field decreases algebraically with time as t^{-2}. The length of the filament increases as $t^{3/2}$, whilst its width increases as $t^{1/2}$

2.4 Three Dimensions

In three dimensions, there are three basic forms for the matrix σ:

$$\sigma^{(3a)} = \begin{pmatrix} \lambda_1 & 0 & 0 \\ 0 & \lambda_2 & 0 \\ 0 & 0 & -\lambda_1 - \lambda_2 \end{pmatrix} \; ; \quad \sigma^{(3b)} = \begin{pmatrix} 0 & 0 & 0 \\ U' & 0 & 0 \\ 0 & U' & 0 \end{pmatrix} \; ; \quad \sigma^{(3c)} = \begin{pmatrix} \lambda & 0 & 0 \\ U' & \lambda & 0 \\ 0 & 0 & -2\lambda \end{pmatrix},$$

with corresponding fundamental matrices

$$\mathcal{T}^{(3a)} = \begin{pmatrix} e^{-\lambda_1 t} & 0 & 0 \\ 0 & e^{-\lambda_2 t} & 0 \\ 0 & 0 & e^{(\lambda_1+\lambda_2)t} \end{pmatrix} \; ; \quad \mathcal{T}^{(3b)} = \begin{pmatrix} 0 & 0 & 0 \\ -U't & 0 & 0 \\ \frac{1}{2}(U't)^2 & -U't & 0 \end{pmatrix} \; ; \tag{40}$$

$$\mathcal{T}^{(3c)} = \begin{pmatrix} e^{-\lambda t} & 0 & 0 \\ -U't\,e^{-\lambda t} & e^{-\lambda t} & 0 \\ 0 & 0 & e^{2\lambda t} \end{pmatrix}. \tag{41}$$

We can assume without loss of generality that $\lambda_1 \geq 0$, $\lambda_1 \geq \lambda_2$, and $U' > 0$, but the sign of λ_2 and λ is arbitrary; however, we must have $\lambda_3 = -\lambda_1 - \lambda_2 \leq 0$. The case of greatest interest to us is (3a). The relevant $\boldsymbol{k}(t)$, corresponding to (16), is

$$\mathcal{J}_t^{(3a)} \cdot \boldsymbol{k}_0 = \left(\mathrm{e}^{-\lambda_1 t} k_{01}, \, \mathrm{e}^{-\lambda_2 t} k_{02}, \, \mathrm{e}^{|\lambda_3| t} k_{03}\right), \tag{42}$$

which is used in (12) to give

$$\theta(\boldsymbol{x}, t) = \int \hat{\theta}_0(\boldsymbol{k}_0) \, \mathrm{e}^{\mathrm{i}\boldsymbol{k}(t) \cdot \boldsymbol{x}} \exp\left\{-\tfrac{1}{2}\kappa\left(\lambda_1^{-1}\left(1 - \mathrm{e}^{-2\lambda_1 t}\right) k_{01}^2\right.\right.$$
$$\left.\left. + \lambda_2^{-1}\left(1 - \mathrm{e}^{-2\lambda_2 t}\right) k_{02}^2 + |\lambda_3|^{-1}(\mathrm{e}^{2|\lambda_3| t} - 1) k_{03}^2\right)\right\} \mathrm{d}^3 k_0. \tag{43}$$

What happens next depends on the sign of λ_2: the question is whether $\mathrm{e}^{-2\lambda_2 t}$ grows or decays for $t \gg |\lambda_2|^{-1}$. If $\lambda_2 > 0$, then we have

$$\theta(\boldsymbol{x}, t) \simeq \int \hat{\theta}_0(\boldsymbol{k}_0) \, \mathrm{e}^{\mathrm{i}\boldsymbol{k}(t) \cdot \boldsymbol{x}} \exp\left\{-\tfrac{1}{2}\kappa\left(\lambda_1^{-1} k_{01}^2\right.\right.$$
$$\left.\left. + \lambda_2^{-1} k_{02}^2 + |\lambda_3|^{-1} \mathrm{e}^{2|\lambda_3| t} k_{03}^2\right)\right\} \mathrm{d}^3 k_0, \tag{44}$$

whilst for $\lambda_2 < 0$

$$\theta(\boldsymbol{x}, t) \simeq \int \hat{\theta}_0(\boldsymbol{k}_0) \, \mathrm{e}^{\mathrm{i}\boldsymbol{k}(t) \cdot \boldsymbol{x}} \exp\left\{-\tfrac{1}{2}\kappa\left(\lambda_1^{-1} k_{01}^2\right.\right.$$
$$\left.\left. + |\lambda_2|^{-1} \mathrm{e}^{2|\lambda_2| t} k_{02}^2 + |\lambda_3|^{-1} \mathrm{e}^{2|\lambda_3| t} k_{03}^2\right)\right\} \mathrm{d}^3 k_0. \tag{45}$$

Both approximations are valid when $t \gg \max(\lambda_1^{-1}, |\lambda_2|^{-1})$. For $\lambda_2 = 0$ the situation is similar to the 2D case (2a):

$$\theta(\boldsymbol{x}, t) \simeq \int \hat{\theta}_0(\boldsymbol{k}_0) \, \mathrm{e}^{\mathrm{i}\boldsymbol{k}(t) \cdot \boldsymbol{x}} \exp\left\{-\frac{\kappa}{2\lambda_1}\left(k_{01}^2 + \mathrm{e}^{2\lambda_1 t} k_{03}^2\right)\right\} \mathrm{d}^3 k_0,$$

valid when $t \gg \lambda_1^{-1}$.

The rest of the calculation is very similar to the 2D case (2a), in going from (22) to (30). In both (44) and (45) the x_3 direction is stabilised, that is we need to blow up the k_{03} integral to remove the time dependence from the exponential and find that the integral is dominated by $k_{03} \simeq 0$. The x_2 direction is also stabilised in (45), so we can set $k_{02} \simeq 0$. We thus find for $\lambda_2 \geq 0$,

$$\theta(\boldsymbol{x}, t) \simeq \mathrm{e}^{-|\lambda_3| t} \, G(x_3; \ell_3) \int \theta_0(\mathrm{e}^{-\lambda_1 t} x_1, \mathrm{e}^{-\lambda_2 t} x_2, \tilde{x}_3) \, \mathrm{d}\tilde{x}_3, \tag{46}$$

and for $\lambda_2 < 0$,

$$\theta(\boldsymbol{x}, t) \simeq \mathrm{e}^{-(|\lambda_2| + |\lambda_3|)t} \, G(x_2; \ell_2) \, G(x_3; \ell_3) \int \theta_0(\mathrm{e}^{-\lambda_1 t} x_1, \tilde{x}_2, \tilde{x}_3) \, \mathrm{d}\tilde{x}_2 \, \mathrm{d}\tilde{x}_3, \tag{47}$$

where $\ell_i := \sqrt{\kappa/|\lambda_i|}$. Contracting directions have their spatial dependence given by a time-independent Gaussian, with an overall exponential decay; stretching directions do just that: they stretch the initial distribution, with no diffusive effect. Solutions of the form (46) are called *pancakes*, and those of the form (47) are called *ropes* or *tubes*.

There is another way of thinking about the asymptotic forms (30), (46), and (47) [15]: contracting directions are stabilised near some constant width ℓ_j, and expanding directions lead to exponential growth of the width of an initial distribution along the direction. Thus, the volume of the initial distribution grows exponentially at a rate given by the sum of λ_is associated with stretching directions, but the total amount of θ remains fixed (the mean is conserved). Hence, the concentration at a point should decay inversely proportional to the volume, which is exactly what (30), (46), and (47) predict.

3 Random Strain Models

In Sect. 2 we analysed the deformation of a patch of concentration field (a "blob") in a linear velocity field. Though this is interesting in itself, it is a far cry from reality. We will now inch slightly closer to the real world by giving a random time dependence to our velocity field.

3.1 A Single Blob

Consider a single blob in a 2D linear velocity field of the type we treated in Sect. 2.2 (case (2a)). Now assume the orientation and stretching rate λ of the straining flow change randomly for every time τ. This situation is depicted schematically in Fig. 3. We assume that the time τ is much larger than a typical stretching rate λ, so that there is sufficient time for the blob to be deformed into its asymptotic form (30) at each period, which predicts that at each period the concentration field will decrease by a factor $\exp(-\lambda^{(i)}\tau)$, where $\lambda^{(i)}$ is the stretching rate at the ith period. The concentration field after n periods will thus be proportional to the product of decay factors:

$$\theta \sim e^{-\lambda^{(1)}\tau}e^{-\lambda^{(2)}\tau}\cdots e^{-\lambda^{(n)}\tau}$$

$$= e^{-(\lambda^{(1)}+\lambda^{(2)}+\cdots+\lambda^{(n)})\tau}. \tag{48}$$

We may rewrite this as

$$\theta \sim e^{-\Lambda_n t}, \tag{49}$$

where $t = n\tau$, and

$$\Lambda_n := \frac{1}{n}\sum_{i=1}^{n}\lambda^{(i)}$$

is the "running" mean value of the stretching rate at the nth period. As we let n become large, how do we expect the concentration field to decay? We might expect that it would decay at the mean value $\bar{\lambda}$ of the stretching rates $\lambda^{(i)}$. This is not the case: the running mean (49) does not *converge* to the

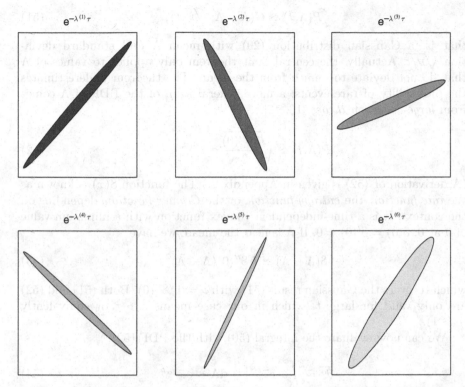

$e^{-\lambda^{(1)}\tau}$ $e^{-\lambda^{(2)}\tau}$ $e^{-\lambda^{(3)}\tau}$

$e^{-\lambda^{(4)}\tau}$ $e^{-\lambda^{(5)}\tau}$ $e^{-\lambda^{(6)}\tau}$

Fig. 3. A single blob being stretched for a time τ by successive random straining flows. The amplitude of the concentration field decays by $\exp(-\lambda^{(i)}\tau)$ at each period

mean $\bar{\lambda}$. Rather, by the central limit theorem its expected value is $\bar{\lambda}$, but its fluctuations around that value are proportional to $1/\sqrt{t}$. These fluctuations have an impact on the decay rate of θ.

The ensemble of variables $\lambda^{(i)}$ is known as a *realisation*. Now let us imagine performing our blob experiment several times and averaging the resulting concentration fields: this is known as an *ensemble average* over realisations. Ensemble-averaging smooths out fluctuations present in each given realisation. We may then replace the running mean Λ_n by a sample-space variable Λ, together with its probability distribution $P(\Lambda, t)$. The mean (expected value) of the αth power of the concentration field is then proportional to

$$\overline{\theta^\alpha} \sim \int_0^\infty e^{-\alpha\Lambda t} P(\Lambda, t) \, d\Lambda. \tag{50}$$

The overbar denotes the expected value. The factor $e^{-\alpha\Lambda t}$ gives the amplitude of θ^α given that the mean stretching rate at time t is Λ, and $P(\Lambda, t)$ measures the probability of that value of Λ occurring at time t.

The form of the probability distribution function (PDF) $P(\Lambda, t)$ is given by the central limit theorem:

$$P(\Lambda, t) \simeq G(\Lambda - \bar{\Lambda}; \sqrt{\nu/t}), \tag{51}$$

that is, a Gaussian distribution (29) with mean $\bar{\Lambda}$ and standard deviation $\sqrt{\nu/t}$. Actually, the central limit theorem only applies to values of Λ that do not deviate too much from the mean. The theorem underestimates the probability of rare events: a more general form of the PDF of Λ comes from *large-deviation theory* [16, 17]:

$$P(\Lambda, t) \simeq \sqrt{\frac{t\, 8''(0)}{2\pi}}\, e^{-t8(\Lambda - \bar{\Lambda})}. \tag{52}$$

(A derivation of (52) is given in Appendix 2.) The function $8(x)$ is known as the *rate function*, the *entropy function*, or the *Cramér function*, depending on the context. It is a time-independent convex function with a minimum value of 0 at 0: $8(0) = 8'(0) = 0$. If Λ is near the mean, we have

$$8(\Lambda - \bar{\Lambda}) \simeq \tfrac{1}{2}\, 8''(0)(\Lambda - \bar{\Lambda})^2, \tag{53}$$

which recovers the Gaussian result (51) with $\nu = 1/8''(0)$. Both (51) and (52) are only valid for large t (which in our case means $t \gg \tau$ or equivalently $n \gg 1$).

We can now evaluate the integral (50) with the PDF (52):

$$\overline{\theta^\alpha} \sim \int_0^\infty e^{-tH(\Lambda)}\, d\Lambda \sim e^{-\gamma_\alpha t}, \tag{54}$$

where we have omitted the nonexponential prefactors and defined

$$H(\Lambda) := \alpha\Lambda + 8(\Lambda - \bar{\Lambda}).$$

Since t is large, the integral is dominated by the minimum value of $H(\Lambda)$: this is the perfect setting for the well-known saddle-point approximation. The minimum occurs at Λ_{sp} where $H'(\Lambda_{\text{sp}}) = \alpha + 8'(\Lambda_{\text{sp}} - \bar{\Lambda}) = 0$ and is unique because 8 is convex and has a unique minimum. The decay rate is then given by

$$\gamma_\alpha = H(\Lambda_{\text{sp}}), \quad \text{with} \quad H'(\Lambda_{\text{sp}}) = 0. \tag{55}$$

There's a caveat to this: for α large enough the saddle point Λ_{sp} is negative. This is not possible: the stretching rates are defined to be nonnegative (the integral (54) involves only nonnegative Λ). Hence, the best we can do is to choose $\Lambda_{\text{sp}} = 0$—the integral (54) is dominated by realisations with no stretching. Thus, in that case $\gamma_\alpha = H(0)$ or

$$\gamma_\alpha = 8(-\bar{\Lambda}). \tag{56}$$

We re-emphasise: for small enough α, the saddle point is positive and the decay rate is given by (55). Beyond that, we must choose zero as the saddle point and the decay rate is given by (56). To find the critical value α_{crit} where we pass from (55) to (56), observe that this happens as the saddle point nears zero. Thus, we may solve our saddle-point equation $H'(\Lambda_{\mathrm{sp}}) = 0$ by Taylor expansion:

$$H'(\Lambda_{\mathrm{sp}}) \simeq \alpha_{\mathrm{crit}} + \mathcal{S}'(-\bar{\Lambda}) + \Lambda_{\mathrm{sp}}\,\mathcal{S}''(-\bar{\Lambda}) = 0. \tag{57}$$

But the saddle point will not be small unless the first terms cancel in (57), that is $\alpha_{\mathrm{crit}} = -\mathcal{S}'(-\bar{\Lambda})$. We may thus recapitulate the result for the decay rate:

$$\gamma_\alpha = \begin{cases} \alpha\Lambda_{\mathrm{sp}} + \mathcal{S}(\Lambda_{\mathrm{sp}} - \bar{\Lambda}), & \alpha < -\mathcal{S}'(-\bar{\Lambda}); \\ \mathcal{S}(-\bar{\Lambda}), & \alpha \ge -\mathcal{S}'(-\bar{\Lambda}). \end{cases} \tag{58}$$

Clearly γ_α is continuous, and it can be easily shown that $\mathrm{d}\gamma_\alpha/\mathrm{d}\alpha$ is also continuous.

As an illustration, we use the Gaussian approximation (53) for the Cramér function, with $\nu = 1/\mathcal{S}''(0)$. The critical α is $\alpha_{\mathrm{crit}} = -\mathcal{S}'(-\bar{\Lambda}) = \bar{\Lambda}/\nu$. The saddle point is positive for $\alpha < \bar{\Lambda}/\nu$, so from (58) we get

$$\gamma_\alpha = \begin{cases} \alpha\left(\bar{\Lambda} - \tfrac{1}{2}\alpha\nu\right), & \alpha < \bar{\Lambda}/\nu; \\ \bar{\Lambda}^2/2\nu, & \alpha \ge \bar{\Lambda}/\nu. \end{cases} \tag{59}$$

This is plotted in Fig. 4. Notice that the solid curve (for a random flow) lies below the dashed line (for a nonrandom flow). This is a general result: if $f(x)$ is a convex function and x a random variable, *Jensen's inequality* says that

$$\overline{f(x)} \ge f(\overline{x}). \tag{60}$$

Now, $\mathrm{e}^{-\alpha t \Lambda}$ is a convex function of Λ, so we have

$$\overline{\mathrm{e}^{-\alpha t \Lambda}} \ge \mathrm{e}^{-\alpha t \bar{\Lambda}},$$

which means that the rate of decay satisfies

$$\gamma_\alpha \le \alpha\,\bar{\Lambda},$$

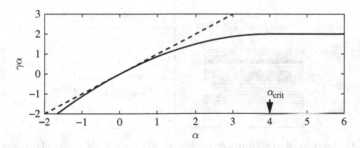

Fig. 4. Decay rate (59) for the concentration of a blob in a Gaussian random stretching flow (*solid curve*). The *dashed line* is for a fixed, nonrandom flow as in Sect. 2.2. Here $\bar{\Lambda} = 1$, $\nu = 1/4$, so $\alpha_{\mathrm{crit}} = \bar{\Lambda}/\nu = 4$

which is exactly what is seen in Fig. 4. Thus, *fluctuations in* Λ *inevitably lead to a slower decay rate* γ_α.

Stronger fluctuations also mean that the decay rate γ_α saturates more quickly with α. Clearly, in the absence of fluctuations we recover the nonrandom result: $\bar{\Lambda}/\nu$ is infinite and only the $\alpha < \bar{\Lambda}/\nu$ case is needed in (59). If there are lots of fluctuations, $\bar{\Lambda}/\nu$ is small, and there is a greater probability of obtaining a realisation with no stretching. For large enough fluctuations this exponentially decreasing probability dominates, and we obtain the second case in (58).

3.2 Many Blobs

In Sect. 3.1 we considered the evolution of the concentration of a single blob of concentration in a random straining field. Now we turn our attention to a large number of blobs, homogeneously and isotropically distributed, with random concentrations. We assume that the mean concentration over all the blobs is zero. A simplified view of this initial situation is depicted in Fig. 5a, with shades of grey indicating different concentrations. If we now apply a uniform straining flow of the type (2a) (see Sect. 2.2), the blobs are all stretched horizontally (the x_1 direction) and contracted in the vertical (x_2) direction, as shown in Fig. 5b. They are pressed together in the x_2 direction until diffusion becomes important (Fig. 5c). The effect of diffusion is to homogenise the concentration field until it reaches a value which is the average of the

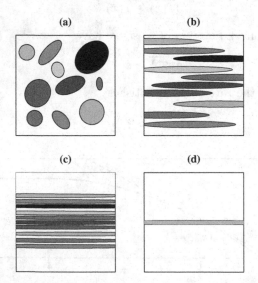

Fig. 5. (a) An initial distribution of blobs with random concentrations (b) is stretched by a constant strain (c) until the blobs reach the diffusive limit in the contracting direction and begin to overlap. (d) Finally, they combine into one very long blob with the average concentration of all the blobs

concentration of the individual blobs. This is depicted by the long grey blob in Fig. 5d, which will itself keep contracting until it reaches the diffusive length ℓ.

Of course, unlike the situation depicted in Fig. 5, here the initial concentration field θ_0 represents the concentration of *all* the homogeneously distributed blobs, so it does not decay at infinity: we must thus use (28) rather than (30). The summing (and hence averaging) over blobs is manifest in (28), which contains an integral over the initial distribution θ_0 in the x_2 direction, windowed by a Gaussian.

In practice, this implies that the expected value of the concentration at a point $\boldsymbol{x}$ on the grey filament is given by

$$\langle \theta(\boldsymbol{x},t)\rangle_{\text{blobs}} \sim \mathrm{e}^{-\Lambda t} \sum_i^N \overline{\theta_0^{(i)}} \longrightarrow 0\,, \tag{61}$$

where $\theta_0^{(i)}$ is the initial concentration of the ith blob, and $\langle \cdot \rangle_{\text{blobs}}$ denotes the expected value of the sum over the overlapping blobs at point $\boldsymbol{x}$ (not the same as spatial integration $\langle \cdot \rangle$). We assume that $N \gg 1$ blobs have overlapped. Equation (61) gives the concentration at a point, summed over N overlapping blobs. Of course, (61) converges to zero for large N, because the blobs average out. Not so for the fluctuations at that point: by the central limit theorem, we have

$$\langle \theta^2(\boldsymbol{x},t)\rangle_{\text{blobs}} \sim \mathrm{e}^{-2\Lambda t} \sum_i^N \overline{\theta_0^{(i)^2}} = N\mathrm{e}^{-2\Lambda t}\, \overline{\theta_0^2}\,, \tag{62}$$

since the initial blobs have identical distributions. The blob-summed fluctuation amplitude $\langle \theta^2 \rangle_{\text{blobs}}$ is thus proportional to the number N of overlapping blobs. But the number of overlapping blobs is proportional to $\mathrm{e}^{\Lambda t}$: as time increases more and more blobs converge to a given $\boldsymbol{x}$ in the contracting direction and overlap diffusively (this can be seen in (28): the width of the windowing region grows as $\mathrm{e}^{\lambda t}$). Assuming the variance of $\theta_0^{(i)}$ is finite, we conclude from (62) that

$$\langle \theta^2(\boldsymbol{x},t)\rangle_{\text{blobs}}^{1/2} \sim \mathrm{e}^{-\Lambda t/2}\,. \tag{63}$$

Compare this to (49) for the single-blob case: the overlap between blobs has led to an extra square root. Thus, the ensemble averages $\overline{\langle \theta^2(\boldsymbol{x},t)\rangle}^{\alpha}_{\text{blobs}}$, for the overlapping blobs are computed exactly as in Sect. 3.1, resulting in (58). Because of the assumption of homogeneity, the point-average is the same as the average over the whole domain (see Sect. 4 for more on this), and we have[3]

$$\langle \theta^2 \rangle^{\alpha} = \overline{\langle \theta^2(\boldsymbol{x},t)\rangle}^{\alpha}_{\text{blobs}} \sim \mathrm{e}^{-\gamma_\alpha t}\,, \tag{64}$$

---

[3] In going from (63) to (64), we've implicitly assumed that the initial concentration field has Gaussian statistics, because we've used the fact that the higher even moments are proportional to powers of the second moment.

with γ_α given by (58). (In (64) the angular brackets denote spatial averaging, not spatial integration, because the total variance is infinite in this case.)

3.3 Three Dimensions

In three dimensions, we will only treat case (3a) (a purely straining flow) of Sect. 2.4. For $\lambda_2 < 0$, where the asymptotic concentration is given by (47) (ropes), the situation is basically identical to the 2D case of Sects. 3.1 and 3.2: the statistics of the stretching direction λ_1 determine γ_α from (58). The contracting directions x_2 and x_3 are stabilised by diffusion.

For $\lambda_2 \geq 0$, the asymptotic concentration is given by (46) (pancakes). We have two fluctuating quantities to worry about (λ_1 and λ_2). But since the decay rate in (46) only depends on λ_3, we can instead focus on its fluctuations only. For a single blob, the average (50) is then replaced by

$$\overline{\theta^\alpha} \sim \int_0^\infty e^{-\alpha|\Lambda_3|t}\, P_3(|\Lambda_3|, t)\, d|\Lambda_3| \sim e^{-\gamma_\alpha t}, \tag{65}$$

where of course Λ_3 is the average of λ_3. This PDF achieves a distribution of the large-deviation form (52). The analysis thus follows exactly as in Sects. 3.1 and 3.2, except the Cramér function for $|\Lambda_3|$ must be used.[4]

4 Practical Considerations

One may rightly wonder if blobs in, a random uniform straining flow, as depicted in Sect. 3, bear any resemblance to reality. The single-blob scenario doesn't, but the many-blobs scenario has a fighting chance, as we will try to justify here. There are two important considerations: where does the ensemble-averaging come from, and what are the stretching rates given by?

The decay rate (58) depends crucially on ensemble-averaging: with that averaging the decay rate fluctuates wildly for a given realisation. At the end of Sect. 3.2 we assumed that homogeneity allowed us to generalise from the average at a point to the average over the whole domain. But the average over the whole domain can actually do a lot more for us: it can provide the ensemble of blobs that we need for averaging! Thus, we can forget about speaking of realisations as if we were running many parallel experiments and instead speak of the moments of the concentration field as given by an average over randomly distributed blobs. The decay rate will then be naturally smoothed out over blobs experiencing different stretching histories. The saturation of the decay rate with α in (58) is due to $\overline{\theta^{2\alpha}}$ being dominated by the fraction of blobs that have experienced no stretching.

What about the stretching rates λ? Luckily, it is not them but their time-average Λ that matters. If we imagine following a blob as it moves through

---

[4] There are a few exceptional cases to consider [15].

the flow, we can see that this time-averaged stretching rate is nothing but the finite-time Lyapunov exponent associated with this blob and its particular initial condition. A given blob will be constantly reoriented as it moves along in the flow, so its finite-time Lyapunov exponent is not just the average of the stretching rates (in fact, it must be strictly less than this average). But in a chaotic system we are guaranteed that, on average, these reorientations do not lead to a vanishing (infinite-time) Lyapunov exponent. This is guaranteed by the celebrated Oseledec multiplicative theorem for random matrices [18].[5] We may thus use for $P(\Lambda, t)$ the distribution of finite-time Lyapunov exponents, which is well known to have the large-deviation form (52) [19].

The result of these considerations is the *local theory* of passive scalar decay. It is called local because of the reliance of such a local concept as the finite-time Lyapunov exponents, which come from a linearisation near fluid element trajectories. In Sect. 4.1 we discuss a specific example. We postpone a discussion of the validity of the local theory until Sect. 4.2, but for now we point out that it is known to be exact at least in some simple model flows [15, 20].

The derivation presented in this section was based on the work of Balkovsky and Fouxon [15], who used a slightly more rigorous approach. Son [21] also obtained the decay rate (58) using path-integral methods. Earlier, Antonsen et al. [8] derived the decay rate for the second moment $\langle \theta^2 \rangle$ in terms of the Cramér function, using a different (and not quite equivalent) approach, though they did not allow for the second case in (58).

4.1 An Example: Flow in a Microchannel

We illustrate how to compute the decay rates γ_α with a practical problem. Specifically, we will use a 3D model of a microchannel. The system is shown in Fig. 6. It consists of a narrow channel, roughly $100\,\mu m$ wide and slightly shallower. These types of channel are widely used in microfluidics applications ("lab-on-a-chip"), and often one wants to achieve good mixing in the lateral cross-section of the channel. This is difficult, since the Reynolds number of the flow varies between 0.1 and 100—far from turbulent. Clever techniques have to be used to induce chaotic motion of the fluid particle trajectories in order to enhance mixing. Stroock et al. [22] used patterned grooves at the bottom of the channel to induce vortical motions and found that the mixing efficiency was dramatically increased. Here we use a variation on this where the bottom is patterned with an electro-osmotic coating, which induces fluid motion near the wall [23]. The effect of the electro-osmotic coating is well approximated by a moving wall boundary condition. The pattern is chosen in a so-called herringbone pattern to maximise the mixing efficiency (though not in a staggered herringbone, which is even better but is more difficult to model). Rather than solving the full equations numerically, we adopt here an analytical model based on Stokes flow in a shallow layer [24]. The longitudinal

---

[5] The reorientations also tend to decrease the correlation time τ [15].

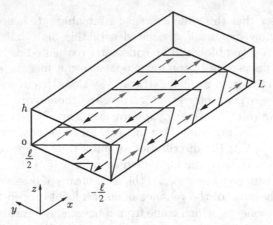

Fig. 6. Microchannel with a periodic patterned electro-osmotic potential at the bottom. The *arrows* indicate the direction of fluid motion at the bottom. The width of the channel is about $100\,\mu m$ and its height $10\text{--}50\,\mu m$, and the period of the pattern is L. A typical mean fluid velocity is $10^2\text{--}10^3\,\mu m/s$

(x) direction is taken to be periodic. The flow is steady, but because it is three dimensional it can still exhibit chaos.

Figure 7 shows two Poincaré sections for the flow. These are taken at two constant x planes, one at $x = 0$ and the other at the midpoint of the x-periodic pattern. The two shades represent two trajectories that have periodically punctured those planes many times over. It is clear from the figures that the flow contains large chaotic regions, as well as smaller regular regions (known as islands). We focus here on the chaotic regions.

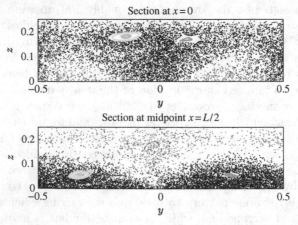

Fig. 7. Poincare sections for the microchannel. The dark dots represent the same trajectory periodically puncturing two vertical planes many times over (with and against the prevailing flow direction). The grey dots show two trajectories in regular, nonmixing regions

Now that we have established (or at least strongly suspect) the existence of chaotic regions, we can compute the distribution of finite-time Lyapunov exponents. There are many ways of doing this: because we are not interested in extremely long times, the most direct route may be used. We have an analytical form for the velocity field, so the velocity gradient matrix is easily computed. This allows us to linearise about trajectories in the standard manner [19, 25]. Each trajectory will thus have a finite-time Lyapunov exponent associated with it, which shows the tendency of infinitesimally close trajectories to diverge exponentially. This is then repeated over many different trajectories within the same chaotic region, and a histogram is made of the finite-time Lyapunov exponents. This histogram changes with time, as shown in Fig. 8. For these relatively early times, it changes dramatically and does not exhibit a self-similar form.

The evolution of the mean and standard deviation of the distribution is shown in Fig. 9. The mean is converging to a constant $\bar{\Lambda} \simeq 0.116$, and the standard deviation is decreasing as $\sqrt{\nu/t}$, with $\nu \simeq 0.168$. These facts taken together are strongly indicative that the distribution is converging to a Gaussian

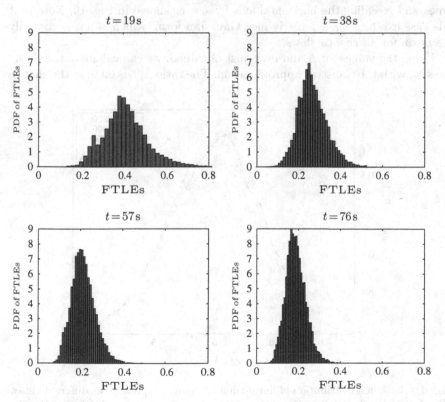

Fig. 8. Evolution of the distribution of finite-time Lyapunov exponents for the microchannel. The average crossing time for particles in the channel is L/U

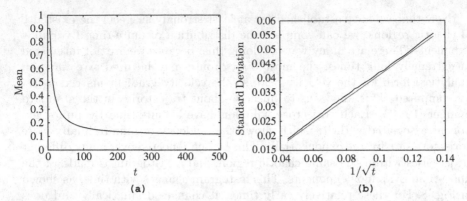

(a)   (b)

Fig. 9. (a) Evolution of the mean $\bar{\Lambda}$ of the distribution of Lyapunov exponents. The mean converges to $\bar{\Lambda} \simeq 0.116$ s^{-1}. (b) Standard deviation of the distribution of Lyapunov exponents versus $1/\sqrt{t}$. The *straight line* represents $\sqrt{\nu/t}$, with $\nu \simeq 0.168$ s^{-1}

of the form (51). This is easily confirmed by plotting the PDFs at different times and rescaling the horizontal axis by $\sqrt{t}$, as shown in Fig. 10. Note that this case exhibits a particularly nice Gaussian form, which is not necessarily the norm for all chaotic flows.

Using the values for $\bar{\Lambda}$ and ν we just obtained, we can calculate the decay rates γ_α with the Gaussian approximation. The ratio $\bar{\Lambda}/\nu$ is 0.69, so the change

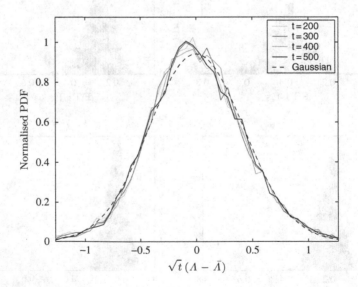

Fig. 10. Rescaled distribution of finite-time Lyapunov exponents at different times. The *dashed line* is the Gaussian form (51), with parameters as in the caption for Fig. 9

in character in (59) occurs at $\alpha \geq 0.69$. Since from (64) the decay of $\langle \theta^2 \rangle^\alpha$ is given by γ_α, this means that moments of order $2\alpha \geq 1.38$ will decay at the same rate. This includes the variance $\langle \theta^2 \rangle$, so we have from (64)

$$\langle \theta^2 \rangle \sim e^{-\gamma_1 t}, \quad \text{with} \quad \gamma_1 = \bar{\Lambda}^2/2\nu \simeq 0.040 \text{ s}^{-1}.$$

The mixing time is thus $\gamma_1^{-1} \simeq 25$ s. This is about a factor of four improvement over the purely diffusive time for, say, DNA molecules ($\kappa \simeq 10^{-10} \text{ m}^2 \text{ s}^{-1}$). This is not spectacular, but can be greatly increased by staggering the herringbone pattern. The mixing time assuming the decay proceeds at the rate of the mean Lyapunov exponent $\bar{\Lambda}$ is roughly 9 m^2 s, so that the fluctuations multiply this by a factor of three!

Of course, we do not know if this is actually a good estimate for the mixing time, since we haven't directly solved the advection–diffusion equation numerically: this is prohibitive in a 3D domain for such a small diffusivity. This is one of the advantages of the local theory: it is usually less expensive to compute the distribution of finite-time Lyapunov exponents than it is to solve the advection–diffusion equation directly. We will say more on the validity of the local theory in Sect. 4.2.

4.2 Limitations of the Local Theory

So is this local theory of mixing correct? Well, certainly not always, even in the Batchelor regime. There are many assumptions underlying the model, some of them difficult to verify. (Do blobs really undergo a series of stretching events as described here? Do correlations between these events matter?) My feeling is that sometimes it will, but most of the time it won't. More experiments and numerical simulations are needed to get to the bottom of this. For a detailed discussion of possible problems with the local theory, see Fereday and Haynes [26]. They make a good case that the theory must break down for long times: the blobs discussed here meet the boundaries of the fluid domain and must begin to fold. The folding forces them to interact with themselves in a correlated fashion. We enter the regime of the *strange eigenmode* [27], which has received a lot of attention lately [14, 26, 28, 29, 30, 31, 32, 33, 34, 35]. Maybe we'll hear more about that in 10 years!

Acknowledgments

I thank Martin Ewart, Bill Young, Andy Thompson, and Emmanuelle Gouillart for their valuable comments.

Appendix 1 The Advection–Diffusion Equation in a Comoving Frame

We start from the advection–diffusion equation (1) and derive its form (8) for a linearised velocity field. We want to transform from the fixed spatial coordinates x to coordinates r measured from a reference fluid trajectory $x_0(t)$. The coordinates r are not quite material (Lagrangian) coordinates, since we follow the trajectory of only one fluid element.

We thus let

$$x = x_0(t) + r, \qquad \frac{dx_0(t)}{dt} = v(x_0(t), t), \tag{66}$$

and write the concentration field as

$$\theta(x, t) = \tilde{\theta}(r, t).$$

The time derivative of θ can be written as

$$\left.\frac{\partial}{\partial t}\right|_x \theta(x, t) = \left.\frac{\partial}{\partial t}\right|_r \tilde{\theta}(r, t) + \nabla_r\tilde{\theta} \cdot \left.\frac{\partial r}{\partial t}\right|_x, \tag{67}$$

where $\partial/\partial t|_x$ denotes a derivative with x held constant. Now from (66)

$$\left.\frac{\partial r}{\partial t}\right|_x = -\frac{dx_0}{dt} = -v(x_0(t), t).$$

Spatial derivatives are unchanged by (66): $\nabla_x\theta = \nabla_r\tilde{\theta}$. Hence, inserting (67) into (1), we find

$$\left.\frac{\partial}{\partial t}\right|_r \tilde{\theta} + \{v(x_0(t) + r, t) - v(x_0(t), t)\} \cdot \nabla_r\tilde{\theta} = \kappa \nabla_r^2\tilde{\theta}. \tag{68}$$

We Taylor expand the velocity field in (68) to get

$$\left.\frac{\partial}{\partial t}\right|_r \tilde{\theta} + r \cdot \sigma(t) \cdot \nabla_r\tilde{\theta} = \kappa \nabla_r^2\tilde{\theta}, \qquad \sigma(t) := \nabla v(x_0(t), t), \tag{69}$$

where we neglected terms of order $|r|^2$. This is only valid if the velocity field changes little over the region we consider (i.e. if it is smooth enough), which is true for large Schmidt number. Equation (69) is the same as (8) and tells us how to find $\sigma(t)$.

Appendix 2 Large-Deviation Theory

In this appendix we will justify the large-deviation form of the PDF, (52), assuming little prior knowledge of probability theory.

First, we define the *generating function* $e^{-s(k)}$ of a random variable x by

$$e^{-s(k)} = \int p(x) e^{-ikx} \, dx,$$

that is, the generating function is simply the Fourier transform of the PDF of x. We have $s(0) = 0$, $\bar{x} = -\mathrm{i}\,s'(0)$, and $\overline{x^2} - \bar{x}^2 = s''(0)$. Now define the random variable X to be the mean of n variables:

$$X_n = \frac{1}{n} \sum_{i=1}^{n} x_i$$

where the x_i are independent and identically distributed with PDF $p(x_i) = p(x)$. How do we find the PDF $P(X_n)$ of X_n, in the limit where n is large? First observe that (from here on we drop the subscript on X_n)

$$P(X) = \int p(x_1) \cdots p(x_n) \, \delta\left(\frac{x_1 + \cdots + x_n}{n} - X\right) \mathrm{d}x_1 \cdots \mathrm{d}x_n$$

since the joint PDF $p(x_1, \ldots, x_n) = p(x_1) \cdots p(x_n)$ by independence of the x_i. The generating function $\mathrm{e}^{-S(k)}$ for $P(X)$ is then

$$\mathrm{e}^{-S(k)} = \int P(X) \mathrm{e}^{-\mathrm{i}kX} \, \mathrm{d}X$$

$$= \int p(x_1) \cdots p(x_n) \delta\left(\frac{x_1 + \cdots + x_n}{n} - X\right) \mathrm{e}^{-\mathrm{i}kX} \, \mathrm{d}x_1 \cdots \mathrm{d}x_n \, \mathrm{d}X.$$

We do the X integral, and then observe that we get a product of n identical x_i integrals, each of which is equal to $\mathrm{e}^{-s(k/n)}$:

$$\mathrm{e}^{-S(k)} = \int p(x_1) \cdots p(x_n) \, \mathrm{e}^{-\mathrm{i}(k/n)(x_1 + \cdots + x_n)} \, \mathrm{d}x_1 \cdots \mathrm{d}x_n = \mathrm{e}^{-ns(k/n)}.$$

Thus the generating function for X is the nth power of the generating function for x. We can invert the Fourier transform to find the PDF $P(X)$:

$$P(X) = \frac{1}{2\pi} \int \mathrm{e}^{-S(k)} \, \mathrm{e}^{\mathrm{i}kX} \, \mathrm{d}k = \frac{n}{2\pi} \int \mathrm{e}^{-n(s(K) - \mathrm{i}KX)} \, \mathrm{d}K, \qquad (70)$$

where $K = k/n$. We let

$$H(K, X) = s(K) - \mathrm{i}KX.$$

For large values of n the integral in (70) is dominated by the stationary points $K_{\mathrm{sp}}(X)$ of $H(K, X)$ (saddle-point approximation):

$$K_{\mathrm{sp}}(X) \quad \text{such that} \quad \frac{\partial H}{\partial K}(K_{\mathrm{sp}}, X) = s'(K_{\mathrm{sp}}) - \mathrm{i}X = 0. \qquad (71)$$

(I will often leave out the X dependence of $K_{\mathrm{sp}}(X)$ to shorten the expressions.) In that case we can approximate the integrand in (70) using

$$H(K, X) = H(K_{\mathrm{sp}}, X) + \tfrac{1}{2} s''(K_{\mathrm{sp}})(K - K_{\mathrm{sp}})^2 + \mathcal{O}\left((K - K_{\mathrm{sp}})^3\right),$$

which allows us to do the integral explicitly:

$$P(X) = \sqrt{\frac{n}{2\pi \, s''(K_{\mathrm{sp}}(X))}} \; \mathrm{e}^{-nH(K_{\mathrm{sp}}(X),X)}, \tag{72}$$

where $K_{\mathrm{sp}}(X)$ is given by (71). As a final step, let us calculate the mean of X using this PDF:

$$\overline{X} = \int X P(X)\,\mathrm{d}X = \int X \sqrt{\frac{n}{2\pi \, s''(K_{\mathrm{sp}}(X))}} \; \mathrm{e}^{-nH(K_{\mathrm{sp}}(X),X)}\,\mathrm{d}X. \tag{73}$$

Again, for large n we can use the saddle-point method to evaluate this integral. The important observation is that the saddle point X_0 of $H_{\mathrm{sp}}(K(X),X)$ satisfies

$$\frac{\mathrm{d}H}{\mathrm{d}X}(K_{\mathrm{sp}}(X_0),X_0) = \frac{\partial H}{\partial K}(K_{\mathrm{sp}}(X_0),X_0)\,\frac{\mathrm{d}K_{\mathrm{sp}}}{\mathrm{d}X}(X_0) - \mathrm{i}\,K_{\mathrm{sp}}(X_0) = 0.$$

The $\partial H/\partial K$ term vanishes because it is evaluated at K_{sp}; hence, $K_{\mathrm{sp}}(X_0) = 0$, which implies $H(K_{\mathrm{sp}}(X_0),X_0) = 0$. Inserting this into the integral (73), we find $\overline{X} = X_0$: *the mean of X and the minimum of H coincide*. This means that it makes sense to define

$$\mathcal{S}(X - \overline{X}) := H(K_{\mathrm{sp}}(X),X), \quad \text{with } \mathcal{S}(0) = 0 \text{ and } \mathcal{S}'(0) = 0, \tag{74}$$

which is the sought-after Cramér function. Note also that $\mathcal{S}''(X - \overline{X}) = 1/s''(K_{\mathrm{sp}}(X))$, and that for large n the nonexponential coefficient in (72) can thus be approximated by evaluating it at the saddle point $K_{\mathrm{sp}}(\overline{X}) = 0$, with $s''(0) = 1/\mathcal{S}''(0)$. The final form of our large-deviation result is thus

$$P(X) = \sqrt{\frac{n\,\mathcal{S}''(0)}{2\pi}} \; \mathrm{e}^{-n\mathcal{S}(X-\overline{X})}, \tag{75}$$

which is the same as (52).

As a simple example (treated in every textbook, see for example [17]), consider a random variable x with PDF

$$p(x) = (1 - \varepsilon)\,\delta(x - x_+) + \varepsilon\,\delta(x - x_-), \tag{76}$$

where $x_+ > x_-$ are constants—this is a *binomial distribution* (or Bernoulli distribution in this case). If we take the mean X of n such variables, what is the PDF of X for large n? First, we compute the generating function for x:

$$\mathrm{e}^{-s(k)} = \int \{(1 - \varepsilon)\,\delta(x - x_+) + \varepsilon\,\delta(x - x_-)\}\,\mathrm{e}^{-\mathrm{i}\,kx}\,\mathrm{d}x$$

$$= (1 - \varepsilon)\,\mathrm{e}^{-\mathrm{i}\,kx_+} + \varepsilon\,\mathrm{e}^{-\mathrm{i}\,kx_-}. \tag{77}$$

We take the logarithm to obtain $s(k)$ and find $K_{\mathrm{sp}}(X)$ by solving the saddle-point equation (71):

$$\frac{\partial H}{\partial K} = s'(K_{\rm sp}) - \mathrm{i}\,X = 0 \quad \Longleftrightarrow \quad K_{\rm sp}(X) = \frac{1}{\mathrm{i}\Delta} \log\left(\frac{1-\varepsilon}{\varepsilon}\frac{x_+ - X}{X - x_-}\right),$$

where $\Delta := x_+ - x_-$ and we restrict $x_- \le X \le x_+$. Inserting this into $H(K_{\rm sp}(X), X)$, we find from (74)

$$\mathcal{S}(X - \overline{X}) = -\frac{X - x_-}{\Delta}\log\left(\frac{1-\varepsilon}{\varepsilon}\frac{x_+ - X}{X - x_-}\right) + \log\left(\frac{x_+ - X}{\varepsilon\Delta}\right).$$

It is easy to verify that, since $\overline{X} = (1-\varepsilon)x_+ + \varepsilon\,x_-$, we have $\mathcal{S}(0) = \mathcal{S}'(0) = 0$ and $\overline{X^2} - \overline{X}^2 = 1/\mathcal{S}''(0) = \varepsilon(1-\varepsilon)\,\Delta^2$.

The binomial distribution (76) is a useful model of stretching of an infinitesimal line segment by a uniform incompressible straining flow in two dimensions, assuming the straining axes of the flow change direction randomly at regular intervals τ. If we set $x_\pm = \pm\lambda\tau = \pm\nu$, where λ is the strain rate, then X is the averaged logarithm of the length ℓ of the segment, i.e. $\ell = \mathrm{e}^{nX}$. Thus, the mth power of the length of the segment will on average grow as

$$\overline{\ell^m} = \overline{\mathrm{e}^{mnX}} = \mathrm{e}^{-S(\mathrm{i}\,mn)} = \mathrm{e}^{-ns(\mathrm{i}\,m)} = \left\{(1-\varepsilon)\,\mathrm{e}^{\nu m} + \varepsilon\,\mathrm{e}^{-\nu m}\right\}^n. \tag{78}$$

We know that $\overline{\ell^{-2}}$ must be constant in a 2D incompressible flow, so that the term in braces in (78) must be unity. We use this to solve for ε,

$$\varepsilon = (1 + \mathrm{e}^{2\nu})^{-1},$$

which then allows us to use (78) to write the growth rate χ_m of line segments as

$$\chi_m = \frac{1}{\tau\,n}\log\overline{\ell^m} = \frac{1}{\tau}\log\left(\frac{\cosh(m+1)\nu}{\cosh\nu}\right).$$

The Lyapunov exponent, which is given by $\mathrm{d}\chi_m/\mathrm{d}m$ at $m = 0$, has a value of $\lambda\tanh\nu$ for this flow: it is less than for a uniform straining flow because of the time taken for the segment to realign with the new straining axis when its direction changes.

References

1. G. K. Batchelor: Small-scale variation of convected quantities like temperature in turbulent fluid: part 1. General discussion and the case of small conductivity. J. Fluid Mech. **5**, 134 (1959)
2. R. H. Kraichnan: Small-scale structure of a scalar field convected by turbulence. Phys. Fluids **11**, 945 (1968)
3. R. H. Kraichnan: Convection of a passive scalar by a quasi-uniform random straining field. J. Fluid Mech. **64**, 737 (1974)
4. Y. B. Zeldovich, A. A. Ruzmaikin, S. A. Molchanov and D. D. Sokoloff: Kinematic dynamo problem in a linear velocity field. J. Fluid Mech. **144**, 1 (1984)

5. E. Ott and T. M. Antonsen, Jr.: Fractal measures of passively convected vector fields and scalar gradients in chaotic fluid flows. Phys. Rev. A **39**, 3660 (1989)
6. T. M. Antonsen, Jr and E. Ott: Multifractal power spectra of passive scalars convected by chaotic fluid flows. Phys. Rev. A **44**, 851 (1991)
7. T. M. Antonsen, Jr, Z. Fan and E. Ott: k spectrum of passive scalars in Lagrangian chaotic fluid flows. Phys. Rev. Lett. **75**, 1751 (1995)
8. T. M. Antonsen, Jr, Z. Fan, E. Ott and E. Garcia-Lopez: The role of chaotic orbits in the determination of power spectra. Phys. Fluids **8**, 3094 (1996)
9. B. I. Shraiman and E. D. Siggia: Lagrangian path integrals and fluctuations in random flow. Phys. Rev. E **49**, 2912 (1994)
10. M. Chertkov, G. Falkovich I. Kolokolov and V. Lebedev: Statistics of a passive scalar advected by a large-scale two-dimensional velocity field: analytic solution. Phys. Rev. E **51**, 5609 (1995)
11. M. Chertkov, I. Kolokolov and M. Vergassola: Inverse cascade and intermittency of passive scalar in one-dimensional smooth flow. Phys. Rev. E **56**, 5483 (1997)
12. B. I. Shraiman and E. D. Siggia: Scalar turbulence. Nature **405**, 639 (2000)
13. G. Falkovich, K. Gawędzki and M. Vergassola: Particles and fields in turbulence. Rev. Mod. Phys. **73**, 913 (2001)
14. J.-L. Thiffeault: The strange eigenmode in Lagrangian coordinates. Chaos **14**, 531 (2004)
15. E. Balkovsky and A. Fouxon: Universal long-time properties of Lagrangian statistics in the Batchelor regime and their application to the passive scalar problem. Phys. Rev. E **60**, 4164 (1999)
16. R. S. Ellis: Entropy, Large Deviations, and Statistical Mechanics. Springer, New York (1985)
17. A. Schwartz and A. Weiss: Large Deviations for Performance Analysis. Chapman & Hall, London (1995)
18. V. I. Oseledec: A multiplicative theorem: Lyapunov characteristic numbers for dynamical systems. Trans. Moscow Math. Soc. **19**, 197 (1968)
19. E. Ott: Chaos in Dynamical Systems. Cambridge University Press, Cambridge, UK (1994)
20. A. Fouxon: Evolution of a scalar gradient's probability density function in a random flow. Phys. Rev. E **58**, 4019 (1998)
21. D. T. Son: Turbulent decay of a passive scalar in the Batchelor limit: exact results from a quantum-mechanical approach. Phys. Rev. E **59**, R3811 (1999)
22. A. D. Stroock, S. K. W. Dertinger, A. Ajdari, I. Mezić, H. A. Stone and G. M. Whitesides: Chaotic mixer for microchannels. Science **295**, 647 (2002)
23. S. Hong, J.-L. Thiffeault, L. Fréchette and V. Modi: International Mechanical Engineering Congress & Exposition, Washington, DC. American Society of Mechanical Engineers, New York (2003)
24. M. A. Ewart and J.-L. Thiffeault: A simple model for a microchannel mixer. in preparation (2005)
25. J.-P. Eckmann and D. Ruelle: Ergodic theory of chaos and strange attractors. Rev. Mod. Phys. **57**, 617 (1985)
26. D. R. Fereday and P. H. Haynes: Scalar decay in two-dimensional chaotic advection and Batchelor-regime turbulence. Phys. Fluids **16**, 4359 (2004)
27. R. T. Pierrehumbert: Tracer microstructure in the large-eddy dominated regime. Chaos Solitons Fractals **4**, 1091 (1994)
28. D. Rothstein, E. Henry and J. P. Gollub: Persistent patterns in transient chaotic fluid mixing. Nature **401**, 770 (1999)

29. D. R. Fereday, P. H. Haynes, A. Wonhas and J. C. Vassilicos: Scalar variance decay in chaotic advection and Batchelor-regime turbulence. Phys. Rev. E **65**, 035301(R) (2002)

30. J. Sukhatme and R. T. Pierrehumbert: Decay of passive scalars under the action of single scale smooth velocity fields in bounded two-dimensional domains: from non-self-similar probability distribution functions to self-similar eigenmodes. Phys. Rev. E **66**, 056032 (2002)

31. A. Wonhas and J. C. Vassilicos: Mixing in fully chaotic flows. Phys. Rev. E **66**, 051205 (2002)

32. A. Pikovsky and O. Popovych: Persistent patterns in deterministic mixing flows. Europhys. Lett. **61**, 625 (2003)

33. J.-L. Thiffeault and S. Childress: Chaotic mixing in a torus map. Chaos **13**, 502 (2003)

34. W. Liu and G. Haller: Strange eigenmodes and decay of variance in the mixing of diffusive tracers. Physica D **188**, 1 (2004)

35. A. Schekochihin, P. H. Haynes and S. C. Cowley: Diffusion of passive scalar in a finite-scale random flow. Phys. Rev. E **70**, 046304 (2004)

Transport of Inert and Reactive Particles: Lagrangian Statistics in Turbulent Flow

G. Boffetta,[1] A. Mazzino,[2] and A. Vulpiani[3]

[1] Dipartimento di Fisica and INFN, Università di Torino, via P. Giuria 1, 10125, Torino, Italy
boffetta@to.infn.it
[2] Dipartimento di Fisica and INFN, Università di Genova, via Dodecaneso 33, 16146, Genova, Italy
mazzino@fisica.unige.it
[3] Dipartimento di Fisica, Università di Roma "la Sapienza" and Center for Statistical Mechanics and Complexity INFM UdR Roma1 Piazzale Aldo Moro 5, I-00185 Roma, Italy
Angelo.Vulpiani@roma1.infn.it

Abstract. In this contribution we review different aspects of passive transport in fluids. Two classes of problems are considered: inert substances and substances that are chemically (or biologically) reactive. Concerning the first issue we discuss in particular the problem of standard and anomalous asymptotic diffusion for single particle statistics and the problem of relative dispersion of particle pairs in chaotic and turbulent flows. For what concerns the issue of reacting transport we study the dependence of the front speed on the flow characteristics, considering the case of reaction that is slow or fast with respect to the typical time scales of the advection.

1 Introduction

Fluid transport of concentration fields is of major importance in various domains ranging from astrophysics to geophysics and to chemical engineering [1, 2]. In several applications, the feedback of the advected fields on the velocity field can be neglected. This is the case of passive transport which will be considered in the following. Under this simplifying assumption, in a given velocity field the most general equation describing the evolution of the concentrations of N species, $\theta_i(\boldsymbol{x}, t)$, can be written as

$$\partial_t \theta_i + \boldsymbol{u} \cdot \boldsymbol{\nabla} \theta_i = D_i \Delta \theta_i + \frac{1}{\tau_i} f_i(\theta_1, \ldots, \theta_N) , \qquad (1)$$

where on the lhs the second term accounts for the transport by an incompressible velocity field; on the rhs the first term represents molecular diffusion (D_i is

G. Boffetta et al.: *Transport of Inert and Reactive Particles: Lagrangian Statistics in Turbulent Flow*, Lect. Notes Phys. **744**, 37–70 (2008)
DOI 10.1007/978-3-540-75215-8_2

the diffusion constant for the ith specie) and the second one takes into account possible chemical or biological processes (with characteristic time scale τ_i) taking place among the different substances.

In this contribution we shall consider separately the problem of inert transport (i.e., $f_i = 0$) and that of reacting transport.

In the case of inert transport, the equation for $\theta_i(x, t)$ decouples and we can consider the advection–diffusion for a single field θ:

$$\partial_t \theta + u \cdot \nabla \theta = D \Delta \theta, \qquad (2)$$

where D is the molecular diffusivity; possibly a source term, which represents the injection of scalar fluctuations, may be added in the rhs. Given the field u the main goal is to understand the dynamical and statistical properties of the field θ. Remarkably, in the last few years, much progress has been reached in this direction and we have now a satisfactory understanding of the statistics of passive fields in terms of the motion of advected passive particles. The interested reader may consult the recent review [3] where an exhaustive discussion on passive fields in turbulent flows can be found.

The problem of transport can be recast in terms of particles motion: indeed (2) is nothing but the Fokker–Planck equation of the stochastic process describing the motion of test particles:

$$\dot{x}(t) = u(x(t), t) + \sqrt{2D}\eta(t) \qquad (3)$$

where $u(x(t), t)$ is the Eulerian velocity field at the particle position $x(t)$, and η is a Gaussian white noise with zero mean and $\langle \eta_i(t)\eta_j(t') \rangle = \delta_{ij}\delta(t - t')$.

After the seminal works of Arnold and Aref it is now well recognized that particle motion can be highly non-trivial even in simple laminar velocity fields due to the so-called Lagrangian chaos [4]. Therefore already the single particle motion presents very interesting features. For instance, the dispersion properties are greatly enhanced by the combined effects of the molecular diffusivity and the advection by the velocity field [1, 4]. Indeed at very large times and scales (with respect to the typical time and length scales of u), the test particle undergoes a Brownian process with an enhanced diffusion coefficient [5], i.e., $\langle (x_i(t) - x_i(0))^2 \rangle \simeq 2\,D_{ii}^{\mathrm{E}}\,t$ where the eddy diffusion coefficient $D_{ii}^{\mathrm{E}} > D$ contains the effect of the velocity field. In terms of the field θ this means that the coarse-grained concentration $\langle \theta \rangle$ (where the average is over a volume of linear dimension larger than the typical velocity length scale) obeys the Fick equation:

$$\partial_t \langle \theta \rangle = D_{ij}^{\mathrm{E}} \partial^2_{x_i x_j} \langle \theta \rangle \qquad i, j = 1, \ldots, d, \qquad (4)$$

where d is the space dimension. To compute D^{E} given the velocity field there are now well-established techniques (see [5, 6]).

Moreover, it is also well known that under certain conditions anomalous diffusion may take place, i.e., $\langle (x(t) - x(0))^2 \rangle \sim t^{2\nu}$ with $\nu \neq 1/2$ [2]. In the following we discuss in detail the necessary conditions to observe anomalous diffusion. In particular, we consider the case of incompressible velocity fields where either standard diffusion ($\nu = 1/2$) or superdiffusion ($\nu > 1/2$) may appear [5].

Though interesting and relevant to many problems, the single particle motion is essentially determined by the large-scale properties of the velocity field, which often hide much more interesting phenomena taking place at small scales. Moreover, the standard or anomalous diffusive properties are asymptotic features that in realistic situations may not be attained. In this sense it is more interesting to study the relative motion of two particles. This is indeed characterized by a variety of behaviors in dependence on the statistics of the velocity field at different length scales. For instance, at very small scales (where the velocity field is smooth) particles separate exponentially, i.e., $\langle \ln |\boldsymbol{x}_2(t) - \boldsymbol{x}_1(t)| \rangle \simeq \lambda t + \text{const}$, where λ is the Lyapunov exponent. On the other hand, in the inertial range of fully developed turbulent flows particles separate as $\langle |\boldsymbol{x}_2(t) - \boldsymbol{x}_1(t)|^2 \rangle \sim t^3$, i.e., the Richardson dispersion law holds, in spite of the diffusive single particle behavior at large scales. Since, as shown in the above example, the relative dispersion properties depend on the behavior of the velocity field at different scales, we introduce a scale-dependent description of particle pairs separation in term of the finite size Lyapunov exponent [7]. This kind of description is also very useful to account for the non-asymptotic properties of dispersion and, e.g., to describe transport in finite systems where finite size effects are present.

The problem of reacting transport is much more difficult, because the presence of the production term $f_i(\theta_1, \ldots, \theta_N)$ in the transport equation (1) makes it to be non-linear, and new phenomena appear. Here we consider the simplest non-trivial case of (1), i.e., a unique scalar field $\theta(\boldsymbol{x}, t)$ evolving according to the advection–diffusion–reaction equation

$$\partial_t \theta + \boldsymbol{u} \cdot \boldsymbol{\nabla} \theta = D\Delta\theta + \frac{1}{\tau} f(\theta), \qquad (5)$$

where θ represents the fractional concentration of the reacting substance with the following glossary: $\theta = 0$ indicates fresh material which has still to react, $0 < \theta < 1$ means coexistence of fresh material and products and $\theta = 1$ means that the reaction is over [8]; equivalently one may think of θ as the concentration of a biological organism which is transported by the flow and grows according to the dynamics $f(\theta)$ [9]. Although very generic forms can be considered for $f(\theta)$ [8, 9] here we mainly discuss the case of $f(\theta)$ which are convex functions ($f''(\theta) < 0$) with $f(0) = f(1) = 0$ and $f'(0) = 1$. A typical example is $f(\theta) = \theta(1 - \theta)$. This class of production terms belongs to the so-called Fisher-Kolmogorov-Petrovsky-Piscounoff (FKPP) type [10, 11]. With this choice $\theta = 0, 1$ are the unstable and stable steady states of the dynamics, respectively. Once at the initial time a small portion of the system is such that $\theta \neq 0$ the reaction starts and a front connecting the unstable and stable states propagates.

In the absence of stirring, $\boldsymbol{u} = 0$, it is known that (5), for FKPP non-linearity, generates a front propagating, e.g., from left to right with asymptotic speed $v_0 = 2\sqrt{D/\tau}$, and the thickness of the reaction region is $\xi = 8\sqrt{D\tau}$ [10, 11]. It is worth recalling that the problem of front propagation has been extensively studied in many different fields [12, 13] such as chemical reaction fronts [8], flames propagation in gases [9] and population dynamics of biological communities [12, 13]. In many of these systems the reaction takes place in moving media, i.e., fluids, so that it is important to understand how the presence of a flow modifies the propagation properties.

As a generic feature, in the presence of a non-zero velocity field the front propagates with an average speed v_f greater than v_0 [14, 15, 16]. However, if $f(\theta)$ is not convex, under certain circumstances, the flow may stop ("quench") the reaction [17].

The front velocity v_f is the result of the interplay among the flow characteristics (i.e., intensity U and length scale L), the diffusivity D and the production time scale τ. Here our major concern will be to discuss the dependence of v_f on such quantities. For instance, introducing the Damköhler number $Da = L/(U\tau)$ (the ratio of advective to reactive time scales) and the Péclet number $Pe = UL/D$ (the ratio of diffusive to advective time scales), one can seek for an expression of the front speed as a dimensionless function $v_f/v_0 = \phi(Da, Pe) \geq 1$. In particular, we study the case of cellular flows. We will see that a crucial role in determining $\phi(Da, Pe)$ is played by the renormalization of the diffusion coefficient and chemical time scale induced by the advection [14]. Moreover, we consider an important limit case, i.e., the so-called geometrical optics limit, which is realized for $(D, \tau) \to 0$ maintaining D/τ constant [18]. In this limit one has a non-zero bare front speed, v_0, while the front thickness ξ goes to zero, i.e., the front is sharp. Physically speaking, this limit corresponds to situations in which ξ is very small compared with the other length scales of the problem. Also in this case we provide a simple prediction for the front speed, which turns out to be expressible as a dimensionless function $v_f/v_0 = \psi(U/v_0)$.

Other interesting questions concern the modification of the front geometry as a consequence of advection. In particular, one may ask if the presence of Lagrangian chaos has a role in front dynamics. We shall briefly discuss this problem in the framework of the geometrical optics limit.

The chapter is organized as follows. In Sect. 2 we discuss statistics of single particle and two particles transported by laminar and turbulent flows. Emphasis is put on the conditions for anomalous diffusion and on the non-asymptotic properties of particles pairs separation. Section 3 is devoted to the study of front propagation in fluid flows. After a brief discussion on some general results which do not depend on the specific properties of the velocity field we shall analyze in detail the case of cellular flows in different regimes.

2 Transport of Inert Substances

As stated in the introduction the dynamical and statistical properties of advected passive fields are tightly related to those of test particles. Therefore, from the study of particle motion one can predict many aspects of the dynamics of advected scalar fields.

In particular, the small-scale features of the scalar field can be understood studying the relative motion of test particles, as the following discussion will clarify. Consider for instance the transport equation for θ:

$$\partial_t \theta + \boldsymbol{u} \cdot \nabla \theta = D\Delta\theta + \Phi, \tag{6}$$

where we added an external source of tracer fluctuations, Φ, which acts at a given length scale L_Φ. The link with particle trajectories is evident by solving (6) with the method of characteristics:

$$\begin{aligned}
\theta(\boldsymbol{x}, t) &= \int_{-\infty}^{t} \mathrm{d}s\, \Phi(\boldsymbol{x}(s;t), s), \\
\dot{\boldsymbol{x}}(s;t) &= \boldsymbol{u}(\boldsymbol{x}(s;t), s) + \sqrt{2D}\,\boldsymbol{\eta}(s), \qquad \boldsymbol{x}(t;t) = \boldsymbol{x};
\end{aligned} \tag{7}$$

the second equation is nothing but (3) where we explicitly fixed the final position to be $\boldsymbol{x}$. By using (7) one can then connect the statistics of particle trajectories to the correlation functions of the scalar field. For instance, let us consider the simultaneous two-point correlations:

$$\langle \theta(\boldsymbol{x}_1, t)\theta(\boldsymbol{x}_2, t) \rangle = \int_{-\infty}^{t} \mathrm{d}s_1 \int_{-\infty}^{t} \mathrm{d}s_2\, \langle \Phi(\boldsymbol{x}_1(s_1;t), s_1)\Phi(\boldsymbol{x}_2(s_2;t), s_2) \rangle, \tag{8}$$

with $\boldsymbol{x}_1(t;t) = \boldsymbol{x}_1$ and $\boldsymbol{x}_2(t;t) = \boldsymbol{x}_2$. With a convenient choice of the correlation function of the forcing, e.g., $\langle \Phi(\boldsymbol{x}_1, t_1)\Phi(\boldsymbol{x}_2, t_2) \rangle = \chi(|\boldsymbol{x}_1 - \boldsymbol{x}_2|)\delta(t_1 - t_2)$, and exploiting space homogeneity, (8) can be further simplified to the form

$$C_2(R) = \langle \theta(\boldsymbol{x}, t)\theta(\boldsymbol{x} + \boldsymbol{R}, t) \rangle = \int_{-\infty}^{t} \mathrm{d}s \int \mathrm{d}\boldsymbol{r}\, \chi(\boldsymbol{r})\, p(\boldsymbol{r}, s|\boldsymbol{R}, t), \tag{9}$$

where $p(\boldsymbol{r}, s|\boldsymbol{R}, t)$ is the probability density function for a pair to be at a separation $\boldsymbol{r}$ at time s, under the condition to have a separation $\boldsymbol{R}$ at time t. It is now clear that the knowledge of $p(\boldsymbol{r}, s|\boldsymbol{R}, t)$ allows to predict the behavior of C_2 and so of the scalar spectrum.

Moreover, assuming that $\chi(r)$ drops to zero for $r > L_f$ and $\chi(0) = \chi_0$, (9) may be approximated as $C_2(R) \approx \chi_0 T(R; L_f)$, where $T(R; L_f)$ is the average time the particles took to reach a separation $O(L_f)$ starting from a separation R (while going backward in time). This also calls for studying relative dispersion in terms of the time $T(R_1; R_2)$ needed to reach a separation R_2 being started from R_1. As we shall see, this is at the core of our approach to relative dispersion.

Similar reasonings can be extended to correlations functions involving more than two points. This leads to consider the relative motion of more than two particles, which may be highly non-trivial [19]. A detailed discussion of these aspects is beyond our aims, therefore we limit our discussion to single and two-particles properties. The interested reader may consult [3] for a discussion on the aspects of more than two particles and their consequences on the scalar field properties.

2.1 Standard and Anomalous Diffusion

Investigating the diffusive properties of single particle motion allows to predict the characteristics of the macroscopic motion of concentration fields (cf. (4)).

In this framework it is important to identify the conditions which may lead to anomalous diffusion that brings as a consequence the failure of the Fickian description of transport. Under these circumstances (4) does not hold anymore. From (3) it is easy to obtain the following relation [20]:

$$\langle (x_i(t) - x_i(0))^2 \rangle = \int_0^t dt_1 \int_0^t dt_2 \langle v_i(\boldsymbol{x}(t_1)) \, v_i(\boldsymbol{x}(t_2)) \rangle \simeq 2\,t \int_0^t d\tau \, C_{ii}(\tau) \,, \tag{10}$$

where

$$C_{ij}(\tau) = \langle \, v_i(\boldsymbol{x}(\tau)) \, v_j(\boldsymbol{x}(0)) \, \rangle \tag{11}$$

is the correlation function of the Lagrangian velocity, $\boldsymbol{v} = \dot{\boldsymbol{x}}$.

From (10) it is not difficult to understand that anomalous diffusion can occur only when one or both of the following conditions are violated:

1. Finite variance of the velocity: $\langle v^2 \rangle < \infty$.
2. Fast enough decay of the auto-correlation function of Lagrangian velocities: $\int_0^t d\tau \, C_{ii}(\tau) < \infty$.

If both $\langle v^2 \rangle < \infty$ and $\int_0^t d\tau \, C_{ii}(\tau) < \infty$ then one has standard diffusion and the effective diffusion coefficients are

$$D_{ii}^{\mathrm{E}} = \lim_{t \to \infty} \frac{1}{2\,t} \langle (x_i(t) - x_i(0))^2 \rangle = \int_0^\infty d\tau \, C_{ii}(\tau) \,. \tag{12}$$

Let us now examine two examples in which the above conditions are violated and anomalous diffusion takes place. It is worth remarking that here we use the term anomalous diffusion to indicate a non-standard diffusion in the asymptotic regime. Sometimes in the literature the term anomalous is used also for long (but non-asymptotic) transient behaviors.

Violation of point 1 can be obtained in the so-called Lévy flight model [21]. The simplest instance is the discrete (in time) one-dimensional case, where the particle position $x(t+1)$ at the time $t+1$ is obtained from $x(t)$ as follows:

$$x(t+1) = x(t) + U(t) \,, \tag{13}$$

and $U(t)$s are independent variables identically distributed according to a α-Lévy-stable distribution, $P_\alpha(U)$, i.e.,

$$\int dU e^{ikU} P_\alpha(U) \propto e^{-c|k|^\alpha} \quad \text{and} \quad P_\alpha(U) \sim U^{-(1+\alpha)} \quad \text{for} \quad |U| \gg 1 \tag{14}$$

with $0 < \alpha \le 2$. An easy computation gives $\langle x(t)^q \rangle = C_q \, t^{q/\alpha}$ if $q < \alpha$, and $\langle x(t)^q \rangle = \infty$ if $q \ge \alpha$. Though $\langle x^2 \rangle = \infty$ for any $\alpha < 2$, one can consider the

Lévy flight as a sort of anomalous diffusion in the sense that $x_{\text{typical}} \sim t^{1/\alpha} \gg t^{1/2}$.

Physically more interesting is the Lévy walk model [22] that is still described by (13) but now $U(t)$ is a random variable with finite variance but non-trivial time correlations, so that point 2 is violated. Let us assume that $U(t)$ can assume the values $\pm u_0$ and maintains its value for a duration T which is a random variable with probability density $\psi(T)$. The origin of the possible anomaly is transferred to the correlation function of the Lagrangian velocity: the idea is that one has to generate a correlation such that $C_{ii}(\tau) \sim \tau^{-\beta}$ with $\beta < 1$. By taking $\psi(T) \sim T^{-(\alpha+1)}$ standard diffusion is realized for $\alpha > 2$, while anomalous (super) diffusion takes place for $\alpha < 2$:

$$\langle x(t)^2 \rangle \sim t^{2\nu} \quad \nu = \begin{cases} 1/2 & \alpha > 2 \\ (3-\alpha)/2 & 1 < \alpha < 2 \\ 1 & \alpha < 1 \end{cases}. \tag{15}$$

Besides the above simplified models, more interesting is the understanding of the anomalous diffusion in incompressible velocity fields or deterministic maps. In this direction Avellaneda, Majda and Vergassola [23, 24] obtained a very important and general result about the asymptotic diffusion in an incompressible velocity field $\boldsymbol{u}(\boldsymbol{x})$. If the molecular diffusivity D is non-zero and the infrared contribution to the velocity field is weak enough, namely

$$\int \mathrm{d}\boldsymbol{k} \, \frac{\langle |\, \hat{\boldsymbol{u}}(\boldsymbol{k}) \,|^2 \rangle}{k^2} < \infty, \tag{16}$$

then one has standard diffusion, i.e., the effective diffusion coefficients D_{ii}^{E}s in (12) are finite. The average $\langle \cdot \rangle$ indicates the time average and $\hat{\boldsymbol{u}}$ is the Fourier transform of the velocity. Then there are two possible causes for the superdiffusion:

1. $D > 0$ and, in order to violate (16), $\boldsymbol{u}$ with strong spatial correlation.
2. $D = 0$ and strong correlation between $\boldsymbol{v}(\boldsymbol{x}(t))$ and $\boldsymbol{v}(\boldsymbol{x}(t+\tau))$ at large τ.

One of the few non-trivial systems for which the presence of anomalous diffusion can be proved rigorously is the $2d$ random shear flow $\boldsymbol{u} = (u(y)\,,\,0)$ where $u(y)$ is a random function [6] such that

$$u(y) = \int_{-\infty}^{\infty} \mathrm{d}k \, \mathrm{e}^{\mathrm{i}ky} \, \hat{u}(k) \quad \langle \hat{u}(k) \, \hat{u}(k') \rangle = S(k) \, \delta(k - k'), \tag{17}$$

$S(k)$ is the power spectrum and the average $\langle \cdot \rangle$ is taken over the field realizations. Matheron and De Marsily [25] showed that the anomalous diffusion in the x-direction occurs if $\int \mathrm{d}k \, S(k)k^{-2} = \infty$. On the contrary, if this integral is finite one has standard diffusion and with an effective diffusivity $D_{11}^{\mathrm{E}} \gg D$. Consider now a spectrum such as $S(k) \sim k^{\zeta}$ for $k \mapsto 0$; it is easy to realize that if $\zeta > 1$ standard diffusion takes place, while if $-1 \leq \zeta \leq 1$ one has a superdiffusion [5]

$$\langle \, | \, x(t) - x(0) \, |^2 \, \rangle \sim t^{2\nu} \quad \nu = \frac{3 - \zeta}{4} \geq \frac{1}{2} \,. \tag{18}$$

The condition for the anomalous diffusion $\int dk \, S(k) \, k^{-2} = \infty$ has the following physical interpretation. Dimensionally $\int dk \, S(k) \, k^{-2} \sim \langle u^2 \rangle \, L^2$ where L is the typical length of the function $u(y)$, i.e., the typical distance between two sequent zeros of $u(y)$. If $\langle u^2 \rangle < \infty$ and $\int dk \, S(k) \, k^{-2} < \infty$ the diffusion process is basically similar to that one characterized by a velocity field given by a sequence of strips of size L and velocity $\pm\sqrt{\langle u^2 \rangle}$, i.e., the transversal Taylor diffusion in channels [26]. The origin of the anomalous diffusion is then due to the fact that a test particle travels in a given direction for a very long time before changing direction and so on.

2.2 Asymptotic Methods to Compute Eddy-Diffusivities

In the previous section we have already seen both the conditions under which standard diffusion is expected to occur and, if any, how to compute eddy diffusivities in terms of Lagrangian trajectories (see (12)). The aim of this section is to show how to arrive at the eddy diffusivities exploiting the Eulerian description. The main ingredients to achieve such a goal is the so-called multiscale expansion [27, 28]. The derivations we are reporting below follow from [5]. Generalizations can be found in [29, 30].

Following [5], let $u(x,t)$ in (2) be an incompressible velocity field, periodic (of zero averages) both in space and time.

Our interest here is in the dynamics of the field θ on *large* scales assumed to be $O(1/\epsilon)$, where $\epsilon \ll 1$ is the parameter controlling the scale separation. Because we expect the scalar field to have a diffusive dynamics, the associated time scale is $O(1/\epsilon^2)$. The presence of the small parameter ϵ naturally suggests to look for a perturbative approach. The perturbation is, however, singular [28] since a constant field is a trivial solution of (2). This is why one needs to use asymptotic methods, like multiscale techniques, to handle secular terms. Following the latter strategy, in addition to the *fast* variables x and t, let us then introduce *slow* variables as $X = \epsilon x$ and $T = \epsilon^2 t$. The prescription of the technique is to treat the two sets of variables as independent. It follows that

$$\partial_i \mapsto \partial_i + \epsilon \nabla_i \,; \qquad \partial_t \mapsto \partial_t + \epsilon^2 \partial_T \,, \tag{19}$$

where ∂ and ∇ denote the derivatives with respect to fast and slow space variables, respectively. The solution is sought as a perturbative series

$$\theta(x,t; X, T) = \theta^{(0)} + \epsilon \theta^{(1)} + \epsilon^2 \theta^{(2)} + \cdots, \tag{20}$$

where the functions $\theta^{(n)}$ depend a priori on both fast and slow variables. By inserting (20) and (19) into (2) and equating terms having equal powers in ϵ, we obtain a hierarchy of equations. The solutions of interest to us are those having the same periodicities as the velocity field. The first equation, corresponding to $O(\epsilon^0)$, is

$$\partial_t \theta^{(0)} + (\boldsymbol{u} \cdot \boldsymbol{\partial}) \, \theta^{(0)} = D_0 \, \partial^2 \theta^{(0)} \, . \tag{21}$$

By using Poincaré inequality, one can easily show that for periodic solutions

$$-\partial_t \int \left(\theta^{(0)} \right)^2 \mathrm{d}V = D_0 \int \left(\boldsymbol{\partial} \theta^{(0)} \right)^2 \mathrm{d}V \geq D_0 \left(\frac{2\pi}{L} \right)^2 \int \left(\theta^{(0)} \right)^2 \mathrm{d}V \, , \tag{22}$$

where L is the spatial periodicity length of $\boldsymbol{u}$ (supposed for simplicity to be the same in all directions) and the integral is over the periodicity box. The inequality (22) implies that the solution will relax to a constant with respect to fast variables, i.e.,

$$\theta^{(0)}(\boldsymbol{x}, t; \boldsymbol{X}, T) = \theta^{(0)}(\boldsymbol{X}, T) \, . \tag{23}$$

It can also be easily checked that the transient has no effect on the large-scale dynamics. The equations at order ϵ and ϵ^2 are

$$\partial_t \theta^{(1)} + (\boldsymbol{u} \cdot \boldsymbol{\partial}) \, \theta^{(1)} - D_0 \, \partial^2 \theta^{(1)} = -\boldsymbol{u} \cdot \nabla \theta^{(0)} \, , \tag{24}$$

$$\partial_t \theta^{(2)} + (\boldsymbol{u} \cdot \boldsymbol{\partial}) \, \theta^{(2)} - D_0 \, \partial^2 \theta^{(2)} = -\partial_T \theta^{(0)} - (\boldsymbol{u} \cdot \nabla) \theta^{(1)} + D_0 \, \nabla^2 \theta^{(0)} + 2 D_0 \boldsymbol{\partial} \cdot \nabla \theta^{(1)} \, . \tag{25}$$

Since (24) is linear, its solution can be written as

$$\theta^{(1)}(\boldsymbol{x}, t; \boldsymbol{X}, T) = \theta^{(1)}(\boldsymbol{X}, T) + \boldsymbol{w}(\boldsymbol{x}, t) \cdot \nabla \theta^{(0)}(\boldsymbol{X}, T) \, , \tag{26}$$

where the first term on the rhs is a solution of the homogeneous equation and the vector field $\boldsymbol{w}$ has a vanishing average over the periodicities and satisfies

$$\partial_t \boldsymbol{w} + (\boldsymbol{u} \cdot \boldsymbol{\partial}) \, \boldsymbol{w} - D_0 \, \partial^2 \boldsymbol{w} = -\boldsymbol{u} \, . \tag{27}$$

Due to the incompressibility of the velocity field, the average over the periodicities of the lhs in (24) and (25) is zero. For the equations to have a solution, the average of the rhs should also vanish (Fredholm alternative). The resulting solvability conditions provide the equations governing the large-scale dynamics, i.e., the dynamics in the slow variables. From (25) we obtain

$$\partial_T \langle \theta^{(0)} \rangle = D_0 \, \nabla^2 \langle \theta^{(0)} \rangle - \langle \boldsymbol{u} \cdot \nabla \theta^{(1)} \rangle \, , \tag{28}$$

where the symbol $\langle \cdot \rangle$ denotes the average over the periodicities. Note that the solvability condition for (24) is trivially satisfied. By plugging (26) into (28) we obtain the diffusion equation

$$\partial_T \theta^{(0)}(\boldsymbol{X}, T) = D_{ij}^{\mathrm{E}} \, \nabla^2 \theta^{(0)}(\boldsymbol{X}, T) \, , \tag{29}$$

where the eddy diffusivity tensor is

$$D_{ij}^{\mathrm{E}} = D_0 \delta_{ij} - \frac{1}{2} \left[\langle u_i w_j \rangle + \langle u_j w_i \rangle \right]. \tag{30}$$

Remark that the structure of the eddy diffusivity tensor will reflect the rotational symmetries of $\boldsymbol{u}$ and is in general non-isotropic.

It is not difficult to show that the eddy diffusivity is a positive definite tensor (see again [5]) thus signaling the fact that small-scale dynamics always enhances the large-scale transport.

2.3 A Remark About the Meaning of *Anomalous*

Let us now discuss in more general terms the anomalous diffusion problem considering moments of arbitrary order of the particle's displacement. Two cases are possible [31]: *weak* anomalous diffusion when a unique exponent is involved,

$$\langle |\, x(t) - x(0)\, |^q \rangle \sim t^{q\nu} \quad \forall q > 0 \text{ and } \nu > \frac{1}{2}; \tag{31}$$

strong anomalous diffusion when

$$\langle |\, x(t) - x(0)\, |^q \rangle \sim t^{q\,\nu(q)} \quad \nu(q) \neq \text{const} \quad \nu(2) > \frac{1}{2} \tag{32}$$

and $\nu(q)$ is a non-decreasing function of q.

In terms of the probability $P(\Delta x, t)$ of observing a displacement $\Delta x = x(t) - x(0)$ at time t, weak anomalous diffusion amounts to the scaling property:

$$P(\Delta x, t) = t^{-\nu} F(\Delta x\, t^{-\nu}), \tag{33}$$

where the function F is not necessarily a Gaussian. On the contrary strong anomalous diffusion is not compatible with the scaling (33).

In the case of weak anomalous diffusion, it is natural to conjecture

$$F(z) \propto e^{-c|z|^\alpha}, \tag{34}$$

where in general α is not determined by ν. However, an argument á la Flory due to Fisher [32] suggests that

$$P(\Delta x, t) \sim t^{-\nu} \exp\left[-c\left(\frac{|\Delta x|}{t^\nu}\right)^{1/1-\nu}\right], \tag{35}$$

i.e., $\alpha = \frac{1}{1-\nu}$. Remarkably the random shear flow examined in the previous section is in agreement with the Fisher's prediction, as shown by Bouchaud et al. [33] indeed for $\zeta = 0$, i.e., when $\nu = 3/4$ one has $F(a) \sim e^{-c|a|^4}$ for $|\,a\,| \gg 1$. While for the properties of dispersion the detailed functional dependence of $P(\Delta x, t)$ is not particularly important, it has a non-trivial role in determining the propagation properties in reactive systems [34].

2.4 *Strong* Anomalous Diffusion in Chaotic Flows

If (16) holds then anomalous diffusion may appear only for $D = 0$ and very strong Lagrangian velocity correlations. The latter condition can be realized, e.g., in time periodic velocity fields in which the Lagrangian phase space has a complicated self-similar structure of island and cantori [35]. In such a case superdiffusion is essentially due to the almost trapping of the ballistic trajectories, for arbitrarily long time, close to the cantori that are organized in complicated self-similar structures.

In this framework an interesting example is the Lagrangian motion in velocity field given by a simple model that mimics the Rayleigh–Bénard convection [36] and is described by the stream function:

$$\psi(x, y, t) = \psi_0 \sin\left[\frac{2\pi}{L}(x + B \sin \omega t)\right] \sin\left[\frac{2\pi}{L}y\right] , \qquad (36)$$

where the velocity is given by $\boldsymbol{u} = (\partial_y \psi, -\partial_x \psi)$, $\psi_0 = UL/2\pi$ (L being the periodicity of the cell, here we use $L = 2\pi$) and U the velocity intensity. The even oscillatory instability is accounted for by the term $B \sin \omega t$, representing the lateral oscillation of the rolls [36]. At fixed B, the control parameter for particle diffusion is $\epsilon \equiv \omega L^2/\psi_0$, i.e., the ratio between the lateral roll oscillation frequency (ω) and the characteristic circulation frequency (ψ_0/L^2) inside the cell. Different regimes take place for different values of ϵ. For instance, at $\epsilon \sim 1$ the synchronization between the circulation in the cells and their global oscillation is a very efficient way of jumping from cell to cell. This mechanism, similar to stochastic resonance, makes the effective diffusivity a structured function of the frequency ω [31] (see Fig. 1). Moreover, in the limit of vanishing molecular diffusivity, anomalous superdiffusion takes place in a narrow window of ω values around the peaks, i.e.,

$$\langle (x(t) - x(0))^2 \rangle \propto t^{2\nu(2)} \quad \text{with} \quad \nu(2) > 1/2 , \qquad (37)$$

as reported in Fig. 2 (left).

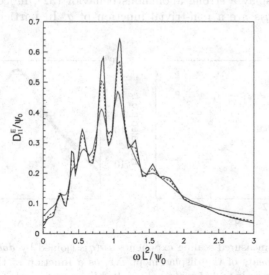

Fig. 1. The turbulent diffusivity $D_{11}^{\mathrm{E}}/\psi_0$ vs the frequency $\omega L^2/\psi_0$ for different values of the molecular diffusivity D/ψ_0. $D/\psi_0 = 3 \times 10^{-3}$ (*dotted curve*); $D/\psi_0 = 1 \times 10^{-3}$ (*broken curve*); $D/\psi_0 = 5 \times 10^{-4}$ (*full curve*)

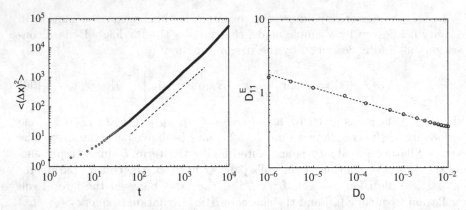

Fig. 2. *Left* Mean-squared displacement vs the time for the flow (36) with $D = 0$ and $\omega = 1.1$. Lengths and times are shown in units of L and L^2/ψ_0, respectively. The best-fit (*dashed*) line corresponds to $2\nu(2) = 1.3$. *Right* The diffusion coefficient D_{11}^{E} as a function of D for the frequency of the roll oscillation $\omega = 1.1$. The diffusivities are reported in units of ψ_0. The best-fit (*dashed*) line has the slope $-\beta = -0.18$

The presence of genuine anomalous diffusion is confirmed by the fact that effective diffusivity diverges as $D_{11}^{\mathrm{E}} \sim D^{-\beta}$ with $\beta > 0$ (see Fig. 2 (right)), as suggested in [5].

The remarkable property of the flow (36) is that moments of the particle displacement display a strong anomalous behavior (32), indeed Fig. 3 (left) shows that $q\nu(q)$s are a non-trivial function of q. In particular, the curve

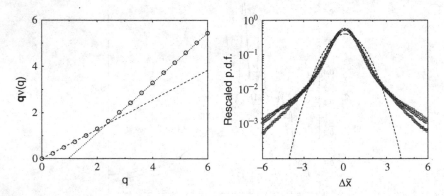

Fig. 3. *Left* The measured scaling exponents $q\nu(q)$s (joined by *dot-dashed straight lines*) of the moments of the displacement Δx, as a function of the order q. The *dashed line* corresponds to $0.65q$ while the *dotted line* corresponds to $q - 1.04$. *Right* The normalized probability distribution function $P(\Delta x(t)/\tilde{\sigma})$ vs $\Delta\tilde{x} \equiv \Delta x/\tilde{\sigma}$ (where $\tilde{\sigma} = \exp\langle\ln|\Delta x(t)|\rangle$) for the three times $t_1 = 500$ (*circles*), $t_2 = 2t_1$ (*diamonds*) and $t_3 = 2t_2$ (*squares*). The *dashed line* represents the Gaussian function

$q\,\nu(q)$ vs q displays a non-linear behavior. A closer inspection shows that two linear regions are present: the first one up to $q \sim 2$, the second elsewhere.

The two linear regions are associated to two different mechanisms in the diffusion process. For small qs, i.e., for the core of the probability distribution function $P(\Delta x, t)$, only one exponent ($\nu_1 \equiv \nu(q) \simeq 0.65$ for $q < 2$) fully characterizes the diffusion process. This means that the typical, i.e., non-rare, events obey a (weak) anomalous diffusion process. Roughly speaking, one can say that at scale l the characteristic time $\tau(l)$ behaves as $\tau(l) \sim l^{1/\nu_1}$. On the other hand, for $q > 2$ the behavior $q\,\nu(q) \simeq q - \text{const}$ suggests that the large deviations are essentially associated to ballistic transport, $\tau(l) \sim l$, basically due to the mechanism of synchronization between the circulation in the cells and their global oscillation.

Strong anomalous diffusion is also highlighted by the normalized probability densities $P(\Delta x, t)$ at different times that do not collapse onto a unique curve (see Fig. 3 (right)) suggesting that the scaling property (33) does not hold.

The above results for the anomalous diffusion in $2d$ time periodic incompressible flow may be encountered also in other systems. For instance, they had been observed in the standard map [37]:

$$\begin{cases} J_{t+1} = J_t + K\sin(\theta_t) \\ \theta_{t+1} = \theta_t + J_{t+1} \qquad \text{mod } 2\pi\,, \end{cases} \qquad (38)$$

where for specific values of K one has the coexistence of many accelerator modes, i.e., ballistic trajectories, and one observes

$$\langle [J_t - J_0]^2 \rangle \sim t^{2\nu(2)} \qquad t \gg 1 \quad \text{with} \quad \nu(2) \neq 1/2\,, \qquad (39)$$

i.e., anomalous diffusion in the action variable. The sticking of the chaotic orbits to stable islands leads to the appearance of blocks of long-range correlation in the sequences of the J_t variable which are responsible for its anomalous behavior [38].

The results discussed in this section, as well as those in [39], suggest that the phenomenon of anomalous diffusion in smooth chaotic dynamics with very long time correlation is possible but rare. For specific values of the parameters, i.e., K or ω, one can have $\nu(2) > 1/2$ but a small change typically restores the standard scenario $\nu(2) = 1/2$.

2.5 Relative Dispersion

So far we limited our discussion to single particle properties, i.e., absolute dispersion. Let us now consider the separation between two particles, $R(t) = x_2(t) - x_1(t)$. The evolution of this separation is ruled by the velocity difference at the scale R:

$$\frac{dR}{dt} = v(x_1(t) + R(t), t) - v(x_1(t), t) = \delta_R v\,. \qquad (40)$$

Since in incompressible flows the separation R typically grows in time [40], from the evolution of the relative separation we can, in principle, extract the contributions of the velocity components at different scales. In this sense the study of relative dispersion provides much information than absolute dispersion which is dominated by the sweeping induced by large-scale flow.

Here we consider both laminar and turbulent flows. A situation of both conceptual and applicative interest, e.g., in geophysical flows, is when the Eulerian velocity field is characterized by two length scales: a small scale l_η below which the velocity is smooth (i.e., $\delta_R v \sim R$ if $R \ll l_\eta$) and a large scale L_0 representing the size of the largest structures present in the flow, i.e., the correlation length of the flow. In a non-turbulent flow, $l_\eta \sim L_0$, while when the flow is turbulent (i.e., the Reynolds number is very high) the two scales are well separated. The interval of scales $l_\eta \ll r \ll L_0$ defines then the so-called inertial range where velocity differences display a non-smooth behavior: $\delta_r v \sim r^h$ with $h < 1$, e.g., $h = 1/3$ in Kolmogorov 1941 turbulence [41]. Moreover, one can also consider the presence of boundaries in the system, e.g., a finite domain of size L_B.

The presence of these scales manifests in different behaviors of the particles separation R. At very small separations $R \ll l_\eta$ the velocity difference in (40) can be reasonably approximated by a linear expansion in R which leads to an exponential growth of the separation of initially close particles, i.e.,

$$\langle \ln R(t) \rangle \simeq \ln R(0) + \lambda t , \tag{41}$$

where the average is taken over many couples with initial separation $R(0)$ and λ is the Lagrangian Lyapunov exponent. The rigorous definition of the Lyapunov exponent requires to take two limits $R(0) \to 0$ and then $t \to \infty$. In physical terms these limits amount to the requirement that the separation, even at very large times, should not exceed l_η. This is a very strict condition, rarely accomplished in real flows, rendering often difficult the experimental observation of the exponential regime (41).

On the opposite limit, for very long times and for separations $R \gg L_0$, the two trajectories $x_1(t)$ and $x_2(t)$ feel two uncorrelated realizations of the velocity and we expect normal diffusion, i.e.,

$$\langle R^2(t) \rangle \simeq 4D^E t , \tag{42}$$

the factor 4 is due to the fact that the two particles are asymptotically independent.

Between the two asymptotic regimes (41) and (42) the behavior of $R(t)$ depends on the particular flow. For instance, as we shall see, if the flow is turbulent with Kolmogorov scaling $\delta_R v \sim R^{1/3}$ an anomalous relative dispersion process takes place with

$$\langle R^2(t) \rangle \sim t^3 , \tag{43}$$

which is the celebrated Richardson dispersion.

In realistic settings, however, the characteristic length scales are not sharply separated, and the description of dispersion in terms of asymptotic quantities is infeasible, and different approaches are needed.

2.6 Scale-by-Scale Description of Dispersion

To treat the difficulties due to the lack of asymptotic let us now introduce a scale-dependent description of dispersion. Consider a particle pairs advected by a smooth velocity field with characteristic length l_η. Assuming that the Lagrangian motion is chaotic, we expect the following regimes:

$$\langle R^2(t)\rangle \simeq \begin{cases} R_0^2 \exp(\Lambda(2)t) & \text{if } \langle R^2(t)\rangle^{1/2} \ll l_\eta \\ 4D^E t & \text{if } \langle R^2(t)\rangle^{1/2} \gg l_\eta \end{cases}, \tag{44}$$

where $\Lambda(2) \geq 2\lambda$ is the second-order generalized Lyapunov exponent [42], D^E is the effective diffusion coefficient for the single particles dispersion. The regimes (44) hold only in the asymptotic limits $t \ll 1/\lambda$ and $t \gg 1/\lambda$, respectively. The presence of fluctuations in the rate of particles separation, which is characterized in terms of the finite time Lyapunov exponents [7], will determine situations such that at the same time different couples are in the two different regimes. Therefore, as consequence of this "contamination", crossovers with spurious regime are generically present (see Fig. 1 in [43]). This comes essentially because we considered the "wrong" variable, i.e., the time instead of the physical one, i.e., the scale.

A possibility for characterizing the dispersion properties in terms of the scale is by introducing the "doubling time" $\tau(\delta)$ at scale δ as follows [42, 43]: given a series of thresholds $\delta^{(n)} = r^n \delta^{(0)}$, one can measure the time $T_i(\delta^{(0)})$ it takes for the separation $R(t)$ to grow from $\delta^{(0)}$ to $\delta^{(1)} = r\delta^{(0)}$, and so on for $T_i(\delta^{(1)})$, $T_i(\delta^{(2)}),\ldots$ up to the largest considered scale. The r factor may be any value larger than 1, properly chosen in order to have a good separation between the scales of motion, i.e., r should be not too large. Strictly speaking, $\tau(\delta)$ is exactly the doubling time if $r = 2$. Then performing the doubling time experiments over N particle pairs separation experiments, the average doubling time $\tau(\delta)$ at scale δ is defined as

$$\tau(\delta) = <T(\delta)>_e = \frac{1}{N} \sum_{i=1}^{N} T_i(\delta), \tag{45}$$

notice that this average is different from the usual time average.

Now we can define the finite size Lyapunov exponent (FSLE) [42, 43] in terms of the average doubling time

$$\lambda(\delta) = \frac{\ln r}{\tau(\delta)}, \tag{46}$$

which quantifies the average rate of separation between two particles at a distance δ. For very small separations (i.e., $\delta \ll l_\eta$) one recovers the Lagrangian

Lyapunov exponent $\lambda = \lim_{\delta \to 0} \ln r / \tau(\delta)$. At large separation the diffusive behavior is signaled by the fact that $\tau(\delta) \sim \delta^2$. Thus the finite size Lagrangian Lyapunov exponent $\lambda(\delta)$ behaves as follows:

$$\lambda(\delta) \sim \begin{cases} \lambda & \text{if } \delta \ll l_\eta \\ 2D^E / \delta^2 & \text{if } \delta \gg l_\eta \end{cases}. \tag{47}$$

One could naively conclude, matching the behaviors at $\delta \sim l_\eta$, that $D^E \sim \lambda l_\eta^2$. This is not always true, since one can have a rather large range for the crossover due to non-trivial correlations which can be present in the Lagrangian dynamics [7, 43]. One can now define the scale-dependent diffusion coefficient as $D(\delta) = \delta^2 \lambda(\delta)$.

One might wonder that the introduction of $\tau(\delta)$ is just another way to look at $\langle R^2(t) \rangle$. This is true only in limiting cases, when the different characteristic lengths are well separated and intermittency is weak. Examples of the usefulness of this indicator may be found in [43, 44, 45].

2.7 Pair Dispersion in Laminar Flows in the Presence of Boundaries

In the case of flows confined in finite domains having size L_B in addition to the two asymptotic regimes (47), another regime appears. For separations close to the average maximal allowed separation $\delta_{\max} \simeq L_B$, the following behavior holds for a broad class of systems [7]:

$$\lambda(\delta) = \frac{D(\delta)}{\delta^2} \propto \frac{1}{\tau_R} \frac{(\delta_{\max} - \delta)}{\delta}, \tag{48}$$

where τ_R is related to the exponential relaxation of the particles' distribution inside the domain to the uniform distribution that is always attained at long times (see [7] for more details).

To exemplify the scale dependence of $\lambda(\delta)$, let us consider again the flow (36) where now, to study the effects of finite boundaries, we confine the tracers' motion in a finite domain. To this aim we slightly modified the oscillating term in (36) as $B \to B \sin(\pi x / L_B)$ with $L_B = 2\pi n/k$ (n is the number of convective cells). In this way the motion is confined in $x \in [-L_B, L_B]$. In Fig. 4 we show $\lambda(\delta)$ for two values of L_B. If L_B is large enough one can distinguish the three regimes: exponential (if δ is much smaller than the cell), diffusive (at scales larger than the unit cell) and saturation (48) (for separation close to the system size). Decreasing L_B, the range of the diffusive regime decreases, and for small values of L_B, it disappears.

Let us now discuss the use of the FSLE for the analysis of experimental Lagrangian data in a convective flow [46]. The experimental apparatus is a rectangular convective tank $L = 15.0$ cm wide, 10.4 cm deep and $H = 6.0$ cm height filled with water. The upper and lower surfaces are kept at constant temperature and the side walls can be considered as adiabatic. Convection

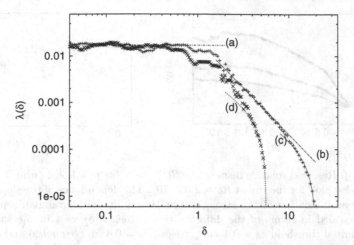

Fig. 4. $\lambda(\delta)$ vs δ for the flow (36) with: $\psi_0 = 0.2$, $B = 0.4$, $\omega = 0.4$. Particle motion has been confined in closed domain with 6 (*crosses*) and 12 (*diamonds*) convective cells, respectively. $\lambda(\delta)$ has been computed with $\mathcal{N} = 2000$ realization. As for the thresholds we used $\delta_n = \delta_0 r^n$ with $\delta_0 = 10^{-4}$ and $r = 1.05$. The *lines* are respectively: (a) Lyapunov regime with $\lambda = 0.017$; (b) diffusive regime with $D^E = 0.021$; (c) saturation regime with $\delta_{max} = 19.7$; (d) saturation regime with $\delta_{max} = 5.7$

is by an electrical circular heater, 0.8 cm in diameter, located in the mid-line of the tank, just above the lower surface (Fig. 5). The heater produces a constant heat flux controlled by a feedback on the power supply. The control parameter of the experiment is the Rayleigh number, Ra, which is varied over a wide range of values. The geometrical configuration constrains the convective

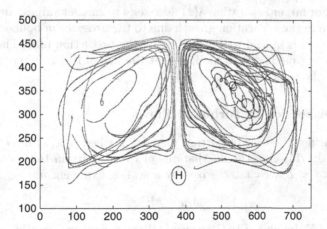

Fig. 5. An example of trajectories reconstructed by the PVT technique (unit in pixels). The *circle on the bottom* represents the heater

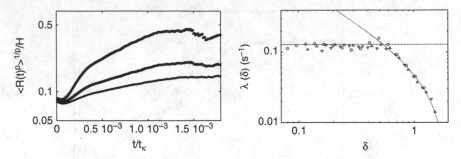

Fig. 6. *Left* Rescaled relative dispersion $\langle R(t)^p \rangle^{(1/p)}$ for $p = 1, 2, 4$ (from *bottom* to *top*) in lin-log plot, for the run at $Ra = 2.39 \times 10^8$. The dependence of the slope on the order p is an indication of the strong intermittency in the Lagrangian separation. Time is rescaled in terms of the diffusion time. *Right* $\lambda(\delta)$ vs δ in the same run. Different initial threshold $\delta_0 = 0.4$ cm (*circles*), $\delta_0 = 0.6$ cm (*triangles*) and $\delta_0 = 0.8$ cm (*inverted triangles*) have been considered to test the statistical robustness of the measure. The *straight line* is the Lyapunov exponent $\lambda = (0.12 \pm 0.01)$s^{-1} and the *curve* represents the saturation regime (48)

pattern to two counter-rotating rolls divided by an oscillating thermal plume (see Fig. 5 for a view of the typical Lagrangian motion). The Eulerian velocity field is thus, basically, two dimensional and time periodic.

Lagrangian data are obtained by particle tracking velocimetry (PTV) technique [47]. In Fig. 6 (left) we show normalized moments of the relative separation between particles, $\langle R^q(t) \rangle^{1/q}$ vs t for the run at $Ra = 2.39 \times 10^8$ computed over about 900 trajectories. As one can see it is very difficult to extract any reliable behavior from this observable. On the other had from the FSLE reported in Fig. 6 (right) it is rather clear what is happening. For small δ, we observe the collapse of $\lambda(\delta)$ to the value of the Lyapunov exponent, independent on δ_0. For larger separation $\lambda(\delta)$ decreases to smaller values, indicating a slowing down in the separation growth due to the presence of boundaries. The behavior of $\lambda(\delta)$ is indeed well described by the prediction (48). The absence of a diffusive regime is due to the fact that the size of Eulerian structure is of the order of the tank size.

2.8 Pair Dispersion in Turbulent Flows

We now consider the case of particle dispersion when $u(x, s)$ in (7) is a turbulent flow. In this case the fluctuations of the velocity in the *inertial range* of scales $\eta < \ell < L$ are characterized by a scaling exponent h:

$$\delta_\ell u \sim \ell^h. \tag{49}$$

According to Kolmogorov 1941 theory [41] the turbulent cascade can sustain a constant energy flux only for $h = 1/3$. Actually, many years of experimental investigations have shown that real turbulent flows are intermittent and it is necessary to introduce a continuous set of scaling exponents [41]. For simplicity, in the following we will first consider the case of non-intermittent turbulence (i.e., $h = 1/3$) and we will discuss the effect of intermittency at the end of the section. Furthermore let us assume that $l_\eta \to 0$, i.e., that the velocity field is turbulent up to the smallest available scale. Of course, this is a very idealized situation which is only approximately true in realistic turbulent flows.

The first important observation is that at variance with smooth velocities (i.e., $\delta_R v \sim R$) where in the limit of zero diffusivity particle trajectories are unique, if v is rough (i.e., $h < 1$) Lagrangian paths are not unique and also initially coincident particles separate. This is exemplified as follows. Let R be the separation between two particles, its temporal evolution (modeled in one dimension for simplicity) is given by $dR/dt = \delta_R v \propto R^{1/3}$ and $R(0) = R_0$. Then for $R_0 > 0$, the solution is given by

$$R(t) = \left[R_0^{2/3} + \frac{2}{3}t \right]^{3/2} . \tag{50}$$

If $R_0 = 0$ two solutions are allowed, i.e., $R(t) = [\frac{2}{3}t]^{3/2}$ and the trivial one $R(t) = 0$: trajectories are not unique. Physically this means that for $R_0 \neq 0$ the solution becomes independent of the initial separation R_0, provided t is large enough. More interestingly one has that trajectories separate anomalously

$$\langle R^2(t) \rangle \sim t^3, \tag{51}$$

as one easily obtains from (50). This is the well-known Richardson law for particles dispersion in turbulent flows [6, 48]. Of course, in realistic flows one should expect that this regime ends at time such that $\langle R(t) \rangle \sim L_0$, and, in the absence of boundaries, the standard diffusive behavior will establish at larger times. It is worth mentioning that the mechanism underlying this "anomalous" diffusive behavior is the same as the one discussed in the context of single particle diffusion: the Lagrangian velocity differences correlation function does not decay when particle separation is within the inertial range so that conditions for the validity of the standard diffusion discussed in Sect. 2.1 are violated [3].

Similar to absolute dispersion, relative dispersion in turbulence can be phenomenologically described in terms of a diffusion equation for the probability density function of pair separation $p(R, t)$ with a space- and time-dependent diffusion coefficient $D(R, t)$. The original Richardson proposal, obtained from experimental data in the atmosphere, is $D(R, t) = D(R) = k_0 \varepsilon^{1/3} r^{4/3}$ [48], where ε has the dimension of energy dissipation (see below) and k_0 is a dimensionless constant. In the d-dimensional isotropic case, this diffusion equation takes the form

$$\partial_t p(R, t) = \frac{1}{R^{d-1}} \left(\frac{\partial}{\partial R} R^{d-1} D(R) \frac{\partial}{\partial R} p(R, t) \right) . \tag{52}$$

Richardson model can be justified by the following argument: the dimensions of D are (time)$^{-1}$ (scale)^2. Using $R/\delta_R v$ and R as the proper time and scale, respectively, one obtains $D(R) \sim R^{4/3}$. Although the above argument is only dimensional, one can prove that if the velocity field is rapidly decorrelating in time (52) is indeed exact [3]. Equation (52) is solved by

$$p(R,t) \sim \frac{R^{d-1}}{t^{3d/2}} e^{-c(R^{2/3}/t)} \tag{53}$$

which gives the Richardson law $\langle R^2 \rangle = \int dR\, R^2 p(R,t) \propto t^3$. Moreover, note that $p(R,t) = t^{-3/2} F(t^{-3/2}R)$ consistently with (33), indeed the assumption of a perfect scale-invariance for the velocity field leads to a "weak" anomalous behavior, i.e., $\langle R^q \rangle \sim t^{3/2q}$. This can be considered a good approximation for two-dimensional turbulent flows in the inverse energy cascade regime [49].

In Fig. 7 we report the results of Boffetta and Sokolov [50] for Richardson diffusion in two-dimensional turbulence. Figure 7 (left) shows the behavior of $\langle R^2(t) \rangle$ vs time. Note that the presence of large crossovers makes the t^3 law not easily detectable. In fact, as follows also from (50) the memory of the initial separation is only asymptotically recovered. On the other hand Fig. 7 (right) reports the computation of the FSLE in the same simulation, as one can see the Richardson law is much more evident as signaled by the scaling $\lambda(\delta) \sim \delta^{-2/3}$. Actually the same qualitative picture persists also in synthetic turbulent flows where the resolution can be enhanced a lot [49]. In realistic turbulent flows, assuming that all the scales can be resolved the following regime for the FSLE should be expected: $\lambda(\delta) = \lambda$ for $\delta \ll l_\eta$; $\lambda(\delta) \sim \delta^{-2/3}$ for $l_\eta \ll \delta \ll L_0$; $\lambda(\delta) \sim \delta^{-2}$ for $\delta \gg L_0$. As an example of realistic turbulent flow, the $\delta^{-2/3}$ has recently been observed in [51] exploiting the

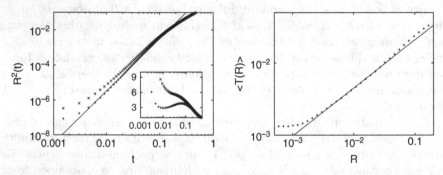

Fig. 7. *Left* Relative dispersion $R^2(t)$ with $R(0) = R_0$ (*pluses*) and $R(0) = 2R_0$ (*crosses*) for separation in the inverse cascade of two-dimensional turbulence. The *line* represents the Richardson law $R^2(t) = g\varepsilon t^3$ with $g = 3.8$. In the *inset* the compensated plot $R^2(t)/(\varepsilon t^3)$ is displayed. *Right* Mean doubling time $\langle T(R) \rangle$ as function of the separation R. The ratio is $\rho = 1.2$ and the average is obtained over about 5×10^5 events. The *line* represents the dimensional scaling $R^{2/3}$. Figures from [50]

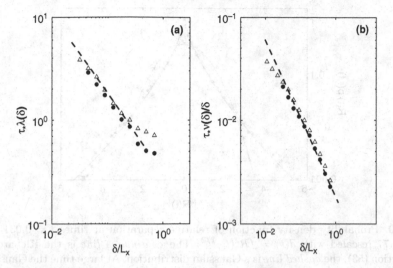

Fig. 8. (a) FSLE at two different resolutions. *Triangles*: 128^3 grid points; *circles*: 96^3 grid points. The *dashed line* corresponds to $\alpha\delta^{-2/3}$ with $\alpha = 0.1$ m^2 s^{-3}. (b) The same as in (a) but for the relative velocity. The *dashed line* has slope $-2/3$

large-eddy simulations of boundary layer flows. In Fig. 8 we can see how the FSLE is able to isolate a nice $\delta^{-2/3}$ behavior despite the small extension of the inertial range of scales (see again [51] and discussions therein).

Finally in Fig. 9 the prediction (53) is compared with the numerical results, showing that at least for the case of two-dimensional Navier–Stokes turbulence the Richardson description is a good approximation even if the velocity field has non-trivial correlations and does not rapidly decorrelate.

For three-dimensional turbulent flows the situation is complicated by the presence of fluctuations in the scaling exponent h in (49) [41]. These fluctuations modify the statistics of relative dispersion and one thus expects an anomalous behavior in the strong sense. Indeed, it is possible to repeat the dimensional argument leading to Richardson scaling taking into account intermittency, for example within the framework of multifractal model [52]. The result is that different moments of separations $\langle R(t)^p\rangle$ grow with exponents different from dimensional predictions $t^{3/2p}$ but the $p = 2$ exponent is not corrected by intermittency.

These results are more simply obtained for the doubling time statistics. The dimensional estimate $T(R) \sim R/\delta_R v$ gives, for an intermittent velocity statistics $\delta_R v$, the prediction [52]

$$\left\langle \frac{1}{T^p(R)} \right\rangle \sim R^{\zeta_p - p}, \tag{54}$$

where ζ_p are the Eulerian intermittent scaling exponents, i.e., $\langle(\delta_R v)^p\rangle \sim R^{\zeta_p}$. Because $\zeta_3 = 1$ [41], the exponent not affected by intermittency is here

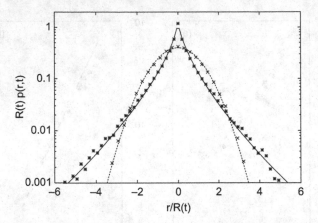

Fig. 9. Probability density function of relative separation at times $t = 0.031$ and $t = 0.77$ rescaled with $d(t) = \langle R^2(t) \rangle^{1/2}$. The *continuous line* is the Richardson prediction (53), the *dashed line* is a Gaussian distribution. At large time the Gaussian behavior is recovered. Figure taken from [50]

$\langle 1/T^3 \rangle \sim R^{-2}$ which is the analogous of Richardson law (51) for doubling times.

Figure 10 is obtained from high-resolution direct numerical simulations of three-dimensional turbulence [53]. Different moments of doubling times (54) are compensated with scaling exponents $\zeta_p - p$ predicted by Eulerian intermittency model. The existence of a scaling range is a direct demonstration of intermittency corrections to Richardson scaling.

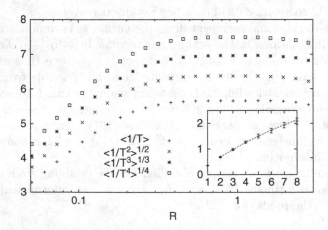

Fig. 10. First moments of the inverse doubling time $\langle (1/T(R))^p \rangle$ compensated with (54) with Eulerian scaling exponents ζ_p obtained from the velocity structure functions (see *inset*). Figure from [53]

3 Transport of Reacting Substances

We consider the case of a unique scalar reactive field $\theta(\boldsymbol{x}, t)$. This model is appropriate in aqueous autocatalytic premixed reactions, and gaseous combustion with a large flow intensity but low value of gas expansion across the flame [8]. The field θ evolves according to the advection–reaction–diffusion (ARD) equation:

$$\partial_t \theta + (\boldsymbol{u} \cdot \boldsymbol{\nabla})\theta = D\Delta\theta + \frac{1}{\tau}f(\theta) , \tag{55}$$

where $f(\theta)$ accounts for the reaction and τ is the reaction characteristic time [8, 14, 15, 16]. For the sake of simplicity, we take $f(\theta) = \theta(1 - \theta)$. However, the results we are going to describe do not depend on the specific form of $f(\theta)$ provided that $f(\theta)$ is convex ($f''(\theta) < 0$) and positive in the interval $(0, 1)$, vanishing at its extremes, and $f'(0) = 1$. This corresponds to the FKPP type of reaction [10, 11].

In Sect. 2 we saw the link between the solution of (2) and Lagrangian trajectories. There exists a similar relation also for (55) [54]:

$$\theta(\boldsymbol{x}, t) = \left\langle \theta(\boldsymbol{x}(0), 0) \exp\left(\frac{1}{\tau}\int_0^t \frac{f(\theta(\boldsymbol{x}(s; t), s))}{\theta(\boldsymbol{x}(s; t), s)}\mathrm{d}s\right)\right\rangle , \tag{56}$$

where the average is performed over all the trajectories $\boldsymbol{x}(s; t)$ that started in $\boldsymbol{x}(0)$ and ended in $\boldsymbol{x}(t; t) = \boldsymbol{x}$ (as in (7)).

Using the maximum principle [54] and noting that $f(\theta)/\theta \le f'(0)$, because of the convexity of $f(\theta)$, one can write an upper bound for $\theta(\boldsymbol{x}, t)$ in terms of the solution, $\theta_\mathrm{L}(\boldsymbol{x}, t)$, of the linearized ARD:

$$\partial_t \theta_\mathrm{L} + (\boldsymbol{u} \cdot \boldsymbol{\nabla})\theta_\mathrm{L} = D\Delta\theta_\mathrm{L} + \frac{f'(0)}{\tau}\theta_\mathrm{L} . \tag{57}$$

In fact, if $\theta(\boldsymbol{x}, 0) \le \theta_\mathrm{L}(\boldsymbol{x}, 0)$ one has [54]

$$\theta(\boldsymbol{x}, t) \le \theta_\mathrm{L}(\boldsymbol{x}, t) . \tag{58}$$

From (56–58) one obtains

$$\theta(\boldsymbol{x}, t) \le \langle\theta(\boldsymbol{x}(0), 0)\rangle \exp\left(\frac{f'(0)}{\tau}t\right) . \tag{59}$$

The rhs of the previous equation is the solution of (57) and, in particular, $\langle\theta(\boldsymbol{x}(0), 0)\rangle$ is nothing but the solution $P(\boldsymbol{x}, t)$ at time t of the passive scalar PDE (2) with initial condition $\theta(\boldsymbol{x}, 0)$ (that we suppose localized around $\boldsymbol{x} = 0$).

In Sect. 2.1 we saw that under general conditions (i.e., spatial and temporal short-range correlations) (2) has the same asymptotic behavior of a Fick equation. As a consequence, we have

$$< \theta(\boldsymbol{x}(0),0) >\equiv P(\boldsymbol{x},t) \sim \frac{1}{\sqrt{2\pi D_{11}^{\mathrm{E}}t}} \exp\left(-\frac{x^2}{4D_{11}^{\mathrm{E}}t}\right) . \tag{60}$$

Equations (59) and (60) imply that, along the x-direction, the field θ is exponentially small until a time t of the order of $x/\sqrt{4D_{11}^{\mathrm{E}}f'(0)/\tau}$, therefore an upper bound for v_f comes out:

$$v_f \le 2\sqrt{D_{11}^{\mathrm{E}}f'(0)/\tau} . \tag{61}$$

The above discussion shows that, if standard diffusion holds, then there is a front propagating with a constant speed, i.e., the solvable case $\boldsymbol{u} = 0$ is recovered with a renormalized diffusion constant. Nevertheless, the analytical determination of v_f, for a given velocity field, $\boldsymbol{u}$, is a rather difficult problem even for simple laminar fields [14, 15, 16, 18].

3.1 Fronts in Cellular Flows

The bound (61) is very general and holds for generic incompressible flows and production terms. Here we numerically investigate the properties of front propagation in the particular case of the cellular flow (36). This kind of flow is interesting because, at variance with shear flows, all the streamlines are closed and, therefore, the front propagation is determined by the mechanisms of contamination of one cell to the other [14, 16]. First we consider the time-independent case, i.e., $B = 0$ in (36). Since we are interested in the propagation in the x-direction we take periodic boundary conditions in y-axis and an infinite extent along the x-axis with boundary conditions $\theta(-\infty, y; t) = 1$ and $\theta(+\infty, y; t) = 0$.

The bulk burning rate [15],

$$v_{\mathrm{f}}(t) = \frac{1}{L}\int_0^L \mathrm{d}y \int_{-\infty}^{\infty} \mathrm{d}x\, \partial_t \theta(x,y;t) , \tag{62}$$

coincides with the front speed when the latter exists, but it is also a well-defined quantity even when the front itself is not well defined. The asymptotic (average) front speed, v_{f}, is determined by

$$v_{\mathrm{f}} = \lim_{T\to\infty} \frac{1}{T}\int_0^T \mathrm{d}t\, v_{\mathrm{f}}(t) .$$

In our discussion, we always suppose that the diffusion time scale is the slowest one and thus $Pe \gg 1$ and $Da \cdot Pe \gg 1$.

At large scales and long times the effects of the velocity field can be modeled in terms of a reaction–diffusion process with renormalized coefficients [14]:

$$\partial_t \theta = D^{\mathrm{E}}\Delta\theta + \frac{1}{\tau_{\mathrm{eff}}}F(\theta) . \tag{63}$$

The renormalized diffusivity D^E accounts for the process of diffusion from cell to cell as a result of the non-trivial interaction of advection and molecular diffusion [6]. The renormalized reaction time τ_{eff} is the time it takes for a single cell to be filled by inert material and depends on the interaction of advection and production. F indicates the functional form of the renormalized chemistry. Therefore, the limiting speed of the front in the moving medium is given by $v_{\text{eff}} \sim \sqrt{D^E/\tau_{\text{eff}}}$ [14, 15]. The problem is now reduced to derive the expressions for the renormalized parameters by means of physical considerations.

In the following sections, using as an interpretative framework the macroscopic model described above, we will present the results of detailed numerical simulations for slow $(Da \ll 1)$ and fast $(Da \gg 1)$ reactions.

3.2 Slow and Fast Reaction Regimes

The renormalized characteristic time can be estimated as follows. At small Da, the reaction is significantly slower than the advection, and consequently the region where the reaction takes place extends over several cells, i.e., the front is distributed. The dependence of D^E on Pe and D is a well-studied problem, the solution of which is [6]

$$\frac{D^E}{D} \sim Pe^{1/2} \qquad Pe \gg 1 . \tag{64}$$

Therefore, in the slow reaction regime, $Da \ll 1$, a single cell is first invaded by a mixture of reactants and products (on the fast advective time scale), and subsequently complete reaction is achieved on the slower time scale $\tau_{\text{eff}} \simeq \tau$ (Fig. 11). In the case of fast reaction, $Da \gg 1$, two sharply separated phases emerge inside the cell and the filling process is characterized by an inward spiral motion of the outer, stable phase (Fig. 11), at a speed proportional to U. Therefore we have

Fig. 11. *Left* Six snapshots of the field θ within the same cell, at six successive times with a delay $\tau/6$ (from *left* to *right*, *top* to *bottom*), as a result of the numerical integration of (55). Here $Da \simeq 0.4, Pe \simeq 315$. *Black* stands for $\theta = 1$, *white* for $\theta = 0$. *Right* The same but for $Da = 4, Pe = 315$, τ is now replaced by $\tau_{\text{eff}} \sim L/U$. Note that a spiral wave invades the interior of the cell, with a speed comparable to U

$$\frac{\tau_{\text{eff}}}{\tau} \sim \begin{cases} 1 & Da \ll 1 \\ Da & Da \gg 1 \end{cases}.$$ (65)

From (64) and (65) we can derive the front speed for a cellular flow. Indeed, recalling that $v_{\text{f}} \sim \sqrt{D^{\text{E}}/\tau_{\text{eff}}}$, we have [14, 15]

$$\frac{v_{\text{f}}}{v_0} \sim \begin{cases} Pe^{1/4} & Da \ll 1,\, Pe \gg 1 \\ Pe^{1/4}Da^{-1/2} & Da \gg 1,\, Pe \gg 1 \end{cases}.$$ (66)

The case of $Pe \ll 1$ is less interesting because the dynamics is dominated by diffusion.

In terms of the typical velocity of the cellular flow, we have $v_{\text{f}} \propto U^{1/4}$ for slow reaction ($U \gg L/\tau$, or equivalently $Da \ll 1$) whereas $v_{\text{f}} \propto U^{3/4}$ for fast reaction ($U \ll L/\tau$, or $Da \gg 1$). The scaling $v_{\text{f}} \propto U^{1/4}$ for slow reaction (i.e., fast advection) is a consequence of $D^{\text{E}} \propto DPe^{1/2}$ [6] in the homogenization limit [14, 15] and has been obtained in [14, 16]. Numerical simulations of (55), with a FKPP production term, confirm these predictions (Fig. 12).

As a remark we mention that, for the class of boundary conditions investigated here, where the region of initially burnt material extends to infinity, no quenching [15] takes place independently of production term used. Numerical simulations show that Arrhenius-type non-linearity gives the same qualitative results as those of FKPP-type reaction presented above, i.e., one has the two scaling laws $v_{\text{f}} \propto U^{1/4}$ and $v_{\text{f}} \propto U^{3/4}$ at fast and slow advections, respectively [14].

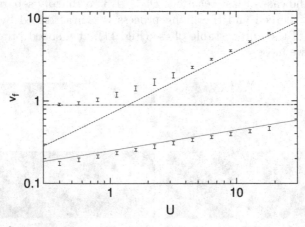

Fig. 12. The front speed v_{f} as a function of U, the typical flow velocity. The *lower curve* shows data at $\tau = 20.0$ (fast advection). The *upper curve* shows data at $\tau = 0.2$ (slow advection). For comparison, the scalings $U^{1/4}$ and $U^{3/4}$ are shown as *dotted* and *dashed lines*, respectively. The *horizontal line* indicates v_0, the front velocity without advection, for $\tau = 0.2$

3.3 Geometrical Optics Limit

When the front thickness and the reaction time are much smaller than the length and time scales of the velocity field fluctuations one has the geometrical optics regime. Mathematically speaking this regime is realized for $(D, \tau) \to 0$ maintaining D/τ constant [18, 55]. In this limit one has a non-zero bare front speed, v_0, while the front thickness ξ goes to zero, i.e., the front is sharp.

The sharp interface separating the reactants from the products is modeled by the G-equation (67) [9, 55]:

$$\frac{\partial G}{\partial t} + \boldsymbol{u} \cdot \boldsymbol{\nabla} G = v_0 |\boldsymbol{\nabla} G| \,. \tag{67}$$

The front is defined by a constant level surface of the scalar function $G(\boldsymbol{r}, t)$.

As far as the cellular flow is concerned, the front border is wrinkled by the velocity field during propagation and its length increases until pockets of fresh material develop [56, 57] (Fig. 13). After this, the front propagates periodically in space and time with an average speed $v_{\rm f}$, which is enhanced with respect to the propagation speed v_0 of the fluid at rest.

Let us now consider the flow (36) with $B = 0$, i.e., the stationary case. The problem addressed here is the dependence of the effective speed $v_{\rm f}$ on the flow intensity, U, and the bare velocity, v_0, that is expected of the form [58]:

$$\frac{v_{\rm f}}{v_0} = \psi\left(\frac{U}{v_0}\right) , \tag{68}$$

where $\psi(\mathcal{U})$ is a function which depends on the flow details.

As far as we know, apart from very simple shear flows (for which $\psi(\mathcal{U}) = 1 + \mathcal{U}$ [59]), there are no methods to compute $\psi(\mathcal{U})$ from first principles. Mainly one has to resort to numerical simulations and phenomenological arguments.

For turbulent flows, by means of dynamical renormalization group techniques, Yakhot [60] proposed

$$\frac{v_{\rm f}}{v_0} = {\rm e}^{(U/v_{\rm f})^\alpha} \,; \tag{69}$$

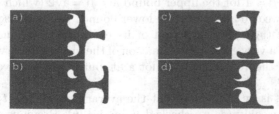

Fig. 13. Snapshot of the front shape with time step $T/8$ (from (**a**) to (**d**)), where T is the period of the front dynamics, for $v_0 = 0.5$, $U = 4.0$ and $L = 2\pi$. Unburnt (burnt) material is indicated in *white (black)*

with $\alpha = 2$, now U indicates the root mean squared average velocity (see also [61, 62]). Therefore, from (69) one has that $v_f \to U/\sqrt{\ln(U)}$ for $U \to \infty$.

For the cellular flow under investigation, albeit the exact form of the function $\psi(U)$ is not known, a simple argument can be given for an upper and a lower bound by mapping the front dynamics onto a one-dimensional problem. The starting point is the following observation. In the optical regime, since the interface is sharp, i.e., $\theta(x, y)$ is a two-valued function ($\theta = 1$ and $\theta = 0$), we can track the farther edge of the interface between product and material $(x_M(t), y_M(t))$, which is defined as the rightmost point (in the x-direction) for which $\theta(x_M, y_M; t) = 1$. Then we can define a velocity

$$\tilde{v}_f = \lim_{t \to \infty} \frac{x_M(t)}{t} , \tag{70}$$

which gives an equivalent value of the standard definition (62). After a transient, in the unit cell $[0, 2\pi]$ (we describe the case $L = 2\pi$) the point $(x_M(t), y_M(t))$ moves to the right along the separatrices of the streamfunction (36), so that $y_M(t)$ is essentially close to the value 0 or π. Along this path one can reduce the edge dynamics to the $1d$-problem:

$$\frac{dx_M}{dt} = v_0 + U\beta |\sin(x_M(t))| , \tag{71}$$

where the second term of the rhs is the horizontal component of the velocity field. We have neglected the y-dependence, replacing it with a constant β which takes into account the average effect of the vertical component of the velocity field along the path followed by (x_M, y_M). By solving (71) in the interval $x_M \in (0, 2\pi)$ one obtains the time, T, needed for x_M to reach the end of the cell. The front speed, as the speed of the edge particle, is then given by $v_f = 2\pi/T$. The final result is

$$\frac{v_f}{v_0} = \psi_\beta(U) = \frac{\pi\sqrt{(U\beta)^2 - 1}}{2\ln\left(U\beta + \sqrt{(U\beta)^2 - 1}\right)} . \tag{72}$$

Note that (72) is valid only for $U\beta \geq 1$.

We have taken $\beta = 1$ for the upper bound and $\beta = 1/2$ (which is the average of $|\cos(y)|$ between 0 and π) for the lower bound. We have also computed the average of $|\cos(y_M(t))|$ in a period of its evolution obtaining $\beta \approx 0.89$ which indeed gives a very good approximation of the measured curve (Fig. 14). We stress that the theoretical curve is not a fit, but it just involves the measured parameter β.

This agreement is an indication that the average of $|\cos(y_M(t))|$ depends on U and v_0 very weakly (as we checked numerically). Previous studies [61, 62] reported an essentially linear dependence of the front speed on the flow intensity, i.e., $v_f \propto U$ for large U which is not too far but different from our result. A rigorous bound has been obtained in [63] by using the G-equation:

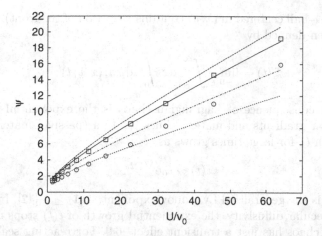

Fig. 14. The measured $\psi(U/v_0)$ as a function of U/v_0 (*squares*), the Yakhot formula (69) with $\alpha = 2$ (*circles*), the function ψ_β for $\beta = 1, 1/2$ (*dashed and dotted lines*) and for $\beta = 0.89$ (*solid line*). The *dashed-dotted line* is the bound (73)

$$v_f \geq U/(\log(1 + U/v_0)) \,. \tag{73}$$

As one can see from Fig. 14, the lower bound (73) seems to be closer to the numerical data than the one obtained with $\beta = 1/2$ in (72). From (72), asymptotically (i.e., for $U \gg v_0$) one has $v_f \sim U/\ln(U)$ which corresponds to (69) for $\alpha = 1$. Expressions as (69) have been proposed for flows with many scales as, e.g., turbulent flows, and in the literature different values of α have been reported [60, 61, 62]. The fact that the simple one-scale vortical flow investigated here displays such a behavior may be incidental. However, we believe that it can be due to physical reasons. Indeed, the large-scale features of the flow, e.g., the absence of open channels (like for the shear flow) can be more important than the detailed multiscale properties of the flow [56].

3.4 Is Chaos Important?

An interesting problem is the thin front dynamics in the presence of Lagrangian chaos generated by the time periodic streamfunction (36). We are mainly interested in addressing the following two issues. First, since trajectories starting near the roll separatrices typically have a positive Lyapunov exponent, it is natural to wonder about the role of Lagrangian chaos on front propagation. Second, as we have shown in Sect. 2.3, we know that for the time-dependent streamfunction (36) the transport properties are strongly enhanced, therefore it is worth to see if similar effects are also reflected in the front speed.

In order to define the instantaneous front length, $\mathcal{L}(t)$, we introduce the variable $\sigma_\epsilon(x, y; t)$ which assumes the value 0 if θ is constant inside a circle of radius ϵ centered in (x, y) at time t, otherwise $\sigma_\epsilon(x, y; t) = 1$ (i.e., $\sigma_\epsilon(x, y; t) =$

1 only if the ϵ-ball centered in (x, y) contains a portion of the front). The front length is then defined by

$$\mathcal{L}(t) = \lim_{\epsilon \to 0} \frac{1}{\epsilon} \int_{-\infty}^{\infty} \mathrm{d}x \int_0^L \mathrm{d}y\, \sigma_\epsilon(x, y; t) \ . \tag{74}$$

A direct consequence of Lagrangian chaos is the exponential growth of passive scalar gradients and material lines [1, 4]: a (passive) material line of initial length ℓ_0 for large times grows as

$$\ell(t) \sim \ell_0 e^{\Lambda(1)t} , \tag{75}$$

where $\Lambda(1)$ is the generalized Lyapunov exponent, $\Lambda(1) \geq \lambda$ [42]. In the presence of molecular diffusivity, the exponential growth of $\ell(t)$ stops due to diffusion, and chaos has just a transient effect [64]. For reacting scalars something very similar happens. Let us compare the evolution of a material line in the passive and reactive cases (Fig. 15). While in the passive case (without molecular diffusivity) structures on smaller and smaller scales develop (due to stretching and folding), in the reactive systems after a number of folding events structures on smaller scales are inhibited as a consequence of the Huygens dynamics: the interface between the two phases merges. This phenomenon is responsible for the formation of *pockets*. Of course, "merging" is more and more efficient as v_0 increases (compare the middle and lower pictures of Fig. 15).

In Fig. 16 we show the time evolution of the line length, $\mathcal{L}(t)$, as a function of t for the passive and reactive material at different values of v_0. While at small times both the passive and reactive scalar lines grow exponentially with a rate close to $\Lambda(1)$, at large time $t > t^*$ (where t^* is a transient time depending on v_0) the reacting ones stop due to merging. Asymptotically, the front length varies periodically with an average value depending on v_0. A rough

Fig. 15. Snapshots at two successive times, $t = 3.6$ and 7.5, of the evolution of passive (*top*) and reactive line of material for two values of v_0 (*middle* $v_0 = 0.7$ and *bottom* $v_0 = 2.1$) for $U = 1.9$, $B = 1.1$ and $\omega = 1.1U$. The initial condition is a straight vertical line

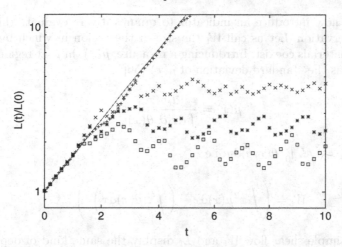

Fig. 16. $\mathcal{L}(t)/\mathcal{L}(0)$ as a function of time for $U = 1.9$, $B = 1.1$ and $\omega = 1.1U$ for the passive (*pluses*) and reactive case: from *top* $v_0 = 0.3$ (*crosses*), 0.5 (*stars*), 0.7 (*squares*). The *straight line* indicates the curve $\exp(\Lambda(1)t)$ with $\Lambda(1) \approx 0.5$, which has been directly measured

argument to estimate t^* is the following: two initially separated parts of the line (e.g., originally at distance ℓ_0) become closer and closer, roughly as $\sim \ell_0 \exp(-\Lambda(1)t)$. When such a distance becomes of the order of $v_0 t$, merging takes place and, hence,

$$t^* \propto \frac{1}{\Lambda(1)} \ln\left(\frac{\Lambda(1)\ell_o}{v_0}\right). \tag{76}$$

In the asymptotic state ($t > t^*$) both the spatial and temporal structures of the flow become periodic.

Let us now switch to the effects of Lagrangian chaos on the asymptotic dynamics of front propagation. An immediate consequence of (76) is that the asymptotic front length (74) behaves as $L_f \sim v_0^{-1}$ for values of small enough v_0. Indeed,

$$L_f \sim L e^{\Lambda(1)t^*} \sim \frac{L^2\Lambda(1)}{v_0}, \tag{77}$$

which is in fairly good agreement with the simulations [56].

It is worth remarking that even if the scaling (77) holds when chaos is present, in general it is not peculiar of chaotic flows. For instance, for the shear flow ($u_x = U \sin(y)$, $u_y = 0$) one has $v_f - U + v_0$. On the other hand, since $v_f \sim L_f v_0$ [9], even if the shear flow is not chaotic $L_f \sim 1/v_0$ for $U/v_0 \gg 1$. From the previous discussion, it seems that the front length dependence on v_0 is not an unambiguous effect of chaos on the asymptotic dynamics. But, looking at Fig. 15 the spatial "complexity" of the front in the presence of Lagrangian chaos is apparent.

Let us now introduce an indicator to quantitatively evaluate this qualitative observation. Let us call W_f the size of the region in which burnt and unburnt materials coexist. Introducing a measure, $\mu(x)$, in that region we can define W_f as the standard deviation of $\mu(x)$ [56]:

$$\mu(x) = \frac{|\partial_x \tilde{\theta}(x)|}{\int dx |\partial_x \tilde{\theta}(x)|},$$ (78)

where $\tilde{\theta}(x) = 1/L \int_0^L \theta(x,y) dy$, i.e.,

$$W_f = \left(\int x^2 \mu(x) dx - \left(\int x \mu(x) dx \right)^2 \right)^{1/2}.$$ (79)

For a simple shear flow W_f and L_f display the same kind of dependence on v_0 (actually they are proportional). In generic chaotic flows there is an increasing of the front length, while chaotic mixing induces a decrease of W_f. This is indeed what one observes in Fig. 17, where we show the ratio L_f/W_f both for the non-chaotic and the chaotic flow. For the latter this ratio diverges for very small v_0 values as a signature of chaos. From a physical point of view the ratio L_f/W_f is an indicator of the spatial complexity of the front. Indeed it indicates the degree of wrinkling of the front with respect to the size of the region in which the front is present. Loosely speaking, we can say that the *temporal* complexity of Lagrangian trajectories converts in the *spatial* complexity of the front.

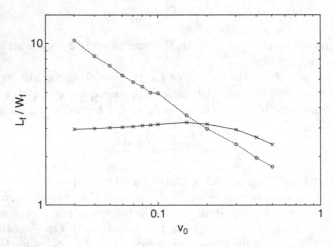

Fig. 17. L_f/W_f as a function of v_0 for the time-dependent (*circles*) and independent (*crosses*) cases with parameters $U = 1.9$, $B = 1.1$, $\omega = 1.1U$ and the time-independent case (*crosses*) with $U = 1.9$

Acknowledgments

We thank M. Abel, P. Castiglione, A. Celani, M. Cencini, G. Lacorata, C. Lopez, R. Mancinelli, P. Muratore-Ginanneschi, R. A. Pasmanter, A. Torcini, M. Vergassola, D. Vergni and E. Zambianchi for fruitful collaborations and interesting discussions during the last years.

References

1. H. K. Moffatt: Rep. Prog. Phys. **46**, 621 (1983)
2. J. P. Bouchaud and A. Georges: Phys. Rep. **195**, 127 (1990)
3. G. Falkovich, K. Gawędzki and M. Vergassola: Rev. Mod. Phys. **73**, 913 (2001)
4. A. Crisanti, M. Falcioni, G. Paladin and A. Vulpiani: *La Rivista del Nuovo Cimento* **14** (12), 1 (1991)
5. L. Biferale, A. Crisanti, M. Vergassola and A. Vulpiani: Phys. Fluids **7**, 2725 (1995)
6. A. J. Majda and P. R. Kramer: Phys. Rep. **314**, 237 (1999)
7. V. Artale, G. Boffetta, A. Celani, M. Cencini and A. Vulpiani: Phys. Fluids **9**, 3162 (1997)
8. J. Xin: SIAM Rev. **42**, 161 (2000)
9. N. Peters: Turbulent Combustion. Cambridge University Press, Cambridge, UK (2000)
10. A. N. Kolmogorov, I. G. Petrovskii and N. S. Piskunov: Moscow Univ. Bull. Math. **1**, 1 (1937)
11. R. A. Fisher: Ann. Eugen. **7**, 355 (1937)
12. E. R. Abraham: Nature **391**, 577 (1998)
13. J. Ross, S. C. Müller and C. Vidal: Science **240**, 460 (1988)
14. M. Abel, A. Celani, D. Vergni and A. Vulpiani: Phys. Rev. E **64**, 046307 (2001)
15. P. Constantin, A. Kiselev, A. Oberman and L. Ryzhik: Arch. Ration. Mech. **154**, 53 (2000)
16. B. Audoly, H. Beresytcki and Y. Pomeau: C. R. Acad. Sci. **328**(Série II b), 255 (2000)
17. N. Vladimirova, P. Constantin, A. Kiselev, O. Ruchayskiy and L. Ryzhik: Combust. Theory Model **7**, 487 (2003)
18. A. R. Kerstein, W. T. Ashurst and F. A. Williams: Phys. Rev. A **37**, 2728 (1988)
19. U. Frisch, A. Mazzino and M. Vergassola: Phys. Rev. Lett. **80**, 5532 (1998)
20. G. I. Taylor: Proc. Lond. Math. Soc. Ser. 2 **20**, 196 (1921)
21. E. Montroll and M. Schlesinger: In: Montroll, E. W., Lebowitz, J. L. (eds.) Studies in Statistical Mechanics, Vol. 11, p. 1. North-Holland, Amsterdam (1984)
22. M. F. Schlesinger, B. West and J. Klafter: Phys. Rev. Lett. **58**, 1100 (1987)
23. M. Avellaneda and A. Majda: Commun. Math. Phys. **138**, 339 (1991)
24. M. Avellaneda and M. Vergassola: Phys. Rev. E **52**, 3249 (1995)
25. G. Matheron and G. De Marsily: Water Resour. Res. **16**, 901 (1980)
26. G. I. Taylor: Proc. R. Soc. A **219**, 186 (1953); Proc. R. Soc. A **225**, 473 (1954)
27. A. Bensoussan, J.-L. Lions and G. Papanicolaou: Asymptotic Analysis for Periodic Structures. North-Holland, Amsterdam (1978)
28. C. M. Bender and S. A. Orszag: Advanced Mathematical Methods for Scientists and Engineers. McGraw-Hill, New York (1978)

29. A. Mazzino: Phys. Rev. E **56**, 5500 (1997)
30. A. Mazzino, S. Musacchio and A. Vulpiani: Phys. Rev. E **71**, 011113 (2005)
31. P. Castiglione, A. Mazzino, P. Muratore-Ginanneschi and A. Vulpiani: Physica D **134**, 75 (1999)
32. M. E. Fisher: J. Chem. Phys. **44**, 616 (1966)
33. J.-P. Bouchaud, A. Georges, J. Koplik, A. Provat and S. Redner: Phys. Rev. Lett. **64**, 2503 (1990)
34. R. Mancinelli, D. Vergni and A. Vulpiani: Eur. Phys. Lett. **60**, 532 (2002); Physica D **85**, 175 (2003)
35. G. M. Zaslavsky, D. Stevens and H. Weitzener: Phys. Rev. E **48**, 1683 (1993)
36. T. H. Solomon and J. P. Gollub: Phys. Rev. A **38**, 6280 (1988)
37. A. J. Lichtenberg and M. A. Lieberman: Physica D **33**, 211 (1988)
38. Y. H. Ichikawa, T. Kamimura and T. Hatori: Physica D **29**, 247 (1987)
39. P. Leboeuf: Physica D **116**, 8 (1998)
40. W. J. Cocke: Phys. Fluids **12**, 2488 (1969)
41. U. Frisch: Turbulence. Cambridge University Press, Cambridge (1995)
42. G. Boffetta, M. Cencini, M. Falcioni and A. Vulpiani: Phys. Rep. **356**, 367 (2002)
43. G. Boffetta, A. Celani, M. Cencini, G. Lacorata and A. Vulpiani: Chaos **10**, 50 (2000)
44. G. Lacorata, E. Aurell and A. Vulpiani: Ann. Geophys. **19**, 121 (2001)
45. M.-C. Jullien: Phys. Fluids **15**, 2228 (2003)
46. G. Boffetta, M. Cencini, S. Espa and G. Querzoli: Europhys. Lett. **48**, 629 (1999); Phys. Fluids **12**, 3160 (2000)
47. G. Querzoli: Atmos. Environ. **16**, 2821 (1996)
48. L. F. Richardson: Proc. R. Soc. **A110**, 709 (1926)
49. G. Boffetta, A. Celani, A. Crisanti and A. Vulpiani: Europhys. Lett. **46**, 177 (1999)
50. G. Boffetta and I. M. Sokolov: Phys. Fluids **14**, 3224 (2002)
51. G. Gioia, G. Lacorata, E. P. Marques Filho, A. Mazzino and U. Rizza: Bound. Layer Meteorol. **113**, 187 (2004)
52. G. Boffetta, A. Celani, A. Crisanti and A. Vulpiani: Phys. Rev. E **60**, 6734 (1999)
53. G. Boffetta and I. M. Sokolov: Phys. Rev. Lett. **88**, 094501 (2002)
54. M. Freidlin: Functional Integration and Partial Differential Equations. Princeton University Press, Princeton, NJ (1985)
55. P. F. Embid, A. J. Majda and P. E. Souganidis: Phys. Fluids **7**(8), 2052 (1995)
56. M. Cencini, A. Torcini, D. Vergni and A. Vulpiani: Phys. Fluids **15**, 679 (2003)
57. W. T. Ashurst and G. I. Shivanshinsky: Combust. Sci. Tech. **80**, 159 (1991)
58. P. D. Ronney: *Some Open Issues in Premixed Turbulent Combustion*, Lect. Notes Phys. **449**, 3–22. Springer, Heidelberg (1995)
59. B. Khouider, A. Bourlioux and A. J. Majda: Combust. Theory Model **5**, 295 (2001)
60. V. Yakhot: Combust. Sci. Tech. **60**, 191 (1988)
61. R. C. Aldredjee: *The Scalar-Field Front Propagation Equation and Its Applications*, Lect. Notes Phys. **449**, 23–35. Springer, Heidelberg (1995)
62. R. C. Aldredge: Combust. Flame **106**, 29 (1996)
63. A. Oberman: PhD Thesis, University of Chicago, 2001
64. A. K. Pattanayak: Physica D **148**, 1 (2001)

Diffusion and Reaction–Diffusion in Steady Flows at Large Péclet Numbers

Y. Pomeau

Laboratoire de Physique Statistique de l'Ecole normale supérieure, 24 Rue Lhomond, 75231 Paris Cedex 05, France
yves.pomeau@lps.ens.fr

Abstract. In many, if not most, geophysical flows molecular diffusion is formally negligible because its relative importance is measured by the inverse of a large Péclet number. This has motivated a number of studies, based in some way or another on various theoretical approaches to turbulence, where the assumed randomness of the turbulent flow plays the center role. If one adds to this complicated situation of the passive scalar the possibility of chemical reactions changing the chemical layout of the fluid, the number of unresolved issues increases dramatically. The chemistry is often such that, without flow, the reaction progresses by a front (the flame front in gaseous combustion) sweeping the system, across which the reaction takes place, from an unburnt (fresh) side to a burnt side. I look at these problems in the case of steady flows made of rolls, as generated by Rayleigh–Bénard instability for instance, certainly an idealization of real turbulence, although it is not clear by how much. The propagation of the chemical reaction in this system depends first on the diffusion of a passive scalar, an interesting question in the limit of a large Péclet number in a flow without open flow lines. The main result there is that the effective diffusion is somewhere in between the molecular diffusion and the "turbulent" diffusion. Once chemical reactions are taken into account, that is when one considers the reaction–diffusion case, one finds that the front speed is the laminar propagation velocity (without flow) times the Péclet number to the power 1/4. I refine this last result and give the behavior of the prefactor in the Zel'dovich limit of a narrow reaction zone.

1 Introduction

If one does not want to be overwhelmed by the problem of statistics of turbulent flows, a way of approaching them is to assume that such a flow has actually a simple structure in space and time and try to do the best of it. For example, this is what is besides the idea of boundary layer per se: the steady fluid equations are solved in the external domain by assuming that the Reynolds number is large and that the flow is almost everywhere potential, but in a narrow layer near the solid where the outside flow has to manage in one way or another the merging with the viscous layer near the solid surface.

Y. Pomeau: *Diffusion and Reaction–Diffusion in Steady Flows at Large Péclet Numbers*,
Lect. Notes Phys. **744**, 71–84 (2008)
DOI 10.1007/978-3-540-75215-8_3

Another extreme example of simplification of turbulent flows is to assume that it has the same spatial structure as near the instability threshold. A wide class of such structures are the cellular flows generated for instance by fluid instabilities like Rayleigh–Bénard thermal convection or Taylor–Couette flow between cylinders.

Below I assume the flow to be "fast" (in a precise sense) but stationary, an assumption that brings an enormous simplification to all the problems considered. Moreover, I shall consider phenomena happening within this steady flow, without changing it. This is the classical "passive" tracer. However, this tracer will not be simply carried by the flow, but it will be submitted to some kind of chemical reaction (the case of a suspension settling by gravity in a fast flow has some interest too, see [1]). That the flow is assumed to be fast is equivalent to say that it carries with it the passive scalar, so that any other effect like molecular diffusion, chemical reaction or gravitational settling is a small perturbation (small but not necessarily irrelevant, as we shall see). Suppose that the flow lines draw a pattern of rolls: in 2D, because of the assumed incompressibility of the fluid, the flow lines are almost all closed, except for a connected network of separatrices. Because of the large flow speed the passive scalar attains very fast an uniform concentration along each flow line, although the transfer across flow lines requires molecular diffusion. As a side remark, the same problem in 3D is far more complicated, and I am unaware of any systematic study of the general structure of incompressible steady 3D flows. Numerically it seems [2] that part of the flow lines remain closed. By analogy with the classical KAM situation in dynamical systems, one expects some flow lines to be chaotic, but they seem to be quite rare. Although the case of reaction–diffusion equations in a shear flow has some interest [3], I shall focus on cellular flows, without mean flow: in such 2D flows all flow lines are closed, but for the separatrices. The network of separatrices being open in some sense, the passive scalar diffuses very fast across the whole system and the effective diffusion coefficient on a large scale is the molecular diffusion coefficient times the (large) square root of the Péclet number, a result recovered in Sect. 2 by the statistical argument in [4] (see also [5]). In Sect. 3, I look at the case studied with Basile Audoly and Henri Berestycki [3], the one where, besides molecular diffusion, some irreversible chemical reaction takes place. A somewhat new result is the reduction of the equations of [3] to a free boundary problem whenever the chemical reaction occurs in a thin zone, as in Zel'dovich theory of combustion.

2 Effective Diffusion in a Fast Cellular Flow

This section is devoted to a short derivation of the effective diffusion coefficient in a cellular flow at large Péclet number. The starting point is the convection–diffusion equation:

$$\frac{\partial u}{\partial t} + A \, \nabla \cdot (q(r)u) = \Delta u \,. \tag{1}$$

In equation (1), $u(r, t)$ is the scalar field, depending a priori on the position r and time t, Δ is the usual space Laplacian, and $q(r)$ the space dependent but steady flow velocity (both $q(r)$ and r are vectors of the usual Euclidan space but no special notation will show it). This equation is written in such a way that the unit length is the width of the cell of the cellular flow, although the time unit is the ratio of this square length to the molecular diffusion constant. This makes formally equal to one the coefficient of the diffusion term on the right hand side of (1). Furthermore the unit velocity is the ratio of the diffusion constant to the unit length. The basic assumption is that the (dimensionless) coefficient A in front of the advection term (1) is large. It is often called the Péclet number, $A = \frac{q\lambda}{D_m}$, λ wavelength of the flow pattern, q typical flow speed and D_m molecular diffusion coefficient. We are going to look at 2D velocity fields, depending on coordinates x and y. For an incompressible flow, one can introduce a stream function $\Psi(x, y)$ such that the Cartesian components of $q(x, y)$ are $q_x = \frac{\partial \Psi}{\partial y}$ and $q_y = -\frac{\partial \Psi}{\partial x}$. The flow lines draw a set of cells wherever they are closed, the boundaries of the cells draw a net of separatrices connecting a network of hyperbolic fixed points. I shall assume that this system is periodic in the two directions of space. In real life, it is usually one dimensional and squeezed between rigid plates. The difference between the two situations may be significant, but I shall not consider it here (see [5, 6]).

The system (1) can be solved by splitting the space into boundary layers near the separatrices and the interior of the cells. However, I shall present below a probabilistic approach to this problem. The first obvious remark is that the molecular diffusion is relevant only to describe the diffusion across the flow lines in the large Péclet number limit. The diffusion across the closed flow lines of a given cell takes a time of order $\frac{\lambda^2}{D_m}$, that is one in our dimensionless units. A diffusing particle will explore the whole cell over this time scale. Imagine now that a particle crosses a separatrix between two cells at some initial time $t = 0$. Its exploration across the flow lines may be seen as similar to the random addition/subtraction of a variable, that denotes the distance across the closed flow lines (in 1D the diffusion equation is the large-scale representation of the addition and subtraction of a random variable). A standard result in the theory of stochastic process states that, starting from zero, and adding or subtracting a random variable, the number of times the sum is equal to zero grows asymptotically like the square root of the number of steps. For the diffusion equation in the cellular flow, that the random sum is zero is equivalent to say that the diffusing particle crosses a separatrix. It does not make sense with a continuous time to state that a dimensionless number of crossing grows like the square root of time, since this square root has a physical dimension and cannot be a number. The relevant dimensionless number is recovered by dividing the square root by another square root time. This time is the travel time between the two end points of the separatrix: then the particle chooses at random to go on one side or the other of the separatrix. It is of order A^{-1} with the scaling of (1). Therefore, during the exploration time of a cell[1] a particle that started on a separatrix shall cross a separatrix (usually again not the one it started on) $N \sim \sqrt{A}$ times. Every time it does so, it goes at random to one side or the other of a separatrix. After N such crossings it will have made N random steps of unit length (the size of an elementary cell). The mean square distance run during this time is therefore N. Since, in our units, this square length is run during an unit time, $N = A^{1/2}$ is also the effective diffusion coefficient, the result of [4]. In ordinary units, the order of magnitude of this effective diffusion coefficient is the geometric average between the turbulent diffusion coefficient $(Aq\lambda)$ and the molecular diffusion coefficient D_m.

The same result follows from a more classical analysis of solutions of (1). It shows that the effective diffusion coefficient is intermediate between the turbulent coefficient (independent of the molecular diffusion) and the molecular coefficient. It is worth pointing out that the turbulent regime (seen here as the limit A tending to infinity) cannot be analyzed simply by looking at the solution of equations without molecular diffusion coefficients (like the heat conductivity, shear viscosity, etc.) simply set to zero: formally the diffusion term is negligible in this limit and could be believed to have no significant effect.

3 Reaction–Diffusion in Fast Cellular Flows

Zel'dovich [7] explained theoretically how a flame propagates in a premixed gas (the situation we are going to look at), as a consequence of the substitution of a metastable state by a stable one, an idea already present in Euler's winning contribution (in Latin!) to the 1738 prize of the Académie des sciences [8] on the nature and the propagation of fire. But, although in many practical applications the effects of hydrodynamic turbulence are very important, they are absent in the classical formulation of the reaction–diffusion equations used by Zel'dovich. Very often the turbulent fluctuations of the hydrodynamic velocity are far bigger than the intrinsic speed of propagation of the flame itself. This section, following a previous Note on the same topic [3], exposes what happens in the same limit (large "fluctuating" fluid velocity), but by assuming, as in the previous section, a frozen velocity field drawing a periodic set of cells bounded by a network of separatrices. The reaction–diffusion equation includes now a non-linear term that represents the change in temperature due to the chemical reaction. This is the $f_r(u)$ term in

$$\frac{\partial u}{\partial t} + A\nabla \cdot (q(r)u) = \Delta u + f_r(u). \tag{2}$$

---

[1] This exploration time may be seen also as the typical time after which a particle starting near the separatrix will reach the center of a cell, usually not one of the two cells split by the separatrix it started from.

The function $f_r(u)$ has positive values and describes the heat release by the chemical reaction (in technical term, one assumes the Lewis number to be one, so that chemistry and heat transfer are one and the same thing). Two different kind of "chemistry" can be considered, depending on the shape of the function $f_r(u)$:

(i) The soft chemistry, where $f_r(u)$ is (for instance) like $ku(u - u_0)$, k positive constant and u_0 temperature of the burned gas, $u = 0$ being that of the fresh reactants.

(ii) A stiff Arrhenius chemistry such that f_r is non-zero only near the maximum of temperature, something like $f_r(u) = \epsilon^{-1} F_A[(u_0(1-\epsilon) - u)(u_0 - u)]$ with $0 < \epsilon \ll 1$, $F_A(.)$ bell shaped, positive and of order 1 vanishes for negative arguments, and $u_1 = u_0(1 - \epsilon)$ is the temperature above which the reaction begins, and that is slightly below the temperature of the burned gas. In the limit $\epsilon \to 0_+$, the thickness of the reaction zone (= the domain where f_r takes non-negligible values) tends to zero proportional to ϵ and flame dynamics reduces itself to the one of a geometrical surface. Solution of the reaction–diffusion (2) are known without streaming term $(A = 0)$. In case (i) (soft chemistry), there is a continuum of planar front solutions in the form $u(x,t) = U_c(x - ct)$ with the boundary conditions $U_c \to 0(/u_0)$ for $(x - ct) \to +(/-)\infty$, the fresh (/burnt) gas being always taken at $+(/-)\infty$.

In case (i) for many initial conditions the relevant solution is the Kolmogoroff solution corresponding to the smallest value of c. On the contrary, in case (ii) there is only one solution of the form $U(x-ct)$ with $c = \frac{1}{u_0}\left(2 \int_{u_0}^{u_1} f_r(u)\, du\right)^{1/2}$ in the limit $\epsilon \to 0$. This stiff reaction case was what was considered by Zel'dovich, and it is relevant for flame propagation in premixed gases. I shall denote as c_L the intrinsic speed (unique for the stiff reaction rate, as given above with an exact vanishing of the reaction when u becomes smaller than u_1 or whenever the Kolmogoroff solution applies in the soft chemistry case). In [3], we examined the case of a fast shear flow (the "Bunsen burner" problem according to the experts in the field): there the efficient speed is the fastest flow velocity toward the fresh gas in the frame of this fresh gas, with some interesting structure near cusps on the flame surface. Below, I shall examine instead the propagation of a front in a periodic cellular flow as described by (2). On long time scales, the state in contiguous cells changes from metastable $(u = 0)$ to stable $(u = u_0)$ and the cell-to-cell propagation occurs at a constant rate, so that on a large scale there is a well defined effective speed of propagation, although there is strictly speaking no solution in the simple form $u(x,y,t) = u(x - ct, y)$. In the limit $A \to \infty$ a rational method of approximation yields a solution of Eq. (2) that represents this propagating front. It relies a lot on the analysis of [6], in particular in its dealing of the vicinity of the separatrices. The main source of difficulty is in the multiconnectedness of the network of separatrices.

Let us assume first that the periphery of a given cell is "lighted," that is that the reaction rate $f_r(u)$ differs significantly from zero there. The invasion of the cell is simple to understand then: a flame propagates inward radially at the laminar velocity c_L: because of the fast mixing along the flow lines, u is almost exactly uniform along a closed flow line, although the propagation across the flow lines depends completely on the reaction and diffusion term, independent of the advection. Therefore the time needed for a cell to burn is just the time for a flame to propagate inward from the periphery to the cell center, a time of order $\frac{\lambda}{c_L}$.

The limit of large A requires one to look at two different cases for the stiff chemistry limit. This can be seen by looking at the time scales for the dynamics near the separatrix. There are two fast time scale there: first the transit time by flow motion from one end of the separatrix to the other, of order A^{-1} (recall that A is large) although the time scale for the chemical reaction is of order ϵ^2, with ϵ small. This time is the time needed for u to change of order ϵ under the effect of the reaction term $f_r(u)$ of order ϵ^{-1} on the right hand side of (2). The two short time scales are independent and each one can be much shorter, of the same order or much larger than the other one (this represents the mathematical possibilities, not necessarily what can be experimentally achieved. In this respect it seems hard to get physically a time scale for the chemistry much longer than the time of transit along the separatrix). The reference [3] deals with the case where the shortest time scale is the flow time, A^{-1} (again perhaps not the most realistic assumption). The opposite case, A large, ϵ^2 small but $1 \gg A\epsilon^2$ is also briefly looked at in the Sect. 3.2 below. If $A\epsilon^2 \gg 1$ one can neglect the chemical reaction in the boundary layer near the separatrix, where everything reduces to linear diffusion ($f_r(u)$ on the right hand side of (2) can be neglected there). The chemical reaction would become dominant in the boundary layer in the opposite limit $1 \gg A\epsilon^2$. Following the ideas of [6], this boundary layer is where a cell-to-cell heat flux takes place. This flux is perpendicular to the separatrix, and as shown in [6] ((7) there), its net effect is to transfer to cell number i a total flux (actually the rate of change of the integral of u over the area of the cell number (i):

$$F_i = s(f_{i+1} + f_{i-1} - 2f_i). \tag{3}$$

In (3) F_i is the sum of the flux from the cells $(i+1)$ and $(i-1)$ toward cell i across the separatrices (the notation f_{i+1}, f_{i-1}, etc. is defined below and it follows the one of [6]. No confusion should arise from the similar notation for the rate of chemical reaction). The flux from one cell to the next occurs across the separatrix and is defined in terms of the average value of u in each cell on a closed flow line far from the boundary layer at the scale of thickness of this boundary layer, but very close to the separatrix on the scale of the cell size. This average is denoted as f_i for cell number i. The intermediate thickness where this average is computed is well defined because the boundary layer has a thickness of order $A^{-1/2}$ although the cell size is the unit length. This yields for the solution of the heat transfer across the boundary layer an asymptotic value f_i, that is defined as the value of u in this matching domain. The order of magnitude of the resulting flux is the length of the separatrix (one in our units) times the gradient of u across the separatrix, namely $\frac{f_{i-1}-f_i}{A^{-1/2}}$ times the molecular diffusion coefficient, that has the value one here. The formula (3) results from the addition of the fluxes from cell $(i-1)$ to i and $(i+1)$ to i. The coefficient s is a kind of effective cell-to-cell diffusion coefficient. It is of order $A^{-1/2}$, from the arguments just given. Its precise expression is derived in [5]:

$$s = \sqrt{\frac{1}{\pi} \int_{y_0}^{y_1} v(y)\mathrm{d}y} \approx A^{1/2}. \tag{4}$$

The integral in (4) is carried over the length of the separatrix, that is along a vertical axis and runs from $y = y_0$ to $y = y_1$ (separatrices are assumed to be straight, this is not necessary but makes notations easier). The velocity there is parallel to the y direction too, and locally the stream function takes the form $\Psi(x,y) \approx -x\, v(y)$, $v(y) = A\, q_y(x = 0, y)$ is the vertical speed on the separatrix.

It remains to show how the flux condition (3) enters into the solution of the reaction–diffusion equation inside cells. For that purpose one writes this reaction-diffusion equation in such a way that the fast uniformization along the flow lines is made more obvious. For that purpose, one uses as a new set of variables of position based on the value of the stream function Ψ or a related variable: following [9] it is convenient to use the area $a(\Psi)$ enclosed inside a closed flow line. This area is a function of Ψ solution of $\frac{\mathrm{d}a}{\mathrm{d}\Psi} = \int \frac{\mathrm{d}l}{q}$, $\mathrm{d}l$ element of length along the flow line and $q(l)$ absolute value of the flow speed q at the point of coordinate l along this line. The reaction–diffusion equation is averaged along the closed flow lines, that is on the fast time scale, with the result, relevant for the slow dynamics:

$$\frac{\partial u_i}{\partial t} = \frac{\partial}{\partial a}\left(D(a)\frac{\partial u_i}{\partial a}\right) + f(u_i). \tag{5}$$

As before the subscript i refers to the cell number. The unknown function u_i depends on the discrete index i and on the stream line index a. Moreover

$$D(a) = \int \mathrm{d}l q(l) \int \frac{\mathrm{d}l}{q(l)}.$$

The function $D(a)$ becomes singular whenever the true no sliding condition applies to the fluid equation (and makes very small the fluid velocity on the solid that bounds the fluid). I shall not consider this and assume free boundary conditions. It remains to impose the boundary condition (3) to solutions of (5). This boundary condition is expressed in two different ways: first, as seen from the point of view of (5), it is a Neuman-like condition for the derivative of $u_i(a_M)$ at $a = a_M$. Here a_M is for the value of a on the streamline bounding the cell. Rigorously this boundary condition should concern a streamline in the matching region between the interior of the cell and the boundary layer near the separatrix, but, as the boundary layer has a thickness of order $1/A$, it does not make any difference to impose the boundary condition right at $a = a_{\mathrm{M}}$. From (5) this flux has the value $-D(a_{\mathrm{M}})\frac{\partial u}{\partial a}$, $D(a_{\mathrm{M}})$ being the value of $D(a)$ on the border of the cell, namely when a reaches its maximum a_{M}, the area of the cell. From the solution of the equation in the boundary layer, we have another expression for the total flux. Equating the two, we obtain

$$-D(a_{\mathrm{M}})\frac{\partial u_i(a)}{\partial a}\Big|_{a=a_{\mathrm{M}}} = s\left(u_{i+1} + u_{i-1} - 2u_i\right)\big|_{a=a_{\mathrm{M}}}. \tag{6}$$

Every relevant quantity has been defined now, and we are ready to pose the problem of flame propagation throughout the structure. This amounts to look for a solution of (5) in the form $u(i - Ct = \zeta, a)$:

$$-C\frac{\partial u}{\partial \zeta} = \frac{\partial}{\partial a}D(a)\frac{\partial u}{\partial a} + f_r(u). \tag{7}$$

This has been derived from (5), by assuming that u changes very little from cell to cell, which allows to consider the variable i as continuous, something that is shown to be legitimate a posteriori in the limit $A \to \infty$. In (7) a changes from zero (the center of the cell) to a_{M}, the latter corresponding to the cell boundary. Now almost all the remaining difficulty lies in the boundary conditions. One boundary condition is to impose that for $\zeta \to -\infty$, u tends to 0 (fresh gas, or metastable state) although it tends to u_0 for $\zeta \to +\infty$ (burnt gas or stable state). The other condition express the exchange flux between the cells (it is derived from (3) by assuming i to be continuous, and that u changes very little from cell to cell):

$$-D(a_{\mathrm{M}})\frac{\partial u(\zeta, a)}{\partial a}\Big|_{a=a_{\mathrm{M}}} = s\frac{\partial^2 u}{\partial \zeta^2}\Big|_{a=a_{\mathrm{M}}}. \tag{8}$$

Even in the simple linear diffusion case (without the reaction term $f_r(u)$) it is not straightforward [5] to impose this last condition. In the present case, we proceed in two steps: first one gets rid of the large parameter A in the equation by rescaling. This yields already a non-trivial result, namely that the propagation speed C is like $A^{1/4}$ at large A. Another byproduct of this reduction is a dimensionless numerical problem (that is without the large parameter A: it has disappeared thanks to the rescaling). This numerical problem (that would yield ultimately the numerical constant in front of the $A^{1/4}$ dependence of C at large A) is rather intricate. However, I shall explain how it reduces to a free boundary problem in the Zel'dovich limit of a very thin flame.

3.1 Rescaling at Large A

In our formulation of the reaction–diffusion problem, the large quantity A appears in the boundary condition (8) only, through the quantity s that scales like $A^{1/2}$. Since it multiplies the second derivative $\frac{\partial^2 u}{\partial \zeta^2}|_{a=a_M}$, one can get rid of A by rescaling ζ as $\overline{\zeta} = \zeta A^{-1/4}$ and $\overline{s} = sA^{-1/2}$. The substitution of $\overline{\zeta}$ in place of ζ in (7) yields on its left hand side $-CA^{-1/4}\frac{\partial u}{\partial \overline{\zeta}}$. There the extra $A^{-1/4}$ is canceled by defining a rescaled C as $\overline{C} = CA^{-1/4}$. This completes the proof (proof in a weak sense for sure) that the effective speed of propagation is of order $A^{1/4}$ at large A. Indeed, this assumes implicitly that the rescaled problem (to be written down shortly) has a solution, that gives a finite value of $\overline{C}$. Notice too at this step that one approximation made on the way is coherent: the effective scale of dependence in ζ is of order $A^{1/4}$, supposing again that the scale of variation with respect to $\overline{\zeta}$ is of order 1. This length scale is much bigger than the cell length, which supports the assumption of a continuous variation with respect to the index i. The physical consequence is that the thickness of the flame is of order $A^{1/4}$ as well. To summarize, the "numerical" problem left to solve is

$$-\overline{C}\frac{\partial u}{\partial \overline{\zeta}} = \frac{\partial}{\partial a}D(a)\frac{\partial u}{\partial a} + f_r(u) \tag{9}$$

with the boundary conditions:

$$- D(a_M)\frac{\partial u(\overline{\zeta}, a)}{\partial a}|_{a=a_M} = \overline{s}\frac{\partial^2 u}{\partial \overline{\zeta}^2}|_{a=a_M} \tag{10}$$

and $\overline{\zeta} \rightarrow -\infty$, u tends to 0 (fresh gas, or metastable state) although it tends to u_0 for $\overline{\zeta} \rightarrow +\infty$ (burnt gas or stable state).

3.2 Free Boundary Problem in the Stiff Reaction Case

This subsection requires first to explain in general how the reaction–diffusion problem, as the one posed in (9) can be reduced to a free boundary problem for the stiff reaction case. We derive first the equation of motion of a flat interface from the reaction-diffusion equation (9), then we relate it to the solution of the set (9, 10) in the limit $\epsilon \rightarrow 0$.

In this limit, one looks for a solution of (9) in the form $u = \tilde{u}(a - \tilde{C}\overline{\zeta})$. Since we are concerned with the cell-to-cell propagation, all the action is near the separatrices, so that one can replace in (9) $D(a)$ by its value on the separatrix, namely $D(a_M)$. The velocity $\tilde{C}$ that is so introduced is a transform of the cell-to-cell velocity of propagation. Putting $b = a - \tilde{a}(\overline{\zeta})$, the time derivative $\frac{d\tilde{a}}{d\overline{\zeta}}$ is found by solving the equation

$$-\overline{C}\tilde{C}\frac{d\tilde{u}}{db} = D(a_M)\frac{d^2\tilde{u}}{db^2} + f_r(\tilde{u}) \tag{11}$$

with appropriate boundary conditions to be discussed later, and with $\tilde{C} = \frac{d\tilde{a}}{d\overline{\zeta}}$.

In the stiff reaction case,

$$f_r(\tilde{u}) = \epsilon^{-1} F_A[(u_0(1 - \epsilon) - \tilde{u})(u_0 - \tilde{u})]$$

with $0 < \epsilon \ll 1$. In this limit, according to Zel'dovich, one splits the solution of (11) into the thin reaction zone where the left hand side of (11) is negligible, and into the diffusive layer where the reaction term is negligible. In the reaction zone, one multiplies the right hand side of (11) by $\frac{d\tilde{u}}{db}$ and integrate from $\tilde{u} = u_0(1 - \epsilon)$ (the fresh side) to $\tilde{u} = u_0$ (the burnt side), where $\tilde{u}$ becomes uniform. This yields:

$$\frac{D_M}{2}(\frac{d\tilde{u}}{db})^2|_{u=u_0(1-\epsilon)} = \int_{u=u_0(1-\epsilon)}^{u_0} f_r(\tilde{u})d\tilde{u} . \tag{12}$$

Although the temperature becomes constant on the burnt side, its knowledge requires the solution of the heat transfer problem on the fresh side. From (12) one gets the value of the normal gradient of $\tilde{u}$ on the interface, on the fresh side:

$$\frac{\partial \tilde{u}}{\partial b}|_{(u=u_{0,f})} = \sqrt{\frac{2}{D_M} \int_{u=u_0(1-\epsilon)}^{u_0} f_r(\tilde{u})d\tilde{u}} . \tag{13}$$

In (13), the subscript $(u = u_{0,f})$ is to mean that the quantity $\frac{\partial \tilde{u}}{\partial b}$ is taken practically for $u = u_0$, but on the fresh side. Furthermore, the ordinary derivative $\frac{d\tilde{u}}{db}$ in the 1D case (11) has been replaced in (13) by the partial derivative $\frac{\partial \tilde{u}}{\partial b}$ to mean that it is the normal gradient of $u(a)$ taken on the fresh side of the flame surface. Therefore, on this fresh side one has to solve (in principle) two conditions, the first condition is given in (13) and represents physically the fact that some heat release takes place on the flame surface (the right hand side) and is balanced by a heat flux (the left hand side). The other condition is that the temperature u is equal to u_0 on the flame surface. Strictly speaking, this u_0 should be $u_0(1 - \epsilon)$ but ϵ is neglected as being small. On the fresh side one has to take into account the heat equation only (since the reaction term is negligible there). This is one more condition than what can be accommodated. The only free parameter left is the flame velocity (here the product $\overline{C}\tilde{C}$, denoted $\hat{C} = \overline{C}\tilde{C}$ later on). For an infinitely extended fresh gas (as shown in Sect. (3.3) below this is not exactly our situation, but this does not change the foregoing analysis in any fundamental way) there is a solution at constant speed, such that u tends to zero at b tending to infinity (minus infinity here, which is possible with a negative $\hat{C}$): $U = u_f exp[-(b\hat{C}/D_M)]$. The multiplicative constant u_f (f for fresh) is found by imposing that $u = u_0$ for $b = 0$ (the location of the flame front), and the speed is found by imposing the gradient condition (13):

$$\hat{C} = \frac{1}{u_0}\sqrt{2D(a_M) \int_{u=u_0(1-\epsilon)}^{u_0} f_r(\tilde{u})d\tilde{u}} . \tag{14}$$

This is a relation between two quantities, $\overline{C}$ and $\tilde{C}$ (remember that $\hat{C} = \overline{C}\tilde{C}$). One more relation is needed. If one assumes (and again this is to be checked a posteriori) that the fresh gas is infinitely extended, at least near the separatrix, the solution in the lighted region is almost exactly a flame propagating at the constant speed $\hat{C}$. This is equivalent to a line drawn in the $(a, \overline{\zeta})$ coordinate system and separating the fresh side (away from the separatrix at $a = a_M$) from the burnt side (in between the flame trajectory and the separatrix). Leaving aside the question of the intersection of this line with the separatrix, since the temperature is constant (in time) and uniform on the burnt side, the boundary condition (10) is automatically satisfied there.

It remains to examine the way the trajectory of the flame merges with the separatrix. The crossing area in the $(a, \overline{\zeta})$ plane is such that the temperature on the separatrix goes from $u_0(1 - \epsilon)$ (on the fresh side) to u_0. Therefore a natural scaling there is to take the variation of u of order ϵ. Putting this into the full reaction–diffusion equation (9), one gets all terms of the same order of magnitude (with respect to ϵ) if a (actually $a - a_M$) scales like ϵ (recall that $f_r(u)$ is of order $\frac{1}{\epsilon}$). The left hand side of (9) scales as the right hand side (that is as $\frac{1}{\epsilon}$) if $\overline{\zeta}$ scales as $\overline{C}\epsilon^2$. Another relation between the scaling of $\overline{\zeta}$ and of a in the transition layer comes from the boundary condition (10). The two sides are of the same order in ϵ if $\overline{\zeta} \sim \epsilon^{1/2}$, $\overline{C} \sim \epsilon^{-1/2}$, the thickness of the boundary layer in the variable a being $a - a_M \sim \epsilon$. This defines the size of the domain in the coordinate plane $(a, \overline{\zeta})$ where the flame trajectory crosses (smoothly) the separatrix at $a = a_M$. Of course, this puts a bound on the domain of validity of the present theory, since the "physical size" of the crossing area has to be much larger than the size of a single cell: otherwise one could not replace the discrete cell index i by a continuous variable. Since $\overline{\zeta}$ scales like $\epsilon^{1/2}$, and ζ like $A^{1/4}\overline{\zeta}$, ζ scales in the region under consideration like $A^{1/4}\epsilon^{1/2}$, the product of a large number $A^{1/4}$ times a small one $\epsilon^{1/2}$. The result must be large with respect to 1, the cell size in our dimensionless units. This requires $\epsilon^2 \gg 1/A$. The physical interpretation of this condition is as follows: $1/A$ is the order of magnitude of the convection time along the separatrix between two cells. It must be much smaller than ϵ, the time scale for the reaction. Otherwise one should resolve the reaction along each separatrix. This would require to replace the inner equation (11) by a more complex inner problem, including an explicit time dependence. In particular, one can think to the opposite limit A large, ϵ small, but $1/A \gg \epsilon$. In this limit, the fastest process is the chemical reaction. Therefore one expects that the effective velocity of propagation becomes independent on ϵ. To get the A dependence of this velocity (again in the limit A large, ϵ small and $1/A \gg \epsilon$), I suggest to do as follows: in this parameter domain the law for the effective flame velocity should merge continuously with the one valid in the opposite limit, that is in the range A large, ϵ small, but $1/A \sim \epsilon$. In this intermediate range, $C = A^{1/4}\overline{C}$, $\overline{C} \sim \epsilon^{-1/2} \sim A^{1/4}$ and $C \sim A^{1/2}$, which yields the scaling of the effective speed of propagation in the range A large, ϵ small and $1/A \gg \epsilon$.

The solution is completed by solving the temperature field on the fresh gas side. There, near the separatrix, the non-linear production term becomes negligible, so that one can use the same type of method as in [6]. This fresh region merges smoothly with the fresh region in front of the advancing flame. Therefore one has to solve the linear diffusion equation in a wedge-like domain: at scales much larger than the scales of the crossing region just considered, one can see the flame as a line of discontinuity, where the temperature is u_0, although the normal gradient is fixed. Moreover, on the separatrix, the boundary condition is the condition (10), that makes a well-defined problem. Notice too that the effective velocity of propagation in the cellular flow is $C = A^{1/4}\overline{C}$, that scales as $C \sim A^{1/4}\epsilon^{-3/2}$ as A becomes large and ϵ tends to zero. Therefore, to derive the effective speed, one cannot use blindly the general formula $C = \sqrt{\frac{D_{\text{eff}}}{\tau}}$ where D_{eff} is the effective diffusivity of a passive scalar in the roll structure and τ a typical time scale for the chemistry. This formula would give correctly the $A^{1/4}$ dependence because D_{eff} is like $D(a_M)A^{1/2}$, but not the ϵ dependence of C: in the stiff reaction case, the flame velocity is ϵ-independent, so that τ must be also ϵ-independent, which does not account for the $\epsilon^{-1/2}$ dependence of C.

3.3 Propagation Inside a Single Cell

This subsection does not concern a central issue of this work. It is only to investigate how a flame propagates inside a cell, once its periphery has begun to burn. The relevant equation of propagation is the 1D equation (5). There the variable of position is the quantity denoted as a and that is a way of indexing the closed streamlines inside a cell. Our formulation scales out any parameter from (5). This has a rather unfortunate consequence, namely the fact that a dimensionless parameter remains hidden: from the derivation of Sect. (3.2) (we look at the stiff reaction case now) the flame has a thermal thickness of order $\frac{D(a_M)}{|\hat{C}|}$, that was assumed implicitly to be of the same order of magnitude as the cell size. Indeed this has no special reason to be true, except for the convenience of the analysis. It makes sense to consider three possible situations: either the cell is much bigger than this length scale, both length scales are of the same order of magnitude or the cell is far smaller than the thermal thickness of the flame. Reference [3] assumed the first case (thermal thickness far smaller than the cell). In this limit, a local approximation is valid: it associates to a flow line of index a a local inward velocity of propagation:

$$C(a) = \frac{1}{u_0}\sqrt{2D(a)\int_{u=u_0(1-\epsilon)}^{u_0} f_r(\tilde{u})\mathrm{d}\tilde{u}}. \tag{15}$$

This formula is of course very close to (14), except that $D(a)$ replaces $D(a_M)$. In this framework in particular there is a well posed free boundary problem in the propagation of the flame through the cell system, since the velocity of propagation near the border of the cell is the limit value of $C(a)$ when a tends to a_M.

The intermediate range does not change the scalings, but now the diffusion problem inside every cell cannot be reduced to a simple exponential like solution of the heat equation. One has to take into account that the flame propagates inside a closed cell, where the full time defendant equation has to be solved. At the scale of the cell thickness the chemical reaction still takes place in a narrow layer of thickness of order ϵ near a closed flow line, but, to be computed, the local speed of this requires, the full solution of the heat transfer problem, including the boundary condition at the center of the cell. However, the scalings with respect to A and ϵ of the physically relevant quantities are the same as in the first case, except of course for the time necessary for a cell to burn completely. In the first case it is "large" and becomes of order 1 in the second case. It is always of order $\frac{\lambda}{C(a_\mathrm{M})}$. The case of a cell much smaller than the thermal thickness of a flame is a priori different. In this case, the heat brought by the chemical reaction shall heat the cell almost uniformly. The rate of heat release is $\lambda\sqrt{2D_m \int_{u_0(1-\epsilon)}^{u_0} \mathrm{d}u f_r(u)}$, where λ is the order of magnitude of the perimeter. This heats up the cell from $u = 0$ to $u = u_0$ in a time $\frac{\lambda u_0}{\sqrt{2D_m \int_{u_0(1-\epsilon)}^{u_0} \mathrm{d}u f_r(u)}}$. This is also the velocity of propagation of the flame inside the cell times the cell diameter, whence the effective speed of propagation inside the cell:

$$C \sim \frac{\sqrt{2D_m \int_{u_0(1-\epsilon)}^{u_0} \mathrm{d}u f_r(u)}}{u_0},$$

the same formula applicable to the stiff reaction case in an infinite system.

4 Summary and Conclusion

This intended to present some recent results concerning solutions of diffusion and reaction–diffusion problems in highly idealized situations, but always in the limit of a large Péclet number, the one relevant for geophysical applications. In this limit, the advection is dominant over the molecular diffusion, but this one cannot be neglected to obtain physically significant results. The speed of propagation of a flame for instance depends on a rather complex interplay of convection effects (the $A^{1/4}$ dependence of the effective flame speed) and of the reaction dynamics (the $\epsilon^{-1/2}$ dependence). Straightforward extensions of the present work can be imagined. For instance, one can use the same line of reasoning for the propagation of a stable thermodynamic phase into a metastable one. No big change in the analysis seems to be expected, at least if the latent heat is neglected, in such a way that there is a finite speed of propagation for the 1D problem without flow. A more complex case is the one of a phase transformation (for instance liquid to vapor or liquid to liquid) with latent heat included. This makes a rather nice exercise in applied math, with not much concrete application perhaps.

References

1. B. Simon and Y. Pomeau: Free and guided convection in evaporating layers of aqueous solutions of sucrose, transport and sedimentation of solid particles. Phys. Fluids **A 3**, 380 (1991)
2. L. de Sèze and Y. Pomeau: Flow lines in an incompressible fluid. J. Phys. (Paris) **39**, C5–95, Colloque C5, (1979)
3. B. Audoly, H. Berestycki and Y. Pomeau: Réaction diffusion en écoulement stationnaire rapide. C.R. Ac. Sci. **328**, Série IIb, 255 (2000)
4. Y. Pomeau. Dispersion dans un écoulement en présence de zones de recirculation. C.R. Ac. Sci. **301**, Série II, 1323 (1985)
5. C. Baudet, E. Guyon and Y. Pomeau: J. Phys. Lett. **46**, 991 (1985); Y. Pomeau: Dispersion at large Péclet number. In: Chaté H. et al. (eds.) Mixing Chaos and Turbulence NATO ASI series, Vol. 373, Kluwer, New York (1999)
6. W. R. Young, Y. Pomeau and A. Pumir: Anomalous diffusion of tracer in convection rolls. Phys. Fluids **A 1**, 462 (1989)
7. *Selected Works of Y.B. Zel'dovich*, Vol. I, Part 2, p. 252 *et sq.* Princeton University press, Princeton (1992)
8. L. Euler: In: De la nature et de la propagation du feu" ASPM Ed. Wasry (52130), France. ISBN2-958902-0-2
9. P. B. Rhines and W. R. Young: J. Fluid Mech. **133**, 133 (1983)

An Introduction to Radiative Transfer for Geophysicists

E. A. Spiegel

Department of Astronomy, Columbia University, New York, NY 10027, USA
eas@astro.columbia.edu

Abstract. To treat the interaction of radiation with matter we describe the radiation as a gas of photons (particles of light) in a manner almost completely free of relativistic and quantum effects. The central aim is to write a kinetic equation describing radiative transfer through matter and to mention some of the simplifying approximations appropriate for a first discussion of the implications of that equation. We discuss the structure of a hydrostatic atmosphere in radiative equilibrium and see how an upper isothermal layer emerges from the simplest treatment of this problem. We conclude by describing the heat equation of a radiating medium and explaining how the diffusion approximation and Newton's law of cooling emerge in suitable limits.

1 Introduction

Seventy years ago, if you looked at what was being written about transport in geophysics and astrophysics, you might have gathered that most of the transport in meteorology and oceanography was by turbulent fluid motions while the astrophysicists' preoccupation was with radiative transport. Since then, astrophysicists have become increasingly concerned with turbulent convection (as they had been in the nineteenth century) while geophysicists have seriously begun to take notice of radiative processes. I offer this oversimplified view of things to rationalize why an astrophysicist has been asked to lecture in this geophysics school: astrophysicists of my generation were all brought up on the theory of radiative transfer. That is why many of the older elementary texts on radiative transfer are by astrophysicists. I shall start by mentioning just a few of these by title and author only, since that should be enough to allow you to find them.

In the astrophysical literature we have the old standard, Chandrasekhar's "Radiative Transfer" now over 50 years old. Chandrasekhar once told me that this was his favorite of the many books he wrote and, in it, you get a nice view of his crisp style. A book written not long after this was "Basic

E. A. Spiegel: *An Introduction to Radiative Transfer for Geophysicists*, Lect. Notes Phys.
744, 85–100 (2008)
DOI 10.1007/978-3-540-75215-8_4   © Springer-Verlag Berlin Heidelberg 2008

Methods in Transfer Problems" by V. Kourganoff, whose aim was pedagogical as is that of the subsequent astrophysical text, "Stellar Atmospheres", by D. Mihalas. (Any book on stellar atmospheres will contain a discussion of radiative transfer in plane-parallel steady atmospheres.) A relevant mathematical reference for the subject is Davisson's book on the related topic of neutron transport theory. There are books on the subject by atmospheric scientists such as Goody's "Atmospheric Radiation" which has been revised into "Atmospheric Radiation: Theoretical Basis" with an added author, Y. L. Yung. Among more recent books with that emphasis, we have "Radiative Transfer in the Atmosphere and Ocean" by G. E. Thomas and K. Stamnes and "An Introduction to Atmospheric Radiation" by K. N. Liou. A useful reference for methods of solution is the two-volume work "Multiple Light Scattering" by H. C. van de Hulst. Finally, I would suggest that if you ever wonder about optical phenomena in our surroundings do not miss Minnaert's "Light and Colour in the Open Air" and, if you like that one, you will enjoy "Color and Light in Nature" by D. K. Lynch and W. Livingston. Where the following account differs from much of the aforementioned work is that time derivatives in the radiative transfer theory are retained even though they are generally not very important for geophysical problems. That is, in geophysics, radiative modes are much faster than the fluid modes, so that the former are enslaved by the latter. This means that we may usually omit the time derivatives in the radiative equations. However, this renders them diagnostic rather than prognostic and I leave it to you to make that transition when you need to. I have not the heart to do that in this general discussion even though all we would be giving up in that case is the retardation time as radiation emitted in one place goes to another place. However, we shall do that explicitly in the section on the radiative heat equation.

My assignment, as I understand it, is to describe how the radiation diffuses through a medium while giving some attention to how it may influence the dynamical processes taking place. To do this, I must decide whether to describe radiation as a rapidly varying electromagnetic field or as a gas of photons. I will adopt the particle picture since it is simpler to take account of the interaction of the radiation with matter from this standpoint. In either case, the radiation is relativistic and quantum mechanical, but I shall cut some corners and concoct a classical version of the theory. Moreover, in all cases of interest here, the photons are so numerous that we need not worry about quantum fluctuations. We shall also reduce the treatment of the photon gas to that of a continuum. In geophysical applications, the material velocities are so small that they are not important for the transfer of radiation so we shall not be worried about the effects of aberration of light and of Doppler shifts caused by motion of the ambient fluid and certainly not about the force that radiation exerts on matter.

In the interests of brevity, I shall not go into the quantum physics of the processes of emission and absorption of radiation except to introduce here the qualitative distinction that is often made between the processes of scattering and of absorption with re-emission. In scattering, the direction of motion of a photon is changed with little or no alteration of its energy. This can happen when a photon bounces off an aerosol or when it excites an atom from a ground state to a state from which it immediately returns to its original configuration. In absorption, the photon is swallowed up and the energy absorbed is later re-emitted in one or more photons, leaving no memory of the original photon. We shall typically treat these as separate processes, though there may sometimes be overlap. If you want to delve into these things, you might look at some old-fashioned approaches first, especially for the molecular transitions. For example, you could begin such a study with E. J. Bowen's "Chemical Aspects of Light" if you do not want to go deeply into modern quantum mechanics.

2 Some Definitions

I have often complained to my mathematical friends that modern math books frequently begin with some 30 pages of unmotivated definitions that I often cannot get through. This practice of mathematicians is what I blame for my limited mathematical knowledge. So if I begin this discussion with some definitions, I must defend myself by saying that their motivation may be clear at least to those who have some smattering of kinetic theory. The idea is that we have a fluid composed of atoms, molecules, droplets and other particles, all of which we shall consider to be microscopic. Interspersed in this complicated fluid is a gas of photons or particles of light. A photon has an associated frequency, ν, and we shall consider only those applications to materials with unit index of refraction so that all our photons move at the constant speed $c \cong 3 \times 10^{10}$ cm/s. The direction of a photon's motion is given by the unit vector $\boldsymbol{\mu}$. The photon's momentum is

$$\mathbf{p} = \frac{h\nu}{c}\boldsymbol{\mu}, \tag{1}$$

and its energy is $h\nu$, where h is the Planck's constant. If you are troubled by the idea of associating a frequency to a light particle, you should read Newton's "Opticks". For him, light consisted of corpuscles subject to fits, to which no doubt a mean frequency can be associated.

At any time, t, we position a photon in a six-dimensional phase space whose coordinates are the three spatial position coordinates and the three components of $\mathbf{p}$. We denote the density of photons in phase space by f so that the probable number of photons in an infinitesimal volume of phase space is given by

$$dN = f(\mathbf{p}, \mathbf{x}, t)d\mathbf{p}\, d\mathbf{x}. \tag{2}$$

If we use spherical coordinates in momentum space with radial coordinate $h\nu/c$, we may write the volume element $d\mathbf{p}$ as $\frac{h^3\nu^2}{c^3}\, d\Omega\, d\nu$ with $d\Omega$ as the element of solid angle. We can then hide phase-space factors by working with the specific intensity

$$I_\nu = \frac{h^4\nu^3}{c^2} f \tag{3}$$

instead of the phase-space density. This quantity describes the properties of what is called a pencil of radiation giving the rate at which energy in a unit frequency interval passes through unit area (normal the pencil's axis) per unit time and per unit solid angle. For a discussion of the formulation in terms of specific intensity, which some people find intuitively more appealing than the phase-space density, see the book by Chandrasekhar, for example. We have included (what may seem to be) an extra factor of $h\nu$ in I_ν so that its dimensions are energy per unit area per unit time per unit frequency per unit solid angle.

As in fluid mechanics, we shall focus on the three lowest moments of the phase-space density. In terms of radiative quantities these are the radiative energy density

$$\mathcal{E} = \frac{1}{c} \int d\Omega \int I_\nu(\boldsymbol{\mu}, x, t) d\nu , \tag{4}$$

the radiative energy flux

$$\boldsymbol{\mathcal{F}} = \int d\Omega \int I_\nu(\boldsymbol{\mu}, x, t) \boldsymbol{\mu} \, d\nu \tag{5}$$

and the radiative pressure tensor

$$\mathbb{P} = \frac{1}{c} \int d\Omega \int I_\nu(\boldsymbol{\mu}, x, t) \boldsymbol{\mu}\boldsymbol{\mu} \, d\nu . \tag{6}$$

And, as in fluid mechanics, we may obtain equations for these macroscopic quantities from an equation for f (or I_ν), to which we turn next.

3 The Transfer Equation

The density in phase space satisfies what is essentially a continuity equation as a density should. We write this as

$$\partial_t f + \mathrm{div}(f\mathbf{V}) = \text{sources} - \text{sinks} , \tag{7}$$

where $\mathbf{V}$ is the velocity in the six-dimensional phase space and the divergence is with respect to the six phase-space coordinates. We consider the dynamics to be Hamiltonian so that the flow in phase space is incompressible ($\mathrm{div}\mathbf{V} = 0$). The transfer equation then becomes

$$\partial_t f + \dot{x}^i \frac{\partial}{\partial x^i} f + \dot{p}^j \frac{\partial}{\partial p^j} f = \text{sources} - \text{sinks}, \tag{8}$$

where repeated indices are summed over.

We have no problem understanding what is meant by $\dot{x}^i$ but when do we encounter $\dot{p}^j$, the acceleration of a photon? There are two cases when the speed of light may vary: when the photons pass through a region where the index refraction varies or when they pass through a strong gravitational field. Neither of these situations arises often in geophysical contexts, so we shall not pay any further attention to them here. (However, we may note in passing that photons move along geodesics and, if we know the metric produced by the gravitational field and the index of refraction, we can then re-express $\dot{\mathbf{p}}$ by using the geodesic equation.) When we assume that the velocity of a photon is a constant between encounters with material particles we may write

$$\dot{x}^i = c\mu^i , \tag{9}$$

where the μ^i are the components of the unit vector in the direction of the photon's motion.

At this point in the development, we switch to the specific intensity as the descriptor of the radiation field because it is more conventionally used than the phase-space density. For many people, the specific intensity is the more intuitive of the two, though I find that phase space makes the introduction to the basic quantities relatively simple. So we multiply (8) by $\frac{h^4\nu^3}{c^2}$, recall (3) and get

$$\partial_t I_\nu + c\mu^i \frac{\partial}{\partial x^i} I_\nu = \text{gains} - \text{losses} , \tag{10}$$

where we have discarded the term in $\dot{\mathbf{p}}$ and renamed the terms on the right to signal the change in outlook. We are working in a fixed inertial frame and that is why we may treat ν and $\boldsymbol{\mu}$ as constants.

From the point of view of the photon gas at a given location in phase space, emission and absorption are evidently to be considered gains and losses, respectively, for the radiation field. And the same is true for scattering into and out of the relevant region of momentum space, respectively. Absorption and scattering are characterized by cross-sections. That is, as a photon moves along, the chance that it interacts with a material particle is expressed in terms of the area that is, in effect, blocked by the particle. That is called the cross-section, α (say). If the number of absorbing and scattering particles per unit volume is $n(\mathbf{x}, t)$, then the number of such particles per unit area in a slab of thickness ds is nds. The fraction of the slab that they present to the photon beam is αnds. The quantity $(\alpha n)^{-1}$ is then a length and is called the mean free path (mfp) of a photon. The mfp depends on the physics of the material particle, the frequency of the photon and the physical state of the matter. (We shall not consider a dependence on scattering angle in this introduction but this is important for some applications such as scattering by non-spherical aerosols.)

We include both absorption and scattering in α but we shall assume that we can separate them out in the formulation (in principle). To do so explicitly would mean going into the quantum mechanics of these processes, and we shall not take that up here. In fact, we cannot make the distinction without having some knowledge of the physical conditions. For example, in a scattering process, a material particle may be caused to change its internal state by a photon. If the particle returns to its original state almost immediately and re-emits an almost identical photon, we may consider that it is the same photon and it will merely have suffered a change in direction and perhaps a slight energy shift. For this to happen, we must assume that the particle has completed the scattering process before its state is modified by collision with another particle. Otherwise it will be sent into another state. In that case, if and when it emits a photon, the new photon will be unrelated to the original one. We see then that one aspect of the separation of the processes of scattering and absorption will depend on the ratio of re-emission lifetime to mean free time between interparticle collisions. There is a ratio of times here that will depend on many things such as the nature of the material particles. Since, as mentioned, there are typically several kinds of material particles, it is often expedient to lump them all together and calculate an averaged cross-section, $\bar{\alpha}$.

For complicated mixtures of particles it is sometimes convenient to speak of a cross-section per unit mass of the ambient medium, rather than per particle. If $\bar{m}$ is the average mass of the material particles, we can write $\varkappa_\nu = \bar{m}\bar{\alpha}$ and, with $\rho = \bar{m}n$, the inverse mean free path becomes $\rho\varkappa_\nu$ where ρ is the mass density. The averaging behind all this is complicated, especially because there are particles that add mass but not much cross-section to the mix at any given frequency.

To keep the distinction between absorption and scattering explicit, let us write the effective cross-sectional area per unit mass as $\varkappa_\nu = k_\nu + \sigma_\nu$ where k_ν is the absorption coefficient per unit mass and σ_ν is the scattering coefficient per unit mass. The expression of frequency dependences of these quantities through subscripts is customary and it also serves to emphasize the problems those dependences present and that we shall avoid.

Those new to the study of radiative transfer often expect to see the specific intensity attenuate as the beam propagates because of the inverse square spreading of radiation. But, as can be seen from the transfer equation, if there are no sources or sinks, the intensity is constant as the photon beam moves through a stationary medium. That is one reason why the intensity is considered a useful quantity to work with. On the other hand, if there is material in the path of the beam, the beam is attenuated by absorption and scattering. A beam of light that traverses a material slab of thickness ds along the s-direction is attenuated through absorption and scattering by an amount $dI_\nu = -I_\nu\rho\varkappa_\nu\,ds$. That is, the loss from the beam is proportional to the intensity of the beam, the density of material in its path, the effective cross-sectional area of this material and the distance traveled. When the beam travels across a finite distance, absorption and scattering cause the intensity to decrease like $I_\nu \propto \exp\left(-\int \rho\varkappa_\nu ds\right)$, where the limits on the integration are at the endpoints of the path. The integral in the exponent is the *optical distance* between the two limits of integration. Optical distance is measured in units of the local mean free path of photons. This is how to express distance when the unit of length varies along the path. Astronomers refer to the optical distance from the surface of an object vertically down to a given depth as optical depth.

Next, we need to include the contributions to the beam as it progresses through the medium. We introduce the emission coefficient, j_ν, such that the rate of emission per unit mass of the medium in all directions and per unit frequency is $4\pi j_\nu$. The local rate of contribution to the beam of radiation is ρj_ν per unit volume per unit frequency and per unit solid angle. The medium also scatters light from all directions at a rate $\int \rho \sigma_\nu I_\nu \, d\Omega$ per unit volume and per unit frequency. The amount per unit solid angle available to any particular beam is $1/(4\pi)$ of this. Normally, σ_ν depends on the scattering angle but we shall consider only the case of isotropic scattering here with σ_ν independent of scattering angle and angle of incidence. Hence we may write that the net rate of gain per unit volume from scattering is $\rho \sigma_\nu J_\nu$ where

$$J_\nu = \frac{1}{4\pi} \int I_\nu \, d\Omega \tag{11}$$

is called the mean intensity. The total rate of gain per unit volume in the beam is then $\rho(j_\nu + \sigma_\nu J_\nu)$.

We wrote the losses and the gains as the amounts subtracted from or added to the beam as it traverses a distance ds along its path. Since the photons move at speed c, a slab of thickness ds is traversed in a time $dt = ds/c$. We are assuming that the index of refraction is unity and that the space is Euclidean, so the beam moves in a straight line. Hence, the net rate of gain of energy in a beam as it moves along is given by $(1/c)dI_\nu/dt$, where this is an Eulerian derivative with $\dot{\mathbf{x}} = c\boldsymbol{\mu}$. We may then write

$$\frac{1}{c}\partial_t I_\nu + \boldsymbol{\mu} \cdot \boldsymbol{\nabla} I_\nu = \rho(j_\nu + \sigma_\nu J_\nu) - \rho(k_\nu + \sigma_\nu) I_\nu . \tag{12}$$

This is known as the equation of transfer.

The calculations of the absorption and scattering coefficients are got from a mixture of quantum mechanical calculations and experiments. We assume that the medium may be described in terms of local thermodynamic quantities such as temperature and pressure. This notion of local thermodynamics is also used to get an approximation for the emission coefficient. Attempts to do better than this have not been overly successful, and one may wonder whether this ought to cause concern for studies of the very high atmosphere.

In equilibrium, the medium is homogenous and steady and the radiation field is isotropic ($I_\nu = J_\nu$). In that case, the radiative intensity is given by Planck's equilibrium expression for the intensity, denoted as B_ν, which is a known function of the temperature. In that ideal case, the equation of transfer boils down to

$$j_\nu = k_\nu B_\nu . \tag{13}$$

This relation is called the Kirchhoff–Planck law and we shall use it to provide an expression for the emission coefficient in general. The transfer equation now becomes

$$\frac{1}{c}\partial_t I_\nu + \boldsymbol{\mu} \cdot \boldsymbol{\nabla} I_\nu = \rho \varkappa_\nu \left(\mathfrak{J}_\nu - I_\nu\right) , \tag{14}$$

where $\varkappa_\nu = k_\nu + \sigma_\nu$ and

$$\mathfrak{J}_\nu = \frac{\sigma_\nu}{\varkappa_\nu} J_\nu + \frac{k_\nu}{\varkappa_\nu} B_\nu \tag{15}$$

is called the source function.

4 Equations of the Radiative Fluid

As one does in going from classical kinetic theory to fluid dynamics, we may here integrate out the momentum variables, frequency and direction, to obtain continuum equations for the radiative flow. Setting

$$I = \int_0^\infty I_\nu \, \mathrm{d}\nu \tag{16}$$

we integrate the transfer equation over frequency and obtain

$$\frac{1}{c}\partial_t I + \boldsymbol{\mu} \cdot \boldsymbol{\nabla} I = \rho \int_0^\infty \varkappa_\nu \left(\mathfrak{J}_\nu - I_\nu\right) \mathrm{d}\nu . \tag{17}$$

Though this is a difficult integro-differential equation, it is far simpler than the full monochromatic transfer equation since we are no longer paying attention to the strong variations in absorption and emission that occur in real conditions. Yet, even if we blur such effects, the transfer equation is not easy to solve and the various techniques for solving even the simplest problems rely on making any simplification that seems qualitatively reasonable. A common simplification that we shall adopt limits attention to scattering processes, as we have defined them, for which the change in frequency induced is not large. Hence we shall here replace σ_ν by a scattering cross-section, σ, that does not depend on frequency. One calls such scattering *coherent*.

A simplification that is then used in studying (17), at least in first approximation, is to replace k_ν in the frequency integral by a mean (over frequency) absorption coefficient, k. The weight function to be chosen to evaluate such a mean will depend on what else is under the integral sign. This could lead to the appearance of more than one mean absorption coefficient in a given problem, but the usual practice is to adopt a single one of them. The choice then depends on what quantities to be calculated seem most important. A difficulty in this approach is that the weight function typically depends on the state of the medium, which is not known in advance. So once you solve the transfer equation, you then have to solve the equations governing the properties of the material medium. That, at least, is the usual order of business and it will clearly call for iteration in most cases. However, if you are simultaneously solving the fluid equations you do not want to have to iterate the whole procedure unless you are ready for some serious computing. That is the bad news. Is there any good news?

Fortunately, in geophysical situations, the motion is usually very subsonic, that is, fluid velocities are less than those of most individual particles. The direct effects of velocities on the radiative problem are then not very great, though fluctuations in the material medium can be a problem. The main effects of motion are typically included in k_ν through its dependence on temperature and the fluid motions do not modify those much except to add a bit of blurring of the spectral lines that favors the use of mean absorption coefficients. (If the fluid velocities did become significant, the mean absorption coefficients would be tensorial, but let us not even think about such things.) So the transfer theory is normally carried out as if the medium were stationary.

Once we have made all the implied simplifications, we write the (frequency) integrated transfer equation as

$$\frac{1}{c}\partial_t I + \boldsymbol{\mu}\cdot\boldsymbol{\nabla}I = \rho\varkappa\,(\mathcal{J} - I)\,, \tag{18}$$

where $\varkappa$ is the mean opacity coefficient and

$$\mathcal{J} = \frac{\sigma}{\varkappa}J + \frac{k}{\varkappa}B\,. \tag{19}$$

The (frequency) integrated mean intensity is related to the radiative energy density through $J = \int_0^\infty J_\nu\,\mathrm{d}\nu = c/(4\pi)\mathcal{E}$. I have used both of these notations because J is the usual astrophysical notation and it is well to know of it. On the other hand, $\mathcal{E}$ is more physically meaningful, so I prefer to use it.

The equilibrium expression for the integrated Planckian intensity is $B = \int_0^\infty B_\nu = \sigma T^4/\pi$ where σ is the Stefan–Boltzmann constant and $a = 4\sigma/c$ is known as the radiation constant. Here, T is a temperature assigned to the radiation field. In the simplest cases, the radiation temperature is almost the same as that of the ambient matter, but life is not always so simple.

To go now to the macroscopic description, we first compute the lowest angular moments of the transfer equation. On integrating over all solid angle, we obtain

$$\frac{\partial\mathcal{E}}{\partial t} + \boldsymbol{\nabla}\cdot\boldsymbol{\mathcal{F}} = 4\pi\rho k(B - J) = \rho k\,c\,(\mathcal{P} - \mathcal{E})\,, \tag{20}$$

where $\mathcal{P} = 4\pi B/c = aT^4$ is the radiative energy density in (local) equilibrium. Here, T is the temperature of the matter. We could use subscripts to distinguish the two temperature fields, matter and radiation, but I did not want to burden you with notation at this stage. Equation (20) corresponds to the continuity equation of fluid dynamics with sources and sinks. If the actual radiative energy density exceeds the Planckian equilibrium value, more radiative energy is absorbed than is emitted by the medium. (Note that k is the mean absorption coefficient; coherent scattering plays no explicit role in the energy balance.) When $\mathcal{P} = \mathcal{E}$ we have the condition of radiative equilibrium in which emission and absorption are in balance and the two implied temperatures are equal.

Next, we multiply the transfer equation by $\boldsymbol{\mu}$ and integrate over angle to obtain

$$\frac{\partial \boldsymbol{\mathcal{F}}}{\partial t} + c^2 \boldsymbol{\nabla} \cdot \mathbb{P} = -\rho \varkappa c \, \boldsymbol{\mathcal{F}} . \tag{21}$$

This is the momentum/energy balance equation and it is $\varkappa$ that appears on the right since both absorption and scattering contribute to the exchange of momentum between matter and radiation. The flow of the radiation through the matter that is driven by the radiative pressure gradient macroscopically resembles fluid flow through a porous medium. For this case, we also have the possibility that the porous medium can move (as in a fluidized bed) but, in geophysical situations, such effects are too small to concern us.

In (20) and (21), we have the equations of motion for the photon fluid. These equations do not offer a complete description and we need to provide an equation for the pressure tensor, $\mathbb{P}$. One way to get such an equation might be to go to a higher moment of the transfer equation, but that way lies infinite regress. Rather, we reach for a simplifying assumption whose validity may be checked against examples where accurate solutions are known. Here we shall use a simple closure relation for the radiative pressure tensor.

If the mean free path of the photons is less than the scale of variation of properties of the medium, radiation coming in from various directions is not so different and, in such cases, we may assume that the radiation field is nearly isotropic. If the intensity does not depend on direction, we readily find that

$$\mathbb{P} = \frac{1}{3} \mathcal{E} \, \mathbb{I} , \tag{22}$$

where $\mathbb{I}$ is the unit tensor. Some call this closure formula the Eddington approximation, though it does follow from a less stringent condition on the intensity field than strict isotropy than Eddington used. Proceeding from such weaker conditions on the intensity is advantageous because the physical boundary conditions are usually posed for the intensity itself rather than on its moments. And, from those conditions, we determine the boundary conditions for the moments as Eddington did. In the simplest cases, we still find (22) in the first approximation. This is sometimes a useful way to proceed. If you deal with cases where the geometry and the boundary conditions are simple you can try to do a better job in treating the frequency dependence of $\varkappa_\nu$.

On the other hand, when the mean free path of photons is long, we need to improve on (22). One approach is to replace the $1/3$ in (22) by something called the Eddington factor. That is too technical for us to pursue here. However, you should be aware that I is not invariant under changes of reference frame. When you want to call I isotropic, say, you ought to stipulate the frame in which that is to be true. If we are to regard the radiation as a fluid, we have to admit that it has a reference frame of its own and that ought to be the frame in which to make simplifications of the radiation field. Such a choice has an influence on the Eddington factor since, even though the medium does not move very fast, the radiative fluid does.

5 The Stationary State

5.1 Radiative Equilibrium

Many branches of geophysics are concerned with thin layers, with such exceptions as studies of the earth's core. For thin layers, we can generally ignore or parameterize the influence of sphericity and confine ourselves to a plane-parallel medium. The classical problem of this kind occurs in a layer in radiative equilibrium whose properties depend only on the vertical coordinate, z. In that case, with $\mathcal{F}$ as the z-component of flux and P the zz-component of $\mathbb{P}$, the moment equations become

$$\frac{\mathrm{d}\mathcal{F}}{\mathrm{d}z} = 0 \tag{23}$$

and

$$\frac{\mathrm{d}P}{\mathrm{d}z} = -\rho\varkappa\,\mathcal{F}\,. \tag{24}$$

We see that $\mathcal{F}$ is a constant and that

$$P = P_0 + \tau\mathcal{F} \tag{25}$$

where P_0 is a constant of integration and

$$\tau = \int_z^{z_{\mathrm{top}}} \rho\varkappa\,\mathrm{d}z' \tag{26}$$

is the optical depth into the medium. Here, z_{top} is the vertical coordinate of the top of the layer, which for an isothermal atmosphere, for example, could be at positive infinity. (Astrophysicists often choose coordinates with z measured downward into a star, starting from 0 at the surface, if there is one, but we have not done this.)

For this exposition we adopt the approximation that $P = \frac{1}{3}\mathcal{E}$ for the case of the plane-parallel layer. Since $\mathcal{E}$ is the radiant energy density, it makes sense to introduce a radiation temperature such that $\mathcal{E} = aT^4$. In the case we are studying here, this temperature is the same as that of the medium and (25) tells us that T^4 is linear in optical depth. However, in the case of many planetary atmospheres, the medium is often modeled as a finite slab with radiation coming in from above and below. There is a considerable literature on this problem, such as you will find in the book on multiple light scattering (1981), by H. C. van de Hulst, by radiative transfer. He uses a method devised by Case in the wake of van Kampen's approach to the Vlasov equation.

5.2 The Hydrostatic State

Even in the simplest case, when we find that T^4 is linear in τ, we would still not know what the medium is really like since we need to find $z(\tau)$ to complete the solution for the structure of the medium. To do this, we introduce the hydrostatic condition

$$\frac{\mathrm{d}p}{\mathrm{d}z} = -g\rho \tag{27}$$

where p is the pressure of the material medium and g is the (often constant) gravitational acceleration. We also need an equation of state and that can be any of a number of things depending on whether we are concerned about (say) sunlight getting in amongst plankton patches or ground radiation coming up through the atmosphere. To have something specific to work with, let us take the equation of state for a perfect gas:

$$p = \Re\rho T . \tag{28}$$

In general, $\Re$ will depend on the chemical composition, the form of water (vapor or droplets), ionization, dissociation and other such local details but here we shall assume that it is constant. Then we have a problem simplified to the extent that the matter and radiation conditions may be solved together.

When we divide (27) by (24) and use the Eddington approximation, we get

$$\frac{\mathrm{d}p}{\mathrm{d}\mathcal{E}} = \frac{gc}{\varkappa\mathcal{F}} . \tag{29}$$

This is not a hard equation to solve but it does require that we specify how $\varkappa$ depends on ρ and T, for instance. For definiteness, and ease of solution, let us then consider the pure isotopic scattering case with $k = 0$ so that $\varkappa = \sigma = $ constant. Then we find that

$$p = \frac{gc}{\sigma\mathcal{F}} (\mathcal{E} - \mathcal{E}_0) , \tag{30}$$

where $\mathcal{E}_0$ is a constant of integration. It is chosen to be the energy density of radiation at the top of the medium where, we assume, the gas pressure vanishes. With $\mathcal{E}_0 = aT_0^4$, this defines the surface temperature of the medium.

On returning to (27) and using the equation of state, we have

$$\frac{\mathrm{d}p}{\mathrm{d}z} = -\frac{gp}{\Re T} . \tag{31}$$

This can be now rewritten as

$$\frac{\mathrm{d}\Theta}{\mathrm{d}z} = -\frac{g}{4\Re T_0} \frac{\Theta^4 - 1}{\Theta^4} , \tag{32}$$

where $\Theta = T/T_0$. Deep into the atmosphere, Θ gets large so we find that Θ is approximately $-g\,z/(4\Re) + $ constant. High in the atmosphere, when Θ is close to unity, we have that $\Theta \approx 1 + \text{const.} \exp[-g z/(4\Re)]$. So the upper atmosphere is essentially isothermal and the lower one (with nearly linear temperature profile nz) is polytropic, that is, the static pressure is proportional to a power of the static density. The "isothermal region" (as it was once called) of our atmosphere was first observed by Teisserenc de Bort who discovered it in 1899. It is now called the stratosphere and is known to display much dynamical activity.

In fact the equation for Θ integrates exactly [1] and it gives

$$-\frac{gz}{4\Re T_0} = \Theta - \frac{1}{2} \tan^{-1} \Theta - \frac{1}{2} \coth^{-1} \Theta \,. \tag{33}$$

If we wanted to include other effects such as non-constant absorption coefficient or conductivity (molecular or turbulent) in this study, we would find using matched expansions that the qualitative structure is robust. In the special case $T_0 = 0$, the surface would be at $z = 0$. Then we would have to deal only with the semi-infinite medium with a linear temperature profile. In that case, the static medium has $p \propto \rho^{4/3}$, which is a case of a pure polytropic atmosphere. When weak disturbances of this polytropic layer are introduced with weak motion of the medium by way of the fluid equations, we can study the normal modes of the medium. When the perturbations behave adiabatically and inviscidly, we may refer to Lamb's study of the normal modes of such atmospheres. Radiative effects have also been included in such studies for cases when the perturbations are optically very thin [2] or very thick [3]. For discussion of the linear problem in the Boussinesq approximation, see the book of Goody. For these various treatments, we need to understand how to include the thermal effects of radiation in the fluid equations. In the next section we introduce that topic.

6 The Radiative Heat Equation

To this point, we have followed the formalism where it led us but sooner or later one needs to see how the transfer theory fits in with the physics of what is going on. In this section we examine how the radiative heat transfer affects the temperature of the material medium. After all, in most geophysical contexts we are not as interested in the solution of the transfer equation as in the role of radiation in heating or cooling the ambient medium. To see into this topic, we may examine the influence of radiative sources and sinks on the thermal budget of the material medium itself. I have already prepared the way a bit by introducing some of the standard formulae from the theory of equilibrium radiation.

For a gas, the radiative heat equation is

$$\frac{de}{dt} - \frac{p}{\rho^2} \frac{d\rho}{dt} = \Omega, \tag{34}$$

where e is the internal energy density and $\mathcal{Q}$ is the net radiative heating (or cooling) rate per unit volume. We could also include other effects such as viscous dissipation and conduction. However, in this section, we shall be concerned only with the radiative effects and will even omit the work term, which brings the adiabatic gradient into play.

To find the rate at which radiation arrives at a point $\mathbf{x}$, we integrate the contributions from all over the medium. At the general point $\mathbf{x}'$ radiation is emitted at a rate $\rho j(\mathbf{x}', t)$ per unit volume and per unit solid angle. This radiation is attenuated on its way to the location of interest by the intervening material and by inverse square spreading. In the direction of $\mathbf{x}' - \mathbf{x}$, radiation from $\mathbf{x}'$ is coming in at a rate

$$\rho j(\mathbf{x}', t) \frac{\exp(-\tau)}{|\mathbf{x}' - \mathbf{x}|^2} \tag{35}$$

per unit volume per unit solid angle where τ is the optical distance between $\mathbf{x}'$ and $\mathbf{x}$. (Here we are neglecting the retardation time that the radiation takes to get from $\mathbf{x}'$ to $\mathbf{x}$.) We integrate this expression along the line from $\mathbf{x}'$ to $\mathbf{x}$ to find the intensity along that line. Then, at each distance from the point of interest, we integrate over the spherical surface surrounding the point at that distance and divide by 4π to obtain

$$J(\mathbf{x}, t) = \int d^3 \mathbf{x}' \, \rho \, j(\mathbf{x}', t) \frac{e^{-\tau}}{4\pi |\mathbf{x}' - \mathbf{x}|^2}. \tag{36}$$

Here we see why the inverse square effect does not show up in the intensity: it is compensated by the volume integral.

The right-hand side of (34) is the negative of the right-hand side of (20) and, with (13), it may be written as

$$\mathcal{Q} = -\rho k c \mathcal{P} + \rho k c \int d^3 \mathbf{x}' \, \rho k \mathcal{P}(\mathbf{x}', t) \frac{e^{-\tau}}{4\pi |\mathbf{x}' - \mathbf{x}|^2}. \tag{37}$$

To gain some insight, we simplify things by putting $\mathcal{P} = aT^4$ and treating ρk as a constant over the medium so that $\tau = k\rho |\mathbf{x}' - \mathbf{x}|$. Then with $de = \rho c_v \, dT$ and guidance from (20) we write

$$c_v \frac{\partial T}{\partial t} = -ckaT^4 + ck \int d^3 \mathbf{x}' \, k\rho aT^4(\mathbf{x}', t) \frac{e^{k\rho |\mathbf{x}' - \mathbf{x}|}}{4\pi |\mathbf{x}' - \mathbf{x}|^2}. \tag{38}$$

We may express this equation for the temperature as

$$\frac{\partial T}{\partial t} = \frac{ka}{c_v} \int d^3 \mathbf{x}' \, T^4(\mathbf{x}', t) \, \mathcal{K}(|\mathbf{x}' - \mathbf{x}|), \tag{39}$$

where

$$\mathcal{K}(|\mathbf{x}' - \mathbf{x}|) = \rho k \frac{e^{k\rho |\mathbf{x}' - \mathbf{x}|}}{4\pi |\mathbf{x}' - \mathbf{x}|^2} - \delta(\mathbf{x}' - \mathbf{x}). \tag{40}$$

For a relatively opaque medium with large $k\rho$, the kernel is sharply peaked. Hence we may expand $T^4(\mathbf{x}', t)$ around the point $\mathbf{x}$ in a Taylor series as

$$T^4(\mathbf{x}') = T^4(\mathbf{x}) + (\mathbf{x}' - \mathbf{x}) \cdot \nabla T^4(\mathbf{x}) + \frac{1}{2}(\mathbf{x}' - \mathbf{x})(\mathbf{x}' - \mathbf{x}) : \nabla\nabla T^4(\mathbf{x}) + \cdots \quad (41)$$

where dyadic notation is used, the colon means a double dot product and we have not explicitly exhibited the time dependence. When we put this into the integral of (39) we find that the antisymmetric terms in the expansion do not contribute because the kernel is symmetric. Thus the heat equation is expanded into

$$\frac{\partial T^4}{\partial t} = \frac{q}{3(k\rho)^2}\nabla^2 T^4 + \frac{q}{5(k\rho)^4}\nabla^4 T^4 + \cdots \quad (42)$$

since the volume integral of $\mathcal{K}$ is zero and where

$$q = \frac{4ackT^3}{c_v} \quad (43)$$

is an inverse time.

If we keep just the leading term on the right, we have a diffusion equation for the radiant energy and the diffusion coefficient,

$$\mathcal{D} = \frac{q}{3(k\rho)^2}, \quad (44)$$

has the desired dimensions of length squared over time. The two terms on the right of (42) together may be represented by a rational approximation for the operators so that

$$\frac{\partial T^4}{\partial t} = \frac{\mathcal{D}}{1 - \frac{3}{5(k\rho)^2}\nabla^2}\nabla^2 T^4 \quad (45)$$

if we treat the Laplacian operator as if it were an algebraic quantity. When the spatial scale of temperature variation is large, ∇^2 is a small operator and this equation reduces to a diffusion equation for T^4. In the limit of small scales of variation, ∇^2 is a large operator and we have

$$\frac{\partial T^4}{\partial t} = -\frac{5}{9}qT^4 \quad (46)$$

which, when suitably linearized, is essentially Newton's law of cooling.

We may multiply out the denominator in (45) to turn that equation into a partial differential equation but this extra step makes for complications when additional terms (such as advection and work terms) are included in the heat equation. In its present form, (45) is well set up for spectral methods. We may get some idea of the dependence of the radiative lifetime on scales variation if we replace ∇^2 by $(\pi/\ell)^2$ to obtain

$$t_{\text{rad}} = \frac{[5(k\rho\ell)^2 + 3\pi^2]}{5\pi^2 \mathcal{D}(k\rho\ell)^2}, \tag{47}$$

where $k\rho\ell$ measures the optical thickness of a perturbation to the uniform state. This formula is useful for gauging the effects of radiation on convective processes and are discussed in [4, 5].

In the case of thermal convection, radiative transfer works in parallel with thermal diffusivity and in series with viscosity, in the terminology of electricians. If the conductive lifetime of a thermal perturbation is t_{cond}, the thermal lifetime of a perturbation under the two effects is

$$t_{\text{therm}} = \frac{1}{\frac{1}{t_{\text{cond}}} + \frac{1}{t_{\text{rad}}}}. \tag{48}$$

Since the viscous effects work in series with the thermal effects, the dissipative lifetime is

$$t_{\text{diss}} = \sqrt{[t_{\text{visc}} \, t_{\text{therm}}]}. \tag{49}$$

The degree of convective instability is then measured by the ratio of the dissipative lifetime to the dynamical time, which is the free fall time under the reduced gravity. That ratio squared is a (radiative) Rayleigh number. Of course, it is not enough to identify the key instability parameter; you need to also know the other ratios of characteristic times just as you need to know the Prandtl number for ordinary thermal convection. Moreover, the boundary conditions play a very significant role in the problem and these depend on various things such as cloud cover which themselves depend on the convection. At that point, the fun begins and my discourse ends.

References

I mention here only a few papers. More important are the books mentioned by title and author in the main text.

1. E. A. Spiegel: Photoconvection. In: Spiegel, E. A., Zahn, J.-P. (eds.) Problems in Stellar Convection. Springer, Berlin (1977)
2. E. A. Spiegel: The effect of radiative transfer on convective growth rates. Astrophys. J. **139**, 959 (1964)
3. O. M. Umurhan: Thesis. Columbia University, New York (2002)
4. E. A. Spiegel: Smoothing of temperature fluctuations by radiative transfer. Astrophys. J. **126**, 202 (1957)
5. W. Unno and E. A. Spiegel: The Eddington approximation in the radiative heat equation. Publ. Astronom. Soc. Jpn **18**, 85 (1966)

Coherent Vortices and Tracer Transport

A. Provenzale,[1] A. Babiano,[2] A. Bracco,[3] C. Pasquero,[4] and J. B. Weiss[5]

[1] ISAC-CNR, Torino, Italy
   a.provenzale@isac.cnr.it
[2] LMD-ENS, Paris, France
   babiano@lmd.ens.fr
[3] EAS and CNS, Georgia Institute of Technology, Atlanta, GA, USA
   abracco@gatech.edu
[4] University of California at Irvine, CA, USA
   claudia.pasquero@uci.edu
[5] ATOC, University of Colorado, Boulder, CO, USA
   jeffrey.weiss@colorado.edu

Summary

Geophysical flows are characterized by the presence of coherent vortices, localized concentrations of energy and vorticity that have a lifetime much longer than the local turbulence time (sometimes called the eddy turnover time).

In the ocean, coherent vortices, or eddies, are ubiquitous features whose size varies between several to a few hundred kilometers, and that account for a large portion of the ocean turbulent kinetic energy [1, 2, 3, 4, 5, 6, 7, 8, 9, 10, 11, 12, 13, 14, 15, 16, 17]. The presence of vortices can be revealed in various ways. Vortices at the ocean surface imprint their signature on the sea surface height and can be tracked by satellite, while floats with looping trajectories can help revealing the presence of vortices at depth.

Coherent vortices significantly affect the dynamics and the statistical properties of ocean flows, with important consequences on transport processes. In this contribution, we shall briefly review some of these issues, focusing on the simplified conceptual model provided by two-dimensional turbulence.

1 Coherent Vortices and Background Turbulence

The dynamics of vortex-dominated geophysical flows can be simulated by adopting the overly simplified configuration of two-dimensional, barotropic turbulence, where the motion is purely horizontal and vertical derivatives vanish. The dynamics of two-dimensional turbulence is described by the vorticity equation

$$\frac{\partial \omega}{\partial t} + \mathbf{u} \cdot \nabla \omega = F + D$$

A. Provenzale et al.: *Coherent Vortices and Tracer Transport*, Lect. Notes Phys. **744**, 101–116 (2008)
DOI 10.1007/978-3-540-75215-8_5

where $\omega(\mathbf{x}, t) = \partial v/\partial x - \partial u/\partial y$ is vorticity, $\mathbf{u} = (u, v)$ is the fluid velocity, $\mathbf{x} = (x, y)$ is space and t is time. The terms F and D represent forcing and dissipation respectively.

The dynamics of two-dimensional turbulence is characterized by the spontaneous emergence, and subsequent dominance, of a population of strong coherent vortices that concentrate most of the energy and vorticity of the flow [6, 19, 18]. In past years we have advocated the view that two-dimensional turbulence can be pictured as a two-component fluid: a sea of coherent vortices immersed into a background turbulence that is quite Kolmogorovian. This two-component view forms the basis of how we interpret Lagrangian (and Eulerian) measurements and how we infer flow properties from them [20, 21, 22].

An important issue is how we identify the two components. Until now, the best way to identify vortices is found to be the direct identification by some vortex census algorithm based on the analysis of local vorticity patches in physical space. A variety of such methods exists [23, 24, 25, 26, 27, 28]; all of them require the knowledge of the full vorticity field. A simplified version of a vortex census, which requires the knowledge of just a few Eulerian time series and provides the gross features of the vortex statistics such as the vortex density and the average vortex size, has also been proposed [22].

Although coherent vortices are local vorticity concentrations, their effects are non-local: The velocity field generated by a coherent vortex is non-local as it extends to large distances from the vortex center, well beyond the region where vorticity is significant. The range where the effect of the vortex on the velocity field is significant depends on the vortex shape and on the degree of baroclinicity: Barotropic vortices extend their influence to far distances, while baroclinic lenses (such as Meddies) have a shorter range of influence. Indeed, the Green's function associated with a barotropic (point) vortex is proportional to $\log(r)$, where r is the distance from the core of the vortex. For a baroclinic (point) vortex, the Green's function goes as $1/r$. Therefore baroclinic vortices have a shorter range of influence than barotropic ones [29]. In terms of the velocity field (and particle dispersion), the two-component view of mesoscale turbulence should not be seen as a purely spatial decomposition of space into separate vortex and non-vortex areas, but rather as the superposition of two dynamical components which can simultaneously act at the same spatial position.

The far-field influence of coherent vortices can be seen in the probability distribution function (PDF) of the velocity. At high Reynolds numbers, when vortices are intense and have sharp profiles, velocity PDFs in barotropic turbulence have non-Gaussian tails indicating that high velocities are more probable than would be the case for a Gaussian field [21, 30]. This non-Gaussianity has been previously discussed in the context of point vortices, which can be thought of as a simplified model of vortex dominated flows at very large Reynolds number [31, 32, 33]. In this context, it has been shown that small velocities have a Gaussian distribution but the PDF has a non-Gaussian tail related to the slow decay with distance of the velocity induced by a single vortex. Convergence to a Gaussian PDF is obtained only in systems with an extremely large number of vortices, orders of magnitude more than exist in the ocean [33].

Float trajectories in the North Atlantic [30] and in the Adriatic Sea [34, 35], indicate that velocity PDFs are non-Gaussian. Typically, they have larger kurtosis than a normal distribution: they have a Gaussian-like core and non-Gaussian tails for high velocities. Similar results have been found from mid-latitude fluid particle trajectories along isobaric surfaces in a simulation of the Atlantic Ocean dynamics at high resolution [36]. Note that we are here referring to either Eulerian or Lagrangian velocity PDFs under the assumption that Lagrangian particles sample the whole domain. In this case, in fact, Lagrangian velocity PDFs in the ocean must converge to the Eulerian ones. This similarity in velocity PDFs between float data, ocean GCMs, simplified turbulence models, and point vortex systems suggests that the non-Gaussian nature of the velocity PDFs is due to the vortex component of ocean mesoscale turbulence.

2 Dynamics of Lagrangian Tracers

The Lagrangian equation of motion for an individual fluid particle moving in a two-dimensional flow is

$$\frac{d\mathbf{X}_i}{dt} = \mathbf{U}_i(t) = \mathbf{u}(\mathbf{X}_i(t), t) \tag{1}$$

where $\mathbf{X}_i(t)$ and $\mathbf{U}_i(t)$ are the Lagrangian position and velocity of the ith particle, and $\mathbf{u}(\mathbf{X}_i, t)$ is the Eulerian velocity at the particle position. In this equation, we do not equate force to mass times particle acceleration, but rather particle velocity to the push of the flow. This happens because the particle is assumed to have negligible size and vanishing inertia with respect to the advecting fluid, i.e., to be a fluid element. When particles have finite size and/or non-vanishing inertia, the equations of motion become more complicated, see e.g. [37, 38] for a discussion of the dynamics of inertial and finite-size particles in vortex-dominated flows.

Numerical simulation of barotropic and of baroclinic (stratified) quasi-geostrophic turbulence and of point-vortex systems indicate that the cores of coherent vortices are associated with islands of regular (non-chaotic) Lagrangian motion that trap particles for times comparable with the vortex lifetime [39], and that vortices are characterized by a strong impermeability to inward and outward particle fluxes, see e.g. Elhmaidi et al. [40] or Provenzale [38] for a review. Particles can have more complex behavior and can eventually migrate from inside to outside of a vortex or vice versa only when highly (and relatively rare) dissipative events take place, as the deformation of a vortex due to the interaction with a nearby vortex, or the formation of a filament.

For this reason, an initially inhomogeneous particle distribution becomes homogeneous only on a very long time scale, which is determined by the typical lifetime of the vortices rather than by their typical eddy turnover time.

The trapping behavior of coherent vortices can be rationalized in terms of potential vorticity (PV) conservation [41]. For an ideal fluid with irrotational external forcing PV is conserved. When some little dissipation and/or rotational forcing is acting on the fluid, as it usually happens, PV is not conserved. If the PV-changing effects are small, PV is quasi-conserved. This means that in regions where PV changes slightly, the particles will be able to shift from one PV surface to another. However, strong PV gradients are much more difficult to overcome, as the change in PV that the particle should achieve to climb (or descend) the gradient may be too large compared to the effect of the forcing and dissipation present in the system. As a result, strong PV gradients can act as transport barriers. This is the main physical reason why intense jets, associated with strong PV gradients, can act as efficient barriers to transport. The same happens for isolated vortices: Vortex edges act as barriers to transport because vortices are regions of anomalous potential vorticity, usually embedded in a background where PV oscillates around a reference value with low variance. The vortex edges are therefore characterized by a large potential vorticity gradient, which fluid particles can rarely cross. This behavior is clearly illustrated by the dynamics of the stratospheric polar vortex over Antarctica [42].

Another important effect of coherent vortices concerns the convergence of Lagrangian time-averages. Lagrangian particles can have a very long memory when coherent structures, whose lifetime is long compared to other time scales in the problem, are present. For instance, if a Lagrangian particle is initially released in the background turbulence outside vortex cores, it will move around without entering any of the vortex cores present in the turbulent flow, until, in a quite rare event such as the formation of a new vortex, the particle will get trapped inside a newly forming vortical structure. From that moment on, the particle will stay inside the vortex for times comparable with the vortex lifetime.

The above example indicates that the temporal convergence of the statistical properties of a set of Lagrangian trajectories can take place on rather long timescales, related to the lifetime of the coherent structures. Of course, ensemble averages over a large number of homogeneously distributed Lagrangian particles do not suffer from this problem and they usually give a more complete picture of the flow. This illustrates the fact that ergodicity (i.e., equivalence of time and ensemble averages) is reached only on very long times, if ever, for Lagrangian statistics of particles moving in vortex-dominated flows, as discussed by Weiss et al. [33] for point vortices and by Pasquero et al. [22] for the vortices of two-dimensional turbulence. An interesting question, then, concerns the trade-off between the number of particles required to provide a meaningful picture of the flow (i.e., a correct estimate of the statistical properties of the flow) and the length of the trajectories. This issue has been discussed in some detail in [22], together with the comparison between Lagrangian and Eulerian second-order statistics (i.e., spectra and decorrelation times).

3 Lagrangian Dispersion in Vortex-Dominated Flows

Lagrangian particles in a Gaussian, homogeneous, stationary and uncorrelated velocity field undergo a Brownian random walk. Under such conditions, the second-order moment of the distribution of particle displacements grows linearly with time:

$$A^2(\tau; t_0) \equiv < (\mathbf{X}_i(t_0 + \tau) - \mathbf{X}_i(t_0))^2 >= 2K\tau \qquad (5)$$

where K is the dispersion (or diffusion) coefficient. Here, $\mathbf{X}_i(t)$ is the position of the ith particle at time t, and the angular brackets denote an ensemble average over all particles. The function $A^2(\tau, t_0)$ measures the absolute (or single-particle) dispersion. For a statistically stationary flow, the absolute dispersion A^2 does not depend on the starting time t_0. Relaxing any of the above assumptions (Gaussianity, homogeneity, stationarity, lack of temporal and spatial correlations) can significantly alter the dispersion law described above.

On short timescales, in particular, the Brownian dispersion law is modified by spatial and temporal correlations in the advecting flow, which induce Lagrangian velocity correlations over a substantial time range. The velocity autocorrelation function for an individual particle (labeled by the index i) is defined as

$$R_i(\tau) = \frac{\overline{(\mathbf{U}_i(t) - \overline{\mathbf{U}}_i) \cdot (\mathbf{U}_i(t + \tau) - \overline{\mathbf{U}}_i)}}{\sigma_i^2}, \qquad (3)$$

where $\mathbf{U}_i(t)$ is the velocity of the ith particle at time t, $\overline{\mathbf{U}}_i$ and σ_i^2 are the mean and variance of the velocity of the ith trajectory, and the overbar indicates an average over time t. Hence, $R_i(0) = 1$ and $R_i(\tau)$ goes to zero for large τ, when the particle velocity loses memory of its initial value. The flow field as a whole is characterized by the ensemble-averaged velocity autocorrelation function, $R(\tau)$, defined by averaging over all trajectories. One simple measure of the memory of Lagrangian particles is the Lagrangian integral time, defined as

$$T = \int_0^\infty R(\tau) d\tau. \qquad (4)$$

Over times much shorter than the Lagrangian integral time, the velocity is almost constant and one observes a ballistic dispersion phase,

$$A^2(\tau) = 2E\tau^2 \qquad (6)$$

where E is the mean kinetic energy of the advecting flow. A standard way of representing absolute dispersion is to define a time-dependent dispersion coefficient, $K(\tau)$, as

$$K(\tau) = \frac{A^2(\tau)}{2\tau} .$$ (7)

In the ballistic phase, $K(\tau) \to E\tau$ as $\tau \to 0$, while in the Brownian dispersion phase $K(\tau) \to K$ for $\tau \to \infty$. The ballistic regime is sometimes visible in the dispersion curves computed from surface drifter data [43]. Subsurface float trajectories are often characterized by a well-defined ballistic regime, associated with very steep Lagrangian spectra at small times [44].

Lagrangian stochastic models (LSM) are employed to reproduce the main statistical properties of particle trajectories in turbulent flows, without resolving the full Eulerian dynamics. Individual trajectories computed by an LSM usually do not have the same characteristics of the particles advected by a realistic flow. The similarity is recovered—if ever!—only statistically, after averaging over particle ensembles and over different realizations of the turbulent flow. Thus, one should not expect an individual stochastic trajectory to resemble an individual float trajectory.

One simple class of stochastic models describes the process of single-particle dispersion. In this case, the spatial correlations of the advecting flow are discarded insofar as they do not translate into temporal correlations of the Lagrangian velocities (see also Rupolo et al. [44] for a discussion of how Eulerian spatial correlations are related to Lagrangian time correlations). A more complex approach deals with particle separation processes, i.e., relative dispersion. In this case, the stochastic model describes the time evolution of the separation of a particle pair, and spatial correlations of the turbulent flow become an essential ingredient of the picture. In the following, we shall consider only single-particle dispersion and the related stochastic descriptions. An exhaustive discussion of the atmospheric applications of Lagrangian stochastic models can be found in the monograph by Rodean [45]; for oceanographic applications see Griffa [46] and Brickman and Smith [47].

The simplest stochastic model for single-particle dispersion is the random walk (or Markoff-0 model). In this approach, the particle displacements are randomly extracted from a Gaussian distribution, and there is no temporal correlation between subsequent displacements. If we assume that there is no mean flow advecting the particles and that the turbulent flow is statistically isotropic, we can write a Lagrangian stochastic differential equation for the random walk as

$$\mathrm{d}\mathbf{X}_i = \sqrt{K}\mathrm{d}\mathcal{W}_i(t).$$ (8)

where $\mathbf{X}$ is the position of the ith particle and the diffusivity, K, is not allowed to vary in space and time. The incremental Wiener random vector, $\mathrm{d}\mathcal{W}_i$, has zero mean and it is δ-correlated in space and time, $\langle \mathrm{d}\mathcal{W}_i(t) \cdot \mathrm{d}\mathcal{W}_j(t') \rangle = \delta_{ij}\,\delta(t-t')\mathrm{d}t$.

The single-particle stochastic description illustrated above can be framed in terms of a deterministic partial differential equation for the time evolution of the probability density function of particle positions, $P(\mathbf{X}|\mathbf{X}(0), t)$. Defining the particle concentration at $\mathbf{x}$ as $\rho(\mathbf{x}, t) = \int P(\mathbf{X} = \mathbf{x}|\mathbf{X}(0), t)\, d\mathbf{X}(0)$, the

Fokker–Planck equation for the evolution of P gives the well-known diffusion equation:

$$\frac{\partial \rho(\mathbf{x}, t)}{\partial t} = \frac{1}{2} K \nabla^2 [\rho(\mathbf{x}, t)]. \tag{9}$$

The assumption of uncorrelated displacements is equivalent to the assumption that the Eulerian fluid velocities decorrelate instantaneously, i.e., that the turbulent structure of the flow has no correlations. In general, this assumption is not appropriate for ocean mesoscale flows, where the temporal correlations of the advecting velocity field cannot be discarded. The simplest way of accounting for a memory in Lagrangian velocities is to consider a Markoff-1 model. In this approach, the time evolution of the Lagrangian velocity of the ith particle, $\mathbf{U}_i$, is described by an Ornstein–Uhlenbeck (OU) process:

$$d\mathbf{U}_i = -\frac{\mathbf{U}_i}{T} dt + \sqrt{\frac{2\sigma^2}{T}} d\mathcal{W}_i. \tag{10}$$

where T is the Lagrangian correlation time and σ^2 is the variance of the Lagrangian velocities. The first term on the r.h.s. is the (deterministic) fading-memory term, and the second term is the "stochastic kick," or random component, of the velocity fluctuation. For this process, the velocity distribution is a Gaussian with zero mean and variance σ^2, and the velocity autocorrelation is an exponential, $R(\tau) = \exp(-\tau/T)$. The (time-dependent) diffusion coefficient can be computed analytically,

$$K(\tau) = \sigma^2 T \left[1 - \frac{T(1 - e^{-\tau/T})}{t} \right], \tag{11}$$

see Griffa [46] for a discussion of this type of stochastic model in the context of oceanographic applications.

In a study of particle dispersion in two-dimensional turbulence, Pasquero et al. [21] showed that the linear Ornstein–Uhlenbeck model provides a good representation of absolute dispersion at short and large times (respectively in the ballistic and Brownian regimes), while at intermediate times it provides estimates of the dispersion coefficient which differ by at most 25% from the values obtained by direct integration of particle dynamics in the turbulent flow. If this discrepancy is acceptable, due for example to uncertain or poorly resolved data, then the use of the Ornstein–Uhlenbeck model is sufficient. To obtain a more precise estimate of the dispersion coefficient, however, a stochastic model that more closely represents the processes of particle dispersion in vortex-dominated mesoscale turbulence is warranted.

Major differences between the Ornstein–Uhlenbeck process and particle dispersion in mesoscale turbulence are related to the facts that the velocity distribution is non-Gaussian [20], the velocity autocorrelation is non-exponential [21], and particles get trapped in vortices for long times [39, 40]. Given these differences, it is indeed surprising that just a 25% discrepancy between the turbulent and the modeled dispersion coefficient is detected.

In an attempt to improve stochastic parameterizations of particle dispersion in mesoscale ocean turbulence, various extensions of the Ornstein–Uhlenbeck model have been proposed. The indications that Lagrangian accelerations in the ocean are correlated in time [44] have stimulated the development of Markoff-2 models where an Ornstein–Uhlenbeck formulation is written for the acceleration $\mathbf{a}$, with $d\mathbf{U} = \mathbf{a}\,dt$ [46]. Higher order models have also been proposed, with the aim of better reproducing other statistical properties of Lagrangian motions such as the sub- or super-diffusive behavior at intermediate times [48]. Superdiffusion has also been obtained by Reynolds [49], using a variation of a Markoff-2 model that includes spin.

Models that include spin have been designed to explicitly describe particle motion in and around coherent structures. In the presence of coherent vortices, particle motion has a rotational component, as evident in the looping trajectories of floats deployed inside mesoscale eddies. The rotational component of the velocity vector along a Lagrangian trajectory is characterized by an acceleration orthogonal to the trajectory. Simple geometrical arguments show that the introduction of the spin in Markoff-1 models corresponds to adding a new term in the stochastic equation for the velocity increment, proportional to the orthogonal velocity component [49, 50]. The individual trajectories produced by these models display spiraling motion, although the ensemble averaged velocity autocorrelation function is not necessarily oscillatory [49]. This model has recently been used to reproduce some statistical properties of Northwest Atlantic float trajectories [51].

On the other hand, it is not clear whether particle spinning inside vortices has any effect on space and timescales larger than those of the vortices themselves. In general, rotational motion inside vortices does not contribute to the large-scale spreading of particles; it is only the motion of the vortex itself that is responsible for particle displacements at large scales. In turn, vortices move because they are advected by other vortices and there is no self-induction of the vortices themselves [33]. As a result, the large-time dispersion properties of Lagrangian particles inside or outside the vortices of two-dimensional turbulence are the same. Thus, for the purpose of understanding particle dispersion at scales larger than the size of the individual vortices, the parameterization of particle motion inside a vortex can probably be neglected. Note, however, that the situation can be very different if the scale of motion of interest are large enough that variations with latitude of the Coriolis parameter, equal to twice the component of the Earth's angular velocity, cannot be neglected [52]. Vortices, indeed, move differently with respect to fluid particles in the background turbulence in presence of differential rotation. Here, significant differences between long-time dispersion properties of particles inside and outside vortices can be detected [52].

In a study of single-particle dispersion in two-dimensional turbulence, Pasquero et al. [21] proposed a parameterization of dispersion in two-dimensional turbulence at scales larger than those of the individual vortices. In doing so, no a priori difference between particles inside and outside vortices is drawn.

The main point of the approach followed in [21] is the observation that the Eulerian velocity at any point is determined by the combined effect of the far field of the vortices and the contribution of the local vorticity field in the background [20]. Thus, even outside vortices, the velocity field induced by the coherent vortices cannot be discarded: on average, 80% of the kinetic energy in the background turbulence outside vortices is due to the velocity field induced by the vortex population. In addition, the non-Gaussian velocities measured in the background turbulence outside vortices are entirely due to the action of the surrounding vortices, which extend their influence far away from their inner cores. This is a signature of the non-locality of the velocity field: a particle moving in a vortex-dominated flow is heavily affected by the vortex dynamics even if it is not located inside them.

In this approach, the stochastic Lagrangian velocity of a particle at the position $\mathbf{X}(t)$ is produced by the sum of two components,

$$\mathbf{U}(\mathbf{X}) = \mathbf{U}_B(\mathbf{X}) + \mathbf{U}_V(\mathbf{X}) \ , \tag{12}$$

where $\mathbf{U}_B(\mathbf{X})$ is the velocity induced by the background turbulence and $\mathbf{U}_V(\mathbf{X})$ is that induced by the vortices. The background-induced velocity is characterized by small energy and slow dynamics (i.e., long temporal correlations), while the vortex-induced component has large energy and it undergoes fast dynamics (whose temporal scale is of the order of the eddy turnover time). In addition, the vortex-induced component is characterized by a non-Gaussian velocity PDF.

A different stochastic equation has then to be used for each of the two components. Since the background-induced velocity component, $\mathbf{U}_B(\mathbf{X})$, has a Gaussian distribution, a standard stochastic OU process can be used to describe it. As for the non-Gaussian, vortex-induced component $\mathbf{U}_V(\mathbf{X})$, a proper description is easily obtained by considering a non-linear Markoff-1 model [21]. In this case, one needs to consider a generalized Langevin equation

$$d\mathbf{U}_V = \mathbf{a}(\mathbf{U}_V)dt + b(\mathbf{U}_V)d\mathcal{W} \tag{13}$$

where the functions $\mathbf{a}$ and b are functions of the velocity $\mathbf{U}_V$. The choice of the function $\mathbf{a}(\mathbf{U}_V)$ is (not uniquely) determined by the corresponding Fokker–Planck equation, with the use of the well-mixed condition [53]. In the end, the model proposed by Pasquero et al. becomes (we omit the particle index i for simplicity of notation):

$$d\mathbf{X} = (\mathbf{U}_B + \mathbf{U}_V)\, dt$$

$$d\mathbf{U}_B = -\frac{\mathbf{U}_B}{T_B}dt + \sqrt{\frac{2\sigma_B^2}{T_B}}d\mathcal{W}_B \tag{14}$$

$$d\mathbf{U}_V = -\frac{2 + |\mathbf{U}_V|/\sigma_V}{(1 + |\mathbf{U}_V|/\sigma_V)^2}\frac{\mathbf{U}_V}{T_V}\, dt + \sqrt{\frac{2\sigma_V^2}{T_V}}d\mathcal{W}_V$$

where $T_B > T_V$, $\sigma_V^2 \gg \sigma_B^2$, and $\mathcal{W}_B$ and $\mathcal{W}_V$ are two independent Wiener processes.

Interestingly, the parameters of the stochastic model depicted above can be obtained from fits to an ensemble of Lagrangian trajectories (i.e., assuming no knowledge of the advecting velocity field). Comparison with particle advection in two-dimensional turbulence shows that this model captures single-particle dispersion with an error of less than 5%, and it does also capture statistical quantities measuring higher-order moments of the dispersion statistics (e.g., the distribution of first-exit times). Note that both the non-linear nature of the vortex-induced velocity and the presence of a low-energy background-induced velocity are essential ingredients of the model. At shorter times, the vortex-induced velocity dominates and it entirely determines statistical properties such as the non-Gaussian velocity distribution. At longer times, the vortex-induced velocity becomes rapidly uncorrelated and the lower-energy background-induced velocity gives a significant contribution to particle dispersion.

One advantage of the model illustrated above is that it has been built from a detailed knowledge of the dynamics of vortex-dominated flows. That is, it is not obtained by ignoring the structure of the flow, but from an attempt to reproduce, in a stochastic framework, some of the essential ingredients of mesoscale turbulence. In particular, this model fully exploits the two-component nature of mesoscale turbulence.

4 Dynamics of Passive and Active Tracers

Transport processes can be approached from an Eulerian perspective, focusing on the advection–diffusion equation for an advected tracer field concentration:

$$\frac{\partial \rho}{\partial t} + \mathbf{u} \cdot \nabla \rho = F_\rho + D_\rho$$

where ρ is the concentration of the advected tracer, and F_ρ and D_ρ are respectively source and sink terms for the tracer.

There is a deep difference between the dynamics of active and passive tracers. A passive tracer does not feed back on the velocity field, as in the case of the concentration of a (dispersed) pollutant or of plankton in the ocean. An active tracer, on the other hand, does feed back on the fluid dynamics, think of temperature in a convecting fluid or vorticity in two-dimensional turbulence (which indeed defines the velocity field by the Biot-Savart law $\omega = \nabla^2 \psi$ and $u = -\partial \psi / \partial y$, $v = \partial \psi / \partial x$). Clearly, to some extent all tracers are "active," either dynamically or thermodynamically. However, it is often assumed that when the feedback is small, or indirect, it can be discarded.

In the absence of sources and sinks, both the spatial average, $\langle \rho \rangle$, and the variance, $\langle (\rho - \langle \rho \rangle)^2 \rangle$, of the tracer concentration are conserved. Without loss of generality, we can put $\langle \rho \rangle = 0$. In a statistically stationary flow, the dynamics of a passive tracer in two-dimensional turbulence is characterized by a direct cascade of tracer variance from large to small scales. In the inertial range, far from the characteristic scales of sources and sinks, dimensional arguments indicate that the tracer variance spectrum, $P_\rho(k)$ where k is the wavenumber, is characterized by a form $P_\rho(k) \propto k^{-1}$ [54].

The situation for vorticity is complicated by the presence of two quadratic invariants when $F = D = 0$: enstrophy, $Z = \langle \omega^2 \rangle$ (again we have made the safe assumption that $\langle \omega \rangle = 0$), analogous to tracer variance, and energy, $E = \langle (u^2 + v^2) \rangle / 2$. The simultaneous conservation of these two quantities induces a direct cascade of enstrophy, analogous to the direct cascade of tracer variance, and an inverse cascade of energy, a specific property of two-dimensional turbulence [55, 56]. In the case that the scales of small-scale dissipation, l_D, of forcing, l_F, and of large-scale boundary effects, L, are sufficiently far from each other, dimensional arguments can be used to determine the form of the spectrum. The inverse energy cascade takes place at scales l that are larger than the forcing scale, $l_F < l < L$, and it is associated with an energy spectrum $E(k) \propto k^{-5/3}$. At scales smaller than the forcing scales, $l_D < l < l_F$, a direct cascade of enstrophy appears, associated with an energy spectrum $E(k) \propto k^{-3}$ and an enstrophy spectrum $Z(k) \propto k^{-1}$.

Direct numerical simulation of forced-dissipated two-dimensional turbulence indicates that the spectrum of passive tracer variance follows with a good approximation the predicted scaling form $P_\rho \propto k^{-1}$. On the other hand, the enstrophy spectrum in the range of the direct enstrophy cascade is usually steeper than the prediction from dimensional arguments.

This difference has recently been explored by Babiano and Provenzale [57], who investigated why the direct cascade is weaker for vorticity than for a passive tracer. The analysis of the vorticity field by means of the local value of the Okubo–Weiss parameter [58, 59], $Q = s^2 - \omega^2$, where $s^2 = (\partial u / \partial x - \partial v / \partial y)^2 + (\partial u / \partial y + \partial v / \partial x)^2$, has shown that the enstrophy cascade is reversed in elliptic regions characterized by dominance of rotation over strain ($Q < 0$). In the cores of the vortices and in small elliptic patches in the background, at finite scales in the enstrophy inertial range one observes an *inverse* enstrophy cascade. In turn, this is associated with gradient-smoothing processes and an inverse energy cascade.

This behavior is consistent with the weaker spectral enstrophy flux, compared to the passive tracer variance flux, and with the steeper logarithmic slope of the enstrophy spectrum. The inversion of the enstrophy cascade in elliptic regions is the main difference between the dynamics of passive tracer and vorticity. In particular, Babiano and Provenzale speculated that the inversion of the enstrophy cascade can be one important mechanism associated with the formation of coherent vortices.

5 Conclusions

Geophysical turbulence is populated with long-lived, energetic structures: vortices, fronts, jets, and waves. Among these, coherent vortices play an especially important role, and affect transport processes in many ways.

As a consequence, transport processes cannot be understood in detail by resorting to simple stochastic parameterizations, but require the development and use of new approaches. In this chapter we have discussed some possible options, that include non-linear stochastic processes and an explicit consideration of the turbulent cascades.

Of course, many issues are still open. One conceptual question is how and why do coherent vortices form. The consideration of the cascades can help address this problem, but much more needs to be done.

Another active topic of research, which has not been discussed here, is the interplay of coherent vortices and the marine ecosystem (see the contribution of Marina Levy in this volume, or Pasquero et al. [60, 61] to discover the view of some of the authors of the present chapter). Mesoscale vortices affect the population dynamics of phyto- and zooplankton, and are associated with secondary currents responsible for localized vertical fluxes of nutrients [62, 63, 64, 65, 66, 67, 68, 69, 70]. The fact that the nutrient fluxes have a fine spatial and temporal detail, generated by the eddy field, has important consequences on primary productivity [60, 65, 71]. Furthermore, vortices can act as shelters for temporarily less-favored planktonic species owing to their trapping properties [72] and can disguise the possible presence of self-sustained oscillations in the plankton system [73]. The horizontal velocity field induced by vortices also plays an important role in determining plankton patchiness [74, 75, 76]. The parameterization of transport in mesoscale turbulence and of its ecological effects [77], needed for properly representing biogeochemical cycles in coarse-resolution climate models, is a key open problem.

Acknowledgments

A large part of the present contribution is based on a chapter published in the volume "Lagrangian Analysis and Prediction of Coastal and Ocean Dynamics," edited by A. Griffa et al. [78].

References

1. M. Arhan, H. Mercier and J. R. E. Lutjeharms: The disparate evolution of three Agulhas rings in the South Atlantic Ocean. J. Geophys. Res. Oceans **104**, 20987–21005 (1999)
2. A. S. Bower, L. Armi and I. Ambar: Lagrangian observations of Meddy formation during a mediterranean undercurrent seeding experiment. J. Phys. Oceanogr. **27**, 2545–2575 (1997)

3. G. R. Flierl: Isolated eddy models in geophysics. Ann. Rev. Fluid Mech. **19**, 493–530 (1987)
4. S. L. Garzoli, P. L. Richardson, C. M. D. Rae, D. M. Fratantoni, G. J. Goni and A. J. Roubicek: Three Agulhas rings observed during the Benguela Current experiment. J. Geophys. Res. Oceans **104**, 20971–20985 (1999)
5. N. G. Hogg and W. B. Owens: Direct measurement of the deep circulation within the Brazil Basin. Deep-sea Res. Part II **46**, 335–353 (1999)
6. J. C. McWilliams: Submesoscale, coherent vortices in the ocean. Rev. Geophys. **23**, 165–182 (1985)
7. D. B. Olson and R. H. Evans: Rings of the Agulhas Current. Deep-sea Res. Part A **33**, 27–42 (1986)
8. D. B. Olson: Rings in the ocean. Annu. Rev. Earth. Planet. Sci. **19**, 133–183 (1991)
9. R. S. Pickart, W. M. Smethie, J. R. N. Lazier, E. P. Jones and W. J. Jenkins: Eddies of newly formed upper Labrador Sea water. J. Geophys. Res. Oceans **101** (C9), 20711–20726 (1996)
10. P. L. Richardson: A census of eddies observed in North-Atlantic SOFAR float data. Progr. Ocean. **31**, 1–50 (1993)
11. P. L. Richardson, G. E. Hufford, R. Limeburner and W. S. Brown: North Brazil current retroflection eddies. J. Geophys. Res. Oceans **99**, 5081–5093 (1994)
12. P. L. Richardson and D. M. Fratantoni: Float trajectories in the deep western boundary current and deep equatorial jets of the tropical Atlantic. Deep-sea Res. Part II **46**, 305–333 (1999)
13. P. L. Richardson, A. S. Bower and W. Zenk: A census of Meddies tracked by floats. Progr. Ocean. **45**, 209–250 (2000)
14. G. I. Shapiro and S. L. Meschanov: Distribution and spreading of Red-Sea water and salt lens formation in the Northwest Indian-Ocean. Deep-sea Res. Part A **38**, 21–34 (1991)
15. D. Stammer: Global characteristics of ocean variability estimated from regional TOPEX/POSEIDON altimeter measurements. J. Phys. Ocean. **27**, 1743–1769 (1997)
16. P. Testor and J. C. Gascard: Large-scale spreading of deep waters in the Western Mediterranean sea by submesoscale coherent eddies. J. Phys. Ocean. **33**, 75–87 (2003)
17. G. Weatherly, M. Arhan, H. Mercier and W. Smethie: Evidence of abyssal eddies in the Brazil Basin. J. Geophys. Res. Oceans **107** (C4), 3027 (2002)
18. J. C. McWilliams: The emergence of isolated coherent vortices in turbulent flow. J. Fluid Mech. **146**, 21 (1984)
19. A. Bracco, J. C. McWilliams, G. Murante, A. Provenzale and J. B. Weiss: Revisiting freely decaying two-dimensional turbulence at millennial resolution. Phys. Fluids **11**, 2931–2941 (2000)
20. A. Bracco, J. LaCasce, C. Pasquero and A. Provenzale: The velocity distribution of barotropic turbulence. Phys. Fluids **12**, 2478–2488 (2000)
21. C. Pasquero, A. Provenzale and A. Babiano: Parameterization of dispersion in two-dimensional turbulence. J. Fluid Mech. **439**, 279–303 (2001)
22. C. Pasquero, A. Provenzale and J. B. Weiss: Vortex statistics from Eulerian and Lagrangian time series. Phys. Rev. Lett. **89**, 284501 (2002)
23. R. Benzi, S. Patarnello and P. Santangelo: On the statistical properties of two-dimensional decaying turbulence. Europhys. Lett. **3**, 811–818 (1987)

24. M. Farge and G. Rabreau: Wavelet transform to detect and analyze coherent structures in two-dimensional turbulent flows. C. R. Acad. Sci. Paris Sér. II **307**, 1479–1486 (1988)
25. M. Farge, K. Schneider and N. Kevlahan: Non-Gaussianity and coherent vortex simulation for two-dimensional turbulence using an adaptive orthogonal wavelet basis. Phys. Fluids **11**, 2187–2201 (1999)
26. J. C. McWilliams: The vortices of two-dimensional turbulence. J. Fluid Mech. **219**, 361–385 (1990)
27. J. C. McWilliams, J. B. Weiss and I. Yavneh: The vortices of homogeneous geostrophic turbulence. J. Fluid Mech. **401**, 1–26 (1999)
28. A. Siegel and J. B. Weiss: A wavelet-packet census algorithm for calculating vortex statistics. Phys. Fluids **9**, 1988–1999 (1997)
29. A. Bracco, J. von Hardenberg, A. Provenzale, J. B. Weiss and J. C. McWilliams: Dispersion and mixing in quasigeostrophic turbulence. Phys. Rev. Lett. **92**, 084501 (2004)
30. A. Bracco, J. H. LaCasce and A. Provenzale: Velocity probability density functions for oceanic floats. J. Phys. Oceanogr. **30**, 461–474 (2000)
31. J. Jiménez: Algebraic probability density tails in decaying isotropic two-dimensional turbulence. J. Fluid Mech. **313**, 223–240 (1996)
32. I. A. Min, I. Mezic and A. Leonard: Levy stable distributions for velocity difference in systems oof vortex elements. Phys. Fluids **8**, 1169–1180 (1996)
33. J. B. Weiss, A. Provenzale and J. C. McWilliams: Lagrangian dynamics in high-dimensional point-vortex systems. Phys. Fluids **10**, 1929–1941 (1998)
34. P. Falco, A. Griffa, P. M. Poulain and E. Zambianchi: Transport properties in the Adriatic Sea as deduced from drifter data. J. Phys. Ocean. **30**, 2055–2071 (2000)
35. A. Maurizi, A. Griffa, P. M. Poulain and F. Tampieri: Lagrangian turbulence in the Adriatic Sea as computed from drifter data: Effects of inhomogeneity and nonstationarity. J. Geophys. Res. Oceans **109**, C04010 (2004)
36. A. Bracco, E. P. Chassignet, Z. D. Garraffo and A. Provenzale: Lagrangian velocity distributions in a high-resolution numerical simulation of the North-Atlantic. J. Atmos. Ocean. Tech. **20**, 1212–1220 (2003)
37. A. Babiano, J. H. E. Cartwright, O. Piro and A. Provenzale: Dynamics of small neutrally buoyant sphere in a fluid and targeting in Hamiltonian systems. Phys. Rev. Lett. **84**, 5764–5767 (2000)
38. A. Provenzale: Transport by coherent barotropic vortices. Annual Rev. Fluid Mech. **31**, 55–93 (1999)
39. A. Babiano, G. Boffetta, A. Provenzale and A. Vulpiani: Chaotic advection in point vortex models and two-dimensional turbulence. Phys. Fluids **6**, 2465–2474 (1994)
40. D. Elhmaidi, A. Provenzale and A. Babiano: Elementary topology of two-dimensional turbulence from a Lagrangian viewpoint and single-particle dispersion. J. Fluid Mech. **242**, 655–700 (1993)
41. M. E. McIntyre: On the Antarctic ozone hole. J. Atmos. Terr. Phys **51**, 29–43 (1989)
42. F. Paparella, A. Babiano, C. Basdevant, A. Provenzale and P. Tanga: A Lagrangian study of the Antarctic polar vortex. J. Geophy. Res. **102**, 6765–6773 (1997)
43. A. Colin de Verdiere: Lagrangian Eddy statistics from surface drifters in the eastern North-Atlantic. J. Marine Res. **41**, 375–398 (1983)

44. V. Rupolo, B. L. Hua, A. Provenzale and V. Artale: Lagrangian velocity spectra at 700m in the western North Atlantic. J. Phys. Oceanogr. **26**, 1591–1607 (1996)
45. H. C. Rodean: Stochastic Lagrangian models of turbulent diffusion. *Meteor. Monogr.* **26**(48), 84 (1996)
46. A. Griffa: Applications of stochastic particle models to oceanographic problems. In: Stochatic Modelling in Physical Oceanography, Adler, R. J., Müller, P., and Rozovskii, R. B. (eds.), Birkhäuser, Boston pp. 114–140 (1996)
47. D. Brickman and P. C. Smith: Lagrangian stochastic modeling in coastal oceanography. J. Atmos. Ocean. Tech. **19**, 83–99 (2002)
48. P. S. Berloff, and J. C. McWilliams: Material transport in oceanic gyres. Part II: Hierarchy of stochastic models. J. Phys. Oceanogr. **32**, 797–830 (2002)
49. A. M. Reynolds: On Lagrangian stochastic modelling of material transport in oceanic gyres. Physica D **172**, 124–138 (2002)
50. B. L. Sawford: Rotation of trajectories in lagrangian stochatsic models of trubulent dispersion. Bound.-Lay. Meteorol. **93**, 411–424 (1999)
51. M. Veneziani, A. Griffa, A. M. Reynolds and A. J. Mariano: Oceanic turbulence and stochastic models from subsurface Lagrangian data for the northwest Atlantic Ocean. J. Phys. Ocean. **34**, 1884–1906 (2004)
52. C. R. Mockett: Dispersion and Reconstruction. In Astrophysical and Geophysical Flows as Dynamical System, WHOI Tech. Rep. WHOI-98-00 (1998)
53. D. J. Thomson: Criteria for the selection of stochastic models of particle trajectories in turbulent flows. J. Fluid Mech. **180**, 529–556 (1987)
54. G. K. Batchelor: Small-scale variation of convected quantity like temperature in turbulent field. J. Fluid Mech. **5**, 113–133 (1959)
55. G. K. Batchelor: Computation of the energy spectrum in homogeneous two-dimensional turbulence. Phys. Fluids Suppl. **12**, II 233 (1969)
56. R. H. Kraichnan: Inertial ranges in two-dimensional turbulence. Phys. Fluids **10**, 1417–1423 (1967)
57. A. Babiano and A. Provenzale: Coherent vortices and tracer cascades in two-dimensional turbulence. J. Fluid Mech. **574**, 429–448 (2007)
58. A. Okubo: Horizontal dispersion of floatable particles in the vicinity of velocity singularities such as convergences. Deep-sea Res. **17**, 445–454 (1970)
59. J. Weiss: The dynamics of enstrophy transfer in two-dimensional hydrodynamics. Physica D **48**, 273–294 (1991)
60. C. Pasquero, A. Bracco and A. Provenzale: Coherent vortices, Lagrangian particles and the marine ecosystem. In: Jirka G. H. and Uijttewaal W. S. J., *Shallow Flows*, Balkema Publishers, Leiden, NL, 399–412 (2004)
61. C. Pasquero, A. Bracco, A. Provenzale and J. B. Weiss: Particle motion in a sea of eddies. In: Griffa A. et al. (eds.) *Lagrangian Analysis and Prediction of Coastal and Ocean Dynamics*. Cambridge University Press, Cambridge (2007)
62. P. G. Falkowski, D. Ziemann, Z. Kolber and P. K. Bienfang: Role of eddy pumping in enhancing primary production in the Ocean. Nature **352**, 55–58 (1991)
63. M. Lévy, P. Klein and A. M. Tréguier: Impact of sub-mesoscale physics on production and subduction of phytoplankton in an oligotrophic regime. J. Marine Res. **59**, 535–565 (2001)
64. M. Lévy: Mesoscale variability of phytoplankton and of new production: Impact of the large-scale nutrient distribution. J. Geophys. Res. Oceans **108** (C11), 3358 (2003)

65. A. P. Martin and K. J. Richards: Mechanisms for vertical nutrient transport within a North Atlantic mesoscale eddy. Deep-sea Res. Part II **48**, 757–773 (2001)

66. A. P. Martin, K. J. Richards, A. Bracco and A. Provenzale: Patchy productivity in the open ocean. Global Biogeochem. Cycles **16**, 1025 (2002)

67. D. J. McGillicuddy and A. R. Robinson: Eddy-induced nutrient supply and new production in the Sargasso Sea. Deep-sea Res. Part I **44**, 1427–1450 (1997)

68. D. J. McGillicuddy, A. R. Robinson, D. A. Siegel, H. W. Jannasch, I. R. Johnson, T. Dickeys, J. McNeil, A. F. Michaels and A. H. Knap: Influence of mesoscale eddies on new production in the Sargasso Sea. Nature **394**, 263–266 (1998)

69. D. Siegel, D. J. McGillicuddy and E. A. Fields: Mesoscale eddies, satellite altimetry, and new production in the Sargasso Sea. J. Geophys. Res. Oceans **104** (C6), 13359–13379 (1999)

70. C. L. Smith, K. J. Richards and M. J. R. Fasham: The impact of mesoscale eddies on plankton dynamics in the upper ocean. Deep-sea Res. II 1807–1832 (1996)

71. C. Pasquero, A. Bracco and A. Provenzale: Impact of the spatiotemporal variability of the nutrient flux on primary productivity in the ocean. J. Geophys. Res. **110**, C07005 (2005)

72. A. Bracco, A. Provenzale and I. Scheuring: Mesoscale vortices and the paradox of the plankton. P. Roy. Soc. Lond. B **267**, 1795–1800 (2000)

73. I. Koszalka, A. Bracco, C. Pasquero and A. Provenzale: Plankton cycles disguised by turbulent advection. Theor. Populat. Biol., doi:10.1016/j.tpb.2007.03.007 (2007)

74. E. R. Abraham: The generation of plankton patchiness by turbulent stirring. Nature **391**, 577–580 (1998)

75. A. Mahadevan and J. W. Campbell: Biogeochemical patchiness at the sea surface. Geophys. Res. Lett. **29**, 1926 (2002)

76. A. P. Martin: Phytoplankton patchiness: the role of lateral stirring and mixing. Progress in Oceanography **57**, 125 (2003)

77. C. Pasquero: Differential eddy diffusion of biogeochemical tracers. Geophys. Res. Lett. **32**, L17603 (2005)

78. A. Griffa, A. D., Jr. Kirwan, A. M. Mariano, T. Ozgokmen, H. Thomas Rossby (eds.): Lagrangian Analysis and Prediction of Coastal and Ocean Dynamics. Cambridge University Press, Cambridge, UK (2007)

Part II

Experiments and Observations

Dispersion and Mixing
in Quasi-two-dimensional Rotating Flows

M. G. Wells, H. J. H. Clercx and G. J. F. van Heijst

Physics Department, Eindhoven University of Technology
P.O. Box 513, 5600MB Eindhoven, The Netherlands
h.j.h.clercx@tue.nl

Abstract. A new rotating-tank experiment has been set up to investigate several aspects of dispersion in forced quasi-two-dimensional turbulence. By superimposing a harmonically varying perturbation on the mean rotation rate the mean flow continually interacts with the no-slip boundaries and forms boundary layers with high-amplitude vorticity twice during the forcing period. By choosing the proper amplitude and frequency of the perturbation it is possible to continuously inject small-scale vorticity in the interior of the flow, either in the form of filamentary structures (detached boundary layers) or as small vortices (after the roll-up of detached boundary layers). We present measurements of the passive scalar spectrum which show good agreement with the k^{-1} spectrum predicted by Batchelor (J. Fluid Mech. 5:113, 1959). Using particle image velocimetry we are able to reconstruct the Lagrangian trajectories of particles. The relative dispersion rates of particle pairs show an initial exponential separation followed by the classical Richardson dispersion, $R^2 \propto t^{3.0\pm0.1}$. The variance of the absolute particle displacement grows as $\sigma \propto t^{1.4}$, similar to the observations in the previous experiments by Solomon et al. (Phys. Rev. Lett. 71:3975, 1993) and Hansen et al. (Phys. Rev. E 58:7261, 1998). Finally, and indicating future directions of research, we present results of a simple chemical reaction in forced quasi-2D turbulence and show how the bulk reaction rate is controlled by the mixing and filamentation processes.

1 Introduction

Strictly two-dimensional (2D) turbulence is an idealisation, since natural flows have a 3D aspect to them. Nevertheless, understanding the simplest 2D case gives a good grasp of more complicated systems that occur in the atmosphere and oceans. Examples of quasi-2D flow where the mixing and dispersion of passive tracers are important are easily found in the atmosphere and in oceanic flows where a combination of geometry, stratification and rotation acts to suppress motion in the vertical direction. In a similar way the magnetic field lines can constrain charged particles in plasma confinement devices and astrophysical flows to quasi-2D behaviour [1]. In the stratosphere the rate that reactive

M. G. Wells et al.: *Dispersion and Mixing in Quasi-two-dimensional Rotating Flows*, Lect.
Notes Phys. **744**, 119–136 (2008)
DOI 10.1007/978-3-540-75215-8_6

chlorine can mix and destroy ozone in the polar vortices is to a large degree controlled by quasi-2D horizontal stirring and mixing. Pierce and Fairliem [2] used output from an atmospheric circulation model to quantify the advection and mixing by subtropical anticyclones within the northern hemisphere winter stratospheric vortex. The deformation of the material lines near the edge of the polar vortex, which then rapidly evolve into elongated filaments as material is drawn around the smaller vortices at the edge of the polar vortex, strongly enhances mixing. The rate of stretching of the material lines was shown to be exponential and the chaotic advection leads to rapid mixing of vortex air with tropical and mid-latitude air, which has important implications for ozone depletion. Chaotic mixing has also been shown to be important in the ocean, as the uptake of carbon dioxide is in part due to the growth of phytoplankton whose growth rate depends on light and nutrient availability, the depth of the mixing layer and lateral stirring of nutrients by fields of quasi-2D eddies. The dynamics of the resulting filamental and patchy fields of phytoplankton has been reviewed by Martin [3]. Such patchiness can lead to enhancement of phytoplankton growth due to the increased horizontal chemical gradients. This is consistent with the increased carbon dioxide uptake in numerical ocean circulation models as found by Martin et al. [4]. Similar enhancement of the bulk reactivity rates was also demonstrated by Paireau and Tabeling [5] in laboratory experiments of mixing of reactants by a quasi-2D chaotic flow in shallow fluid layers.

The dispersion characteristics of passive Lagrangian tracers in models of the ocean have been shown by Bracco et al. [6] to be very similar to what is found in idealised 2D models. They numerically studied the dynamics of passive Lagrangian tracers in 3D quasi-geostrophic turbulence and compared the behaviour with that of 2D barotropic turbulence. Despite the different Eulerian properties of the two flows, they found that the Lagrangian dynamics of passively advected tracers in 3D quasi-geostrophic turbulence is very similar to that of barotropic turbulence, with an initial exponential separation followed by the relative dispersion growing as Richardson's law of $R^2 \propto t^3$ (with R the particle separation and t the time). Similar results were found for the initial relative dispersion of floating surface drifters in the Gulf of Mexico [7] where the mean square pair separation grew exponentially in time from the smallest resolved scale, which was 1 km in this particular study, to approximately 50 km, with an e-folding time of 2–3 days. Thereafter, the dispersion exhibits a power-law dependence on time with an exponent between 2 and 3 up to scales of several hundred kilometres. These oceanographic and atmospheric examples serve to motivate the study of mixing and dispersion in 2D turbulence.

This chapter is structured as follows; in Sect. 2 we review the important theoretical predictions of the dispersion and mixing in 2D flows. Only a limited number of different laboratory experiments have been used to study the properties of these 2D flows and in Sect. 3 we discuss briefly the use of thin electrolytic layers, soap films, density stratified fluids and rotating fluids to make flows quasi-2D. In Sect. 3.1 we describe a rotating laboratory experiment that is able to continuously inject vortices from the boundary into the interior of the flow. The observed turbulent flow field appears to behave similar to globally forced 2D turbulence. Experimental observations of the vortices and their interactions are presented in Sect. 4 along with the observed scalar spectrum (Sect. 4.2) and dispersion rates (Sect. 4.3). Finally, and as an illustration, it is shown in Sect. 4.4 how this experiment can be used to investigate chemical reactions in 2D turbulence.

2 Passive Scalar Dispersion in 2D Turbulence

Stirring and mixing of two distinct bodies of water with certain amounts of passive tracer is accomplished in three stages. Initially there are distinct interfaces separating the two water masses, with possibly a different passive tracer concentration in each water parcel. During the stirring process, the second stage, the water masses are mechanically swirled and folded, and molecular diffusion is unimportant as long as tracer concentration gradients are not too large. The final stage is when mixing occurs: the gradients suddenly disappear and the fluid becomes homogeneous; molecular diffusion is responsible for the sudden mixing. This final stage of mixing is reached when filaments of the fluid have been strained to small enough scales that diffusion can quickly act. In a chemical reaction, molecules of different species must come into contact for the reaction to occur. Thus, when the species are initially separated, the reaction will not begin until the final mixing stage is reached. Thus in many chemically reacting systems the bulk reaction rates are set not only by the chemical kinetics but also by the rates at which different patches of fluid are stretched and folded so that spatial scales are reduced to those where diffusion takes over.

Amongst the few firm results in the theory of turbulence is Batchelor's prediction about the form of the passive scalar spectrum $E_s(k)$ in the convective–dissipative range of wave numbers. The spatial scales of the convective–dissipative regime occur in the second stage of mixing that we have just described. In this wave number range, no eddies are present and velocity fluctuations are strongly damped, but the straining is still strong enough to produce stretching and folding of the passive tracer field (and diffusion of the passive tracer is not important yet). The theory proposed by Batchelor [8] predicted that the stretching and folding leads to an intensity spectrum of the dye or temperature fluctuations that scale as $E_s(k) \propto k^{-1}$.

The Lagrangian dynamics of mixing in a turbulent fluid was first investigated by Richardson [9] who predicted the "four-thirds law" which expresses that the turbulent eddy diffusivity increases as the scale of the flow increases as $K(r) = k_0 \epsilon^{1/3} r^{4/3}$, where ϵ is the rate of dissipation of turbulent kinetic energy and k_0 a constant. His first experiments consisted of releasing pairs of balloons and recording the approximate time and place they were later recovered on the ground. The resulting data showed that the relative separation increased non-linearly with time. In a later experiment with Henry Stommel they threw floating white disks cut from parsnips from a small pier into Loch Long, and by measuring their position with time they confirmed that the rate of spreading increased as the distance between the pairs of parsnips increased, consistent with $R^2 \propto t^3$ [10]. This idea was later analyzed more generally by Batchelor [11]. At very small scales in 2D turbulence, the flow is dominated by straining and an exponential separation of particles has been observed [12].

Two-dimensional turbulent fields are characterised by the presence of coherent vortices. The trapping of particles in the cores of coherent structures changes the dispersion statistics of particles [13]. With such long-lived coherent vortex structures there are two distinct regions of the flow: those regions that mix rapidly, with a particle separation rate that initially grows exponentially and subsequently grows like $R^2 \propto t^3$, and those regions within the vortex cores where particles are trapped. These two different mixing regions can be distinguished by the relative importance of the vorticity field with respect to the strain field (stretching plus shearing deformation). This is done using the so-called Okubo–Weiss parameter, defined as $Q = S^2 - \omega^2$, where ω^2 is the square of the vorticity and S^2 is the total squared strain [14, 15]. When $Q < 0$, the vorticity field is stronger than the strain field and eddy-like structures are present (elliptic regions), while for $Q > 0$ the strain field dominates (hyperbolic regions) where dispersion is stronger. In chemically reacting systems the strain-dominated regions act to increase the heterogeneity of chemical species within the flow.

The energy spectrum $E(k)$ of forced 2D turbulence consists of two distinct power-law regimes, one for wave numbers larger and another for wave numbers smaller than an injection wave number k_i [16, 17]. For $k < k_i$ the spectrum is characterised by an inverse cascade of energy from small scales to large scales with $E(k) \propto k^{-5/3}$, while for large wave numbers $k > k_i$ the energy spectrum decreases more rapidly, $E(k) \propto k^{-3}$, characteristic of the direct enstrophy cascade. The transfer of energy from small spatial scales to large spatial scales is the opposite of 3D turbulence where spectral energy of both active and passive tracers are transferred from larger to smaller scales. In 2D turbulence only passive tracers are advected to small scales, so we could expect the passive scalar spectrum to have a slope of $E_s(k) \propto k^{-1}$ for the same wave number range that corresponds to the k^{-3} direct enstrophy cascade. The theoretical predictions assume that turbulence is isotropic, unbounded and continuously forced with a well-defined wave number k_i at which energy is injected into the system. However, by necessity laboratory experiments are bounded, and turbulence is anisotropic due to the spatial variability in forcing or interactions with no-slip boundaries that act as vorticity sources.

3 Laboratory Experiments of Quasi-2D Turbulence

In recent reviews the various experimental techniques to study 2D flows have been summarised [18, 19]. Quasi-2D turbulent flows are usually generated in shallow fluid layers, soap films, rotating fluids and density stratified fluids. In the first case the turbulence is generated in a relatively thin layer of conducting fluid (or electrolyte), with a spatially varying magnetic field applied perpendicular to the fluid layer. With two electrodes at opposite sidewalls of the container a temporally varying electric field is applied. The resulting Lorentz force acts on the charge carriers and drives the fluid motion. The first such experiment was performed by Sommeria [20] with turbulence generation by steady forcing. If the electromagnetic forcing is turned on and off, rapidly decaying 2D turbulence is produced and vortices are seen to merge together over time to form larger structures [21]. With the second technique, quasi-2D turbulence is generated in a rapidly flowing soap film that is penetrated by a comb [22]. By measuring the short-time fluctuations in film thickness, here considered as a passive tracer, Amarouchene and Kellay [23] found a k^{-1} spectrum over a decade of wave numbers in a fast-flowing soap film. A special arrangement of two combs set in a V-shape allowed Rutgers [24] to continuously inject vortices into the flow field and make simultaneous measurements of both an inverse energy cascade with characteristic $k^{-5/3}$ spectrum and a forward enstrophy cascade with a k^{-3} spectrum. Rotation has also long been known to lead to a 2D flow field [25], with Welander [26] making some of the earliest observations of the mixing and filamentation of a passive tracer in decaying 2D turbulence. Finally, strongly stratified flows that limit vertical motion have been used by a number of investigators to study confined quasi-2D flows. By towing a vertical comb through a two-layer stratification Maassen, Clercx and van Heijst [27, 28, 29] were able to examine the long-time evolution of decaying 2D turbulence and its interaction with the no-slip boundaries in circular, square and rectangular containers.

3.1 Quasi-2D Turbulence by Oscillating Spin-Up

We will now present results of mixing and dispersion in a new experiment to generate forced quasi-2D turbulence. Laboratory experiments in rotating fluids of oscillating spin-up have been conducted where the production of small-scale vorticity near the no-slip sidewalls of the container leads to the formation of quasi-2D vortices. The decay of the vortices is due to either the interaction with other vortices (vortex stripping) or the effects of Ekman pumping at the bottom of the tank. The effects of Ekman damping are minimised by increasing the depth of the fluid.

In the laboratory experiments, the flow is made quasi-2D by a steady background rotation. A small sinusoidal perturbation to the background rotation leads to the periodic formation of eddies in the corners of the tank by the roll-up of vorticity generated along the sidewalls. When the oscillation period is greater than the time scale needed to advect a full-grown corner vortex along a sidewall to a neighbouring corner, dipole structures are observed to form. These dipoles migrate away from the walls, and the interior of the tank is continually filled with new vortices. After several forcing periods a sea of vortices emerges in the interior of the tank. This system is clearly not isotropic, so it is of interest to see how some of the properties of forced 2D turbulence still emerge in this quasi-2D turbulent flow and how the coherent vortex structures evolve.

The most important effect of no-slip boundaries on the evolution of a (turbulent) flow field is that if there is a continued input of energy into the system, then dissipation at the boundaries naturally occurs and the energy of the flow does not grow without limit. The injection of filaments of vorticity and small vortices (after the roll-up of viscous boundary layers) represents a source of vorticity whose scale is independent of the forcing scale itself, but depends on the Reynolds number of the flow (i.e. the boundary-layer thickness). In simulations of a decaying initial distribution of vortices, this source can clearly be seen in spectra where there is a change in the slope of the energy at the wave number corresponding to the thickness of the viscous boundary layer, k_δ. For $k < k_\delta$ the spectrum is characterised by a $k^{-5/3}$ spectrum, characteristic of the inverse energy cascade, and for $k > k_\delta$ there is a k^{-3} spectrum, characteristic of the direct enstrophy cascade [30]. Similarly, Clercx, Maassen and van Heijst [31] found important influences of the boundary upon the energetics of decaying turbulence.

3.2 Experimental Design

The laboratory experiments were performed in a tank of square cross-section, with dimensions $100 \times 100 \times 30$ cm^3 (length $\times$ width $\times$ depth). This tank is mounted on a rotating table (see Fig. 1) and the flow is made quasi-2D by a steady background rotation of 1 rad/s to which a small sinusoidal perturbation is applied:

$$\Omega(t) = \Omega_0(1 + A\sin(ft)), \qquad (1)$$

where Ω_0 is the mean rotation rate, A is the (dimensionless) amplitude of the perturbation and f is the frequency. In a typical experiment the tank is spun up from rest, and through the action of Ekman pumping the flow becomes quasi-2D after approximately 20 min. The perturbation amplitude is $A = 0.06$ and the forcing frequency f is in the range 0.0157–0.126 rad/s.

In the rotating frame of reference the fluid oscillates back and forth around the axis of the tank. A similar forcing of the fluid was made in experiments performed by van Heijst [32] and van de Konijnenberg et al. [33] where the rotation rate Ω_0 of a fluid spinning in solid-body rotation was suddenly increased by a factor $\Delta\Omega_0$. In the rotating frame of reference (with angular velocity $\Omega_0 + \Delta\Omega_0$) this is equivalent to suddenly changing the relative vorticity of the flow by an amount $-2\Delta\Omega_0$. In this case an anticyclonic flow arises in the tank with maximum velocities along the sidewalls. Similar phenomena are observed in the present experiment, in which the background rotation rate is changed continuously. In this case the maximum flow velocity, its absolute value denoted by U, occurs near the sidewalls of the tank and scales with

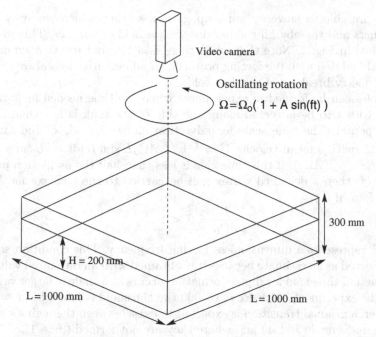

Fig. 1. A sketch of the experimental set-up: a camera is mounted above the table and rotates in the same reference frame to record the movement of particles or dye. The tank has a width $L = 100$ cm and height of 30 cm. The fluid layer has a depth $H = 20$ cm

$(\Omega(t) - \Omega_0) \times L$ so that $U = A\Omega_0 L$, where L is the half-width of the tank. The oscillation acts to provide a forcing to the vorticity field: the overall relative vorticity changes as $\omega = -2A\Omega_0 \sin(ft)$ in the rotating frame of reference.

The induced oscillating flow in the tank is strongly affected by the vertical sidewalls, which imply the presence of viscous boundary layers that contain strong vorticity. Near the corners of the tank the flow can separate and the vorticity produced in the boundary layer will accumulate in the eddies in the corners of the tank, as observed for spin-up flows [32, 34]. These eddies increase in radius with time as the vorticity produced in the viscous boundary layers is continuously advected towards the corners. The size and strength of the eddies is affected by both the strength of the vorticity in the boundary layer (or, stated differently, by the boundary-layer thickness) and the time that the vortex is able to form. Thus as the amplitude of the sinusoidal forcing $A\Omega_0$ is decreased or the frequency f is increased smaller and weaker vortices will form in the corners of the tank. In experiments with an oscillating flow on every forcing cycle vortices will be formed in the corners of the tank. When the forcing changes direction, newly formed vortices detach from the corners and travel with the mean flow. If these vortices reach the next corner they may couple with opposite-signed vortices and vortex dipoles are formed.

Dipoles are able to move by self-propagation, so they will move away from the corners and thereby fill the interior with a field of vortices. This process is sketched in Fig. 2. Note that while there is a fourfold rotation symmetry inherently related with the forcing protocol, small perturbations always result in a symmetry breaking of the flow field.

Dipoles can be formed in experiments when the time needed for 'a newly formed vortex to be advected along one side of the tank is less than half a forcing period. The time scale for advection along one side of the tank of length $2L$, with a mean velocity $\bar{U} = \frac{f}{\pi} \int_0^{\pi/f} A\Omega_0 L \sin(ft)\mathrm{d}t = 2U/\pi$, will be $\tau = 2L/\bar{U} = \pi/(A\Omega_0)$. If this timescale is less than half the oscillation period $T = 2\pi/f$, then a detached vortex will be carried to the next corner. Thus dipoles form if

$$\frac{f}{A\Omega_0} = \mathcal{F} < 1 , \tag{2}$$

where $\mathcal{F}$ represents a dimensionless forcing frequency. This condition should be considered as an estimate because it is obtained with an estimated value for τ. The actual threshold for dipole formation occurs at a slightly larger value of $\mathcal{F}$ (for the experiments with $Re = 15,000$ the threshold is found for $\mathcal{F} \approx 1.7$). Moreover, a gradual transition is expected to exist between the regimes where dipoles are formed ($\mathcal{F} \ll 1$) and where they are not formed ($\mathcal{F} \gg 1$).

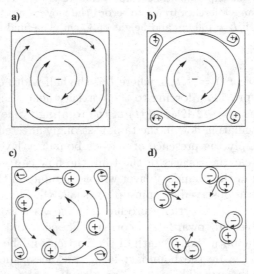

Fig. 2. A schematic cartoon of the formation of dipolar structures in the oscillating flow. (**a**) Initially the interior flow can be well described as having uniform negative vorticity (–). (**b**) The no-slip boundaries lead to the accumulation of positive (+) vorticity in each of the corners, resulting in four vortices. These vortices are then advected by the flow as the forcing changes sign (**c**) and can form dipolar structures by pairing with the negative vortices. These dipoles can then self-propagate into the interior (**d**)

The Reynolds number of the forcing is defined as

$$Re = \frac{UL}{\nu} = \frac{A\Omega_0 L^2}{\nu} \,, \tag{3}$$

where $U = A\Omega_0 L$ and ν is the kinematic viscosity of the fluid. For typical laboratory experiments $Re \approx 15,000$. This Reynolds number should not be confused with a micro-scale Reynolds number based on the ratio of the circulation Γ_v of the vortices to the molecular viscosity ν as $Re_m = \Gamma_v/\nu$. As the total circulation in a bounded flow must be zero, the changes in uniform interior vorticity due to the variable rotation rate must be balanced by thin layers with strong vorticity of the opposite sign. Thus the circulation Γ_v in each of these four vortices will be a quarter of the magnitude of the circulation corresponding to the uniform vorticity over the tank, averaged over half the forcing cycle, or $\Gamma_v = \frac{f}{\pi} \int_0^{\pi/f} 2A\Omega_0 L^2 \sin(ft)\mathrm{d}t = \frac{4}{\pi}A\Omega_0 L^2$. This implies that initially $Re_m \approx 1.3Re$. However, the dissipation during the advection of vorticity from the thin boundary layer, containing steep vorticity gradients, to the corner will result in weaker vortices than expected. Subsequent interactions between vortices and damping processes (lateral diffusion, Ekman damping) will often yield values much smaller than Re for the individual vortices. In our experiments (with $Re \approx 15,000$) initial values of Re_m were measured as high as 4000, with the older vortices having lower values, $Re_m \sim O(10^2 - 10^3)$.

4 Laboratory Results

The formation of dipoles in experiments with $\mathcal{F} < 1$ is most easily visualised by the rapid stirring of dye, as shown in Fig. 3. In this experiment $f = 0.031$ rad/s ($T = 200$ s), $A\Omega_0 = 0.06$ rad/s, so that $\mathcal{F} = 0.52$, corresponding to a regime in which strong dipole formation is expected. A small amount of fluorescein dye was initially injected near the bottom left corner. After $t/T = 0.025$ the flow has sheared and stretched this dye along one wall of the tank, and several vortices have started to stir at smaller scales. A subsequent image at $t/T = 0.125$ (Fig. 3b) shows that the dye has been sheared around the complete perimeter of the tank, and two dipolar structures are beginning to transport dye into the interior. Once the dye is in the interior of the tank, it is rapidly mixed by the field of vortices. In Fig. 3c, about 15 circular structures in the dye streaks can be observed, which indicate the presence of localised vorticity. These vortices result in almost complete mixing of the dye by $t/T = 0.325$ after its injection (Fig. 3d).

To understand the importance of the dipole formation process in determining the amount of mixing in the tank, it is useful to look at an experiment where strong dipoles did not form. The experiment illustrated by the image shown in Fig. 4 had a forcing frequency of $f = 0.126$ rad/s, $A\Omega_0 = 0.06$ rad/s, so that $\mathcal{F} = 2.09$. Dye has been injected into the flow in a similar manner as

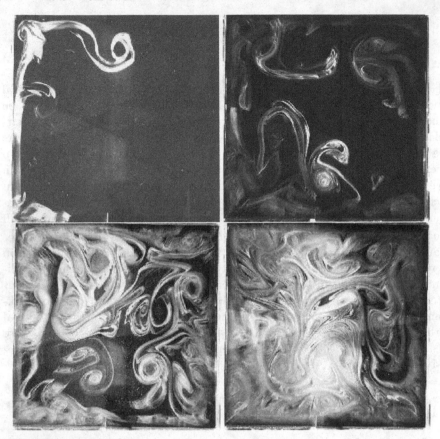

Fig. 3. A sequence of laboratory photographs of a dye tracer showing the turbulent field of vortex structures. The following snapshots are shown: (**a**) $t/T = 0.025$ after dye has been injected in the corner of the tank, (**b**) $t/T = 0.125$, (**c**) $t/T = 0.175$, and by $t/T = 0.325$ the dye has been well mixed in the tank, and the vortex structures are clearly visible (**d**). In this experiment $\mathcal{F} = 0.52$, $f = 0.031$ rad/s (so that $T = 2\pi/f = 200$ s) and $Re = 15,000$

in Fig. 3. After $t/T = 3.6$, the dye has been well mixed around the perimeter of the tank, but the central portion of the container remains free of dye. Flow separation has still occurred in the corners, and the resulting vortices can be seen in the dye filaments, but due to the high oscillation frequency these vortices were unable to form dipoles which would have rapidly left the wall region.

4.1 Particle Tracking

During an experiment, the motion of tracer particles was monitored by a digital camera in order to determine the velocity field. The images were recorded

Fig. 4. A photograph taken approximately four oscillation periods after dye has been released for an experiment with $\mathcal{F} = 2.09$ ($Re = 15,000$). In contrast to the previous figure the dye is not well mixed and has only been transported around the perimeter of the tank. This is due to the forcing parameters having $\mathcal{F} > 1$ so that dipoles could not efficiently form

onto a computer and after the experiment, the Lagrangian trajectories of the particles were determined using the particle tracking algorithm developed by Bastiaans, van der Plas and Kieft [35]. An SMD-1M15 CCD camera with 1024×1024 pixels and 12 bit grey-scale resolution allowed high-quality images to be obtained at a rate of 5 frames per second. After using the particle tracking algorithm typically 9000 particles could be tracked to allow interpolation of the velocity field to a 80×80 grid with about 1 cm^2 resolution. Figure 5a and b shows velocity vectors (indicating the stream lines) and vorticity contour plots from an experiment with $A\Omega_0 = 0.06$ rad/s and $f = 0.031$ rad/s. This implies a forcing scale Reynolds number of $Re = 15,000$ using (3). In the vorticity contour plot, Fig. 5b, one observes individual vortices which are advected by the mean background flow. In these vortices the peak value ω_{max} of the vorticity is around 0.3–0.5 s^{-1} and their radius is $\rho/L \simeq 0.15$. For these strongest vortices the micro-scale Reynolds number can be estimated as $Re_m = \Gamma_v/\nu \approx \frac{1}{2}\omega_{max}\pi\rho^2/\nu \sim 4000$; the other weaker vortices visible in Fig. 5b would have $Re_m \sim 10^2 - 10^3$.

4.2 Passive Scalar Spectra

In Fig. 6a a 1D scalar intensity spectrum $F(k) = \int_k^{k+1} F(\kappa_x, \kappa_y)\mathrm{d}\kappa$ is shown, using a 2D Fourier transform of the pixel intensity of images of fluorescein

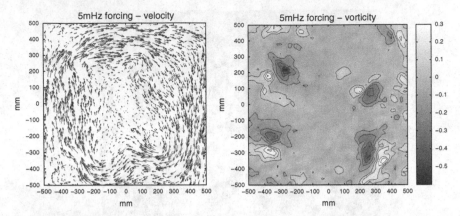

Fig. 5. Graphs showing the flow field in the form of velocity vectors (**a**) and vorticity distribution (**b**) obtained by particle tracking velocimetry in an experiment with $Re = 15,000$ and $\mathcal{F} = 0.52$

dye in a turbulent field. To reduce the influence of the non-periodic boundary conditions of the tracer concentration in these images, the mean is subtracted from the data and a Hanning window is applied before the Fourier transforms were performed. An example of the images used for this spectrum is shown in Fig. 6b at $t/T = 1$ after introduction of the dye. To illuminate the tank we used four slide projectors placed around the square tank, thus illuminating a 1 cm surface layer. In the following analysis we focus on the central 512×512 pixels of the images, which represent a quarter of the total area of the

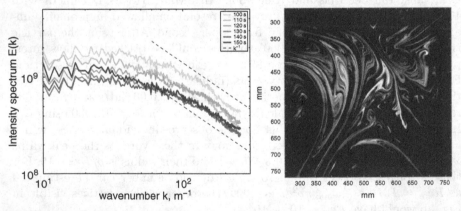

Fig. 6. A sequence of measured intensity spectra of fluorescein dye in the turbulent flow field is plotted in (**a**) from images similar to that shown in (**b**). The spectra in (**a**) and the image in (**b**) are obtained from measurements in the central 50 cm $\times$ 50 cm of the tank in an experiment with $Re = 15,000$ and $\mathcal{F} = 1.05$. The spectra have a slope similar to k^{-1} at high wave numbers

tank. Pragmatic reasons force us to focus on this region in order to minimise any parallax errors in viewing the vertically aligned dye-sheets in the rotating flow. The data shown in Fig. 6 were taken from a sequence of images $t/T = 0.1$ apart in an experiment with $\mathcal{F} = 1.05$ ($A\Omega_0 = 0.06$ rad/s and $f = 0.063$ rad/s, a forcing period of $T = 100$ s). After the experiment had run for 10 min, 5 ml of fluorescein dye was injected into the surface layer of water in one corner of the tank. The striking feature of Fig. 6a is the k^{-1} power-law behaviour of the spectra, consistent with the predictions of Batchelor [8]. The overall intensity can be seen to be decreasing, due to the Ekman pumping slowly transporting dye, that was initially in the visible surface layer, downward. Theoretically, a k^{-1} power law of the scalar spectrum should occur in the same spectral range as the enstrophy cascade of the energy spectrum. It was impossible to resolve a high wave number enstrophy cascade from the velocity measurements, but we note that the k^{-1} power law occurs at wave numbers $k > 80$ m^{-1}, which is in the region where such an enstrophy cascade is indeed expected.

4.3 Relative Dispersion

Using the velocity vectors from experiments similar to those shown in Fig. 5a a fourth order Runge–Kutta technique has been used to integrate trajectories of 400 particles forward in time. These Lagrangian trajectories are shown in Fig. 7. The relative dispersion shown in Fig. 8a shows exponential growth at early times, when the initial particle separation is small, as predicted by Lin [12]. For larger times and larger particle separation the growth follows the classical Richardson scaling of $\langle R^2 \rangle \propto t^{3.0 \pm 0.1}$ (with $\langle \cdot \rangle$ denoting an ensemble average). These results are almost identical to those obtained from numerical simulations of dispersion in quasi-geostrophic turbulence by Bracco et al. [6]. Computation of the growth with time of the particle displacement variance $\sigma^2 \sim \langle (\Delta r)^2 \rangle \propto t^{2\gamma}$ yields $\sigma \propto t^{1.4 \pm 0.1}$. As the exponent γ is between 1 and 2, this indicates the so-called "anomalous" diffusion. This growth rate of variance is similar to previous observations by Solomon, Weeks and Swinney [36] in a rotating annulus experiment forced by an unstable zonal jet. They found that the variance of the displacement grows with time as $\sigma \propto t^{1.65 \pm 0.15}$. Similar results for dispersion were found for freely decaying 2D turbulence in shallow fluid layers by Hansen, Marteau and Tabeling [37] where they noted the presence of Lévy flights and a particle displacement variance that grows in time as $\sigma \propto t^{1.4 \pm 0.1}$.

4.4 Chemical Reactions in Forced 2D Turbulence

In addition to the study of dispersion of passive tracers briefly discussed in the previous sections, preliminary experiments with reactive tracers have been conducted to illustrate the role of stirring and straining of reactive tracers. The experimental parameters are: $Re = 15,000$, $A\Omega_0 = 0.06$ rad/s

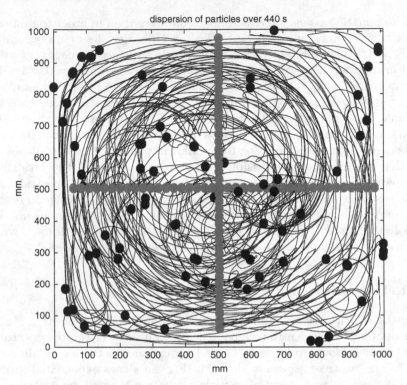

Fig. 7. Using measured velocity fields from a laboratory experiment shown in Fig. 5 a fourth order Runge–Kutta method was used to numerically integrate the position of the particles forward in time. From the initial ordered cross, the particles are rapidly seen to disperse until they fill the full area of the tank

and $f = 0.0157$ rad/s (thus $\mathcal{F} = 0.26$). The difference with the mixing experiments shown in Fig. 3 is the forcing frequency, which is two times smaller in the present experiment with reactive tracers resulting in a larger eddy size. An example of an experiment with a simple acid–base chemical reaction is shown in Fig. 9, where dilute solutions of hydrochloric acid and sodium hydroxide are added to our oscillating flow experiment. The neutralisation in aqueous solution of a strong acid by a strong base, that is the reaction $H^+ + OH^- \rightarrow H_2O$, proceeds very rapidly when well mixed, with a half-life of 10^{-6} s for the typical concentrations we use. In our quasi-2D turbulent flow, the rate-limiting step for this reaction is in bringing the filaments containing the acid and base into contact. The experimental images of passive tracers in Figs. 3 and 6 show that distinct filamental structures remain for several minutes before the tracers have been strained to small enough scales that diffusion smears out the filaments; hence the mixing time scale is several minutes. In Fig. 9 bromothymol blue and methyl red indicators have been added to the aqueous solution, so that a blue-green

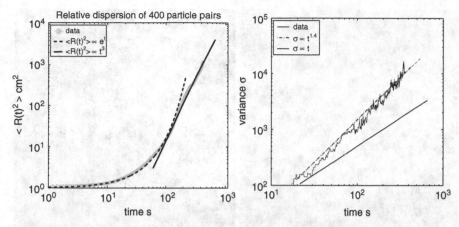

Fig. 8. Using the synthetic particle trajectories of Fig. 7 the relative dispersion (a) and the total variance (b) can be calculated. Graph (a) reveals that at early times the particles' relative separation grows exponentially and then later follows Richardson's scaling for relative dispersion of $\langle R^2 \rangle \propto t^{3.0 \pm 0.1}$. The growth with time of the particles' variance $\sigma^2 \propto \langle (\Delta r)^2 \rangle$ is $\sigma \propto t^{1.4 \pm 0.1}$ indicating anomalous diffusion, see graph (b)

colour represents pH > 7, yellow represents a pH between 6 and 7 and red represents a pH < 5. In the first image a 10 ml acidic solution of pH=5 is added to one side of the tank, and 10 ml of a basic solution of pH=8 is added to the other side. Due to the stirring and straining these initial blobs are rapidly stretched out in the tank. The reaction between the acidic and basic solution proceeds very rapidly once fluid parcels containing acid and base, respectively, come into contact, and on a time scale of several minutes (note that in this experiment $T = 2\pi/f = 400$ s) the initial solutions have been mixed throughout the tank and have reacted completely.

The role of lateral stirring on setting the filamentation and patchiness has been shown to lead to enhancement of phytoplankton growth and increased carbon dioxide uptake in numerical models of ocean circulation by Martin et al. [4]. This enhancement is expected when a chemical reaction with non-linear kinetics takes place in a chaotically mixed system. The reaction will proceed more rapidly in high-concentration regions and more slowly in low-concentration regions, so that the non-linear kinetics give a global enhancement of the reaction rate, in comparison with pure homogeneous mixing. This enhancement of the bulk chemical reactivity rates was demonstrated in quasi-2D electromagnetically forced flows in the experiments performed by Paireau and Tabeling [5]. They noted that the chaotic mixing had a much stronger effect upon the bulk reaction rates of a second-order reaction compared to a first-order chemical reaction.

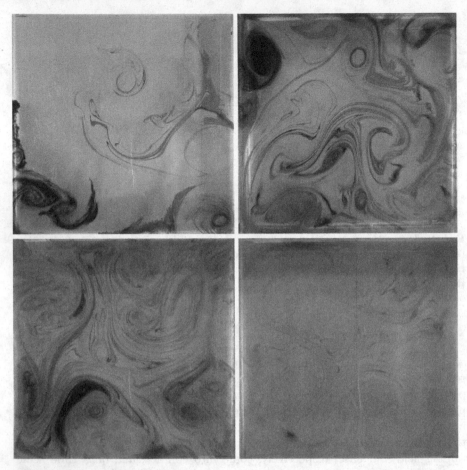

Fig. 9. An experimental visualisation of a chemical reaction in forced 2D turbulence ($Re = 15,000$ and $\mathcal{F} = 0.26$). The snapshots are taken at (**a**) $t/T = 0.075$, (**b**) $t/T = 0.225$, (**c**) $t/T = 0.375$ and (**d**) $t/T = 0.675$, respectively, after the introduction of 10 ml of acid and base to the tank. In these images black blobs and filaments correspond to acidic solution, grey to basic solution and light-grey to the neutral background

5 Conclusion

We have conducted experiments of a quasi-2D forced flow with the boundary layers near the lateral no-slip walls acting as the sole source of vorticity. The average vortex size in the present experiments is controlled by the boundary-layer thickness and the "roll-up" process of these layers. In laboratory experiments we found that the degree of turbulence within the tank is determined by the frequency and amplitude of the forcing. When $A\Omega_0 > f$ (or $\mathcal{F} < 1$) the interior becomes well mixed by a field of dipoles resulting from flow separation in the corners. When $A\Omega_0 < f$ (or $\mathcal{F} > 1$) the flow is not turbulent because the tank oscillates too fast to allow the formation of dipolar structures to occur, so that any vortices formed remain near the walls. For these weakly turbulent cases there is a strong difference in mixing rates between the centre of the tank (where dispersion is very slow) and near the walls, where dispersion occurs rapidly along the periphery of the tank.

The intensity spectrum of the dye, after it had been mixed by the turbulent field, reveals a k^{-1} power-law behaviour at high wave numbers, consistent with the prediction of Batchelor [8]. The laboratory observations of a k^{-1} scalar spectrum are consistent with the presence of the k^{-3} enstrophy cascade.

The relative dispersion shows exponential growth at early times, as predicted by Lin [12], and follows the classical Richardson scaling for larger times. Additionally, anomalous diffusion is conjectured from data of the particle displacement variance with a similar exponent as found by previous studies in different experimental set-ups. An experimental study has been started on the mixing of reactive tracers in rotating quasi-2D turbulent flows. The role of stretching and folding on the chemical reaction has been illustrated, clearly showing the importance of the formation of filamentary structures after which diffusion can facilitate the chemical reaction. This has direct relevance to geophysical applications in the atmosphere (mixing of chemical constituents) and the oceans where similar phenomena are important for dispersion and growth of phytoplankton patches. These mixing phenomena will be investigated in more detail in future laboratory experiments.

Acknowledgments

We would like to thank David Molenaar and Ruben Trieling for helpful discussions and Ad Holten and Gert van der Plas for technical assistance during the laboratory experiments.

References

1. J. Sommeria and R. Moreau: J. Fluid Mech. **118**, 507 (1982)
2. R. B. Pierce and T. D. A. Fairliem: J. Geophys. Res. **98**, 18589 (1993)
3. A. P. Martin: Prog. Oceanogr. **57**, 125 (2003)
4. A. P. Martin, K. J. Richards, A. Bracco and A. Provenzale: Global Biogeochem. Cycles **16**, Art. ID 1025 (2002)
5. O. Paireau and P. Tabeling: Phys. Rev. E **56**, 2287 (1997)
6. A. Bracco, J. von Hardenberg, A. Provenzale, J. B. Weiss and J. C. McWilliams: Phys. Rev. Lett. **92**, 084501 (2004)
7. J. H. LaCasce and C. Ohlmann: J. Mar. Res. **61**, 285 (2003)
8. G. K. Batchelor: J. Fluid Mech. **5**, 113 (1959)
9. L. F. Richardson: Proc. R. Soc. Lond. A **110**, 709 (1926)
10. L. F. Richardson and H. Stommel: J. Atmos. Sci. **5**, 238 (1948)

11. G. K. Batchelor: Proc. Camb. Phil. Soc. **48**, 345 (1952)
12. J. T. Lin: J. Atmos. Sci. **29**, 394 (1972)
13. A. Provenzale: Annu. Rev. Fluid Mech. **31**, 55 (1999)
14. A. Okubo: Deep Sea Res. **17**, 445 (1970)
15. J. B. Weiss: Physica D **48**, 273 (1991)
16. R. H. Kraichnan: Phys. Fluids **10**, 1417 (1967)
17. G. K. Batchelor: Phys. Fluids Suppl. II **12**, 233 (1969)
18. S. D. Danilov and D. Gurarie: Phys. Uspekhi **43**, 863 (2000)
19. H. Kellay and W. I. Goldburg: Rep. Prog. Phys. **65**, 845 (2002)
20. J. Sommeria: J. Fluid Mech. **170**, 139 (1986)
21. P. Tabeling, S. Burkhart, O. Cardoso and H. Willaime: Phys. Rev. Lett. **67**, 3772 (1991)
22. Y. Couder: J. Phys. Lett. **45**, 353 (1984)
23. Y. Amarouchene and H. Kellay: Phys. Rev. Lett. **93**, 214504 (2004)
24. M. A. Rutgers: Phys. Rev. Lett. **81**, 2244 (1998)
25. G. I. Taylor: Proc. R. Soc. Lond. A **100**, 114 (1921)
26. P. Welander: Tellus **7**, 141 (1955)
27. S. R. Maassen, H. J. H. Clercx and G. J. F. van Heijst: Europhys. Lett. **46**, 339 (1999)
28. S. R. Maassen, H. J. H. Clercx and G. J. F. van Heijst: Phys. Fluids **14**, 2150 (2002)
29. S. R. Maassen, H. J. H. Clercx and G. J. F. van Heijst: J. Fluid Mech. **495**, 19 (2003)
30. H. J. H. Clercx and G. J. F. van Heijst: Phys. Rev. Lett. **85**, 306 (2000)
31. H. J. H. Clercx, S. R. Maassen and G. J. F. van Heijst: Phys. Fluids. **11**, 611 (1999)
32. G. J. F. van Heijst: J. Fluid Mech. **206**, 171 (1989)
33. J. A. van de Konijnenberg, H. I. Andersson, J. T. Billdal and G. J. F. van Heijst: Phys. Fluids **6**, 1168 (1994)
34. G. J. F. van Heijst, P. A. Davies and R. G. Davis: Phys. Fluids A **2**, 150 (1990)
35. R. J. M. Bastiaans, G. A. J. van der Plas and R. N. Kieft: Exp. Fluids **32**, 346 (2002)
36. T. H. Solomon, E. R. Weeks and H. L. Swinney: Phys. Rev. Lett. **71**, 3975 (1993)
37. A. E. Hansen, D. Marteau and P. Tabeling: Phys. Rev. E **58**, 7261 (1998)

Quantifying Inhomogeneous, Instantaneous, Irreversible Transport Using Passive Tracer Field as a Coordinate

N. Nakamura

Department of Geophysical Sciences, University of Chicago
nnn@bethel.uchicago.edu

Abstract. Long-lived chemicals in the atmosphere and ocean often reveal surprisingly inhomogeneous distributions, contrary to the intuition that mixing homogenizes them. This is because mixing itself is inhomogeneous: coherent structures in the flow separate regions of fast mixing with semipermeable barriers to transport, and concentrated gradients tend to be found in the latter. The fluxes of materials across the barrier region are of particular interest, since that region is an interface between two fluid masses with distinct chemical characteristics. Yet these fluxes are difficult to measure because the barrier shape is often unsteady and irregular. With spatially fixed (Eulerian) coordinates, it is easy to confuse reversible undulation of the barrier with true, irreversible exchange of matter across it. Particle tracking methods, on the other hand, do not give useful flux–gradient relationships. Given these and other problems with the traditional formalisms, we will advocate a different diagnostic approach to mixing, utilizing a passive tracer field under advection and diffusion. We demarcate a mass of fluid by the isosurfaces of the tracer and ask how much mass is being exchanged through these surfaces. This essentially reduces the problem to 1D mass transport in the tracer coordinate. The purpose of this chapter is to demonstrate how this formalism helps understand the synergy between deformation and diffusion that enhances mixing, quantify inhomogeneous structure of mixing (such as barriers), and extract the instantaneous, irreversible part of transport. We will also touch on the techniques for partitioning fluxes into two opposing directions (and hence quantifying asymmetry in transport) and discuss the present diagnostic in relation to the Eulerian eddy diffusivity and to the probability density function (PDF).

1 Introduction

One of the recurring themes at Aosta School 2004 was the advection–diffusion problem of a passive tracer. For an incompressible fluid, this is governed by

$$\frac{\partial}{\partial t}q^* + \nabla \cdot (\mathbf{v}\, q^*) = D\nabla^2 q^*, \tag{1}$$

N. Nakamura: *Quantifying Inhomogeneous, Instantaneous, Irreversible Transport Using Passive Tracer Field as a Coordinate*, Lect. Notes Phys. **744**, 137–164 (2008)
DOI 10.1007/978-3-540-75215-8_7 © Springer-Verlag Berlin Heidelberg 2008

where q^* is the tracer under advection–diffusion, $\mathbf{v}$ is the advecting flow velocity, and D is the molecular diffusion coefficient. Advection and diffusion lie at the heart of mixing in the atmosphere and ocean.

1.1 Mixing Homogenizes

As pointed out elsewhere in this volume, the essence of mixing may be summarized by the mean tracer variance equation. Multiplying (1) by q^* and taking the domain average, denoted by the angular bracket, and assuming that the boundary fluxes vanish, one obtains

$$\frac{\mathrm{d}}{\mathrm{d}t} \frac{\langle q^{*2} \rangle}{2} = -\langle D |\nabla q^*|^2 \rangle. \tag{2}$$

For a nonzero D the rhs of (2) is negative, so the domain-average tracer variance decays with time; in other words, the process homogenizes the tracer irreversibly (in the thermodynamic sense). Although the flow velocity does not enter (2) explicitly, it acts as a stirring agent by enhancing the mean tracer gradients on the rhs, thereby accelerating homogenization. This is achieved through *deformation*, namely stretching and shearing of tracer surfaces. Deformation is most efficient when the flow is unsteady, as it continuously reorients principal axes of stain and keeps the tracer geometry from equilibrating. Ultimately, diffusion is what destroys the tracer variance, but its efficiency hinges on how fast stirring produces small scales in the tracer.

1.2 Mixing Is Inhomogeneous

Although (2) encapsulates the role of deformation and diffusion in homogenizing a tracer globally, the rate of mixing is not uniform in the atmosphere and ocean. This is evident in the familiar experience that turbulence is encountered only occasionally during an otherwise smooth air ride.

In large-scale flows, coherent structures separate regions of fast mixing with semipermeable barriers in between, and the tracer field tends to develop concentrated gradients in the barrier regions. One such example is found in Fig. 1, which shows the mixing ratio of nitrous oxide (N_2O) on the 830-K isentropic surface ($\sim$30 km above sea level and in the stratosphere) in the Northern Hemisphere. It was observed by the CLAES instrument onboard NASA's *Upper Atmospheric Research Satellite* [1] on 31 December 1992. Nitrous oxide is a chemically inert trace gas in the lower stratosphere, and its mixing ratio is controlled by two processes: a slow, diabatically driven vertical motion, and relatively fast, quasi-horizontal mixing on the isentropic surface [2, 3, 4]. The decrease in N_2O mixing ratio with increasing latitude is caused by the former: an upwelling in the tropics and downwelling in the extratropics convert vertical gradients into meridional gradients. (Since the source of N_2O is at the ground, its mixing ratio decays with altitude in the stratosphere. See also Fig. 4 in Sect. 2.)

UARS CLAES N₂O 830 K 31 December 1992

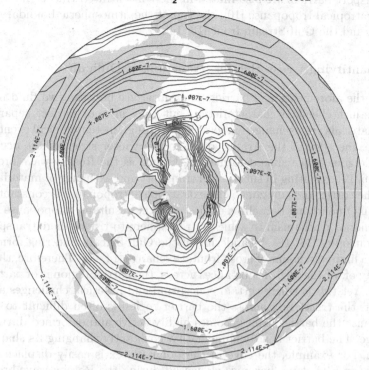

Fig. 1. Azimuthal equidistance plot of N_2O mixing ratio on the 830-K isentropic surface, observed by the CLAES instrument onboard NASA's Upper Atmosphere Research Satellite on 31 December 1992. The center of the plot is the North Pole and the edge is the Equator. Notice there is a small region that lacks data near the pole. The contour interval is 10 ppbv. See text for details

While the diabatic circulation maintains the meridional gradients of N_2O against horizontal mixing, the gradients are not uniform at all. Three regions of weak gradients are separated by bands of concentrated gradients. These regions are (a) the high-latitude air with very low values of N_2O mixing ratio, (b) a pool of midlatitude air with intermediate values, and (3) the tropics with high values of N_2O. These regions are often referred to as the"polar vortex," "surf zone," and "tropical reservoir," respectively [5, 6]. It turns out that latitudinal variation in the diabatic circulation alone cannot explain this kind of inhomogeneity. Rather, the highly nonuniform N_2O gradients stem from inhomogeneity in isentropic mixing: dynamical barriers created by the circulation of the winter stratosphere hamper mixing between these three regions, resulting in the concentrated N_2O gradients there [7, 8, 9]. Although the advection–diffusion equation (1) describes the isentropic mixing of N_2O reasonably well, the rich structure in mixing is all but lost in the globally averaged variance equation (2).

Transport barriers are also hinted in the concentrated tracer gradients at the extratropical tropopause [10], the top of the atmospheric boundary layer [11, 12], and the Gulf Stream front [13, 14].

1.3 Quantifying Cross-Barrier Transport is Difficult

Each of the aforementioned barriers is an interface between two fluid masses with distinct chemical characteristics, but they are not perfect separators. Some materials are exchanged across them through meteorological events. For example, the edge of the Arctic vortex in the winter stratosphere occasionally undergoes irreversible deformation, known as the Rossby-wave breaking [5, 15, 16, 17], resulting in substantial transport and exchange of materials between the vortex and surf zone. The extratropical tropopause is constantly deformed by underlying baroclinic eddies, and the resultant "tropopause folds" are the major mechanism to achieve stratosphere-to-troposphere transport in the extratropics [18, 19]. Meandering of the Gulf Stream and ring formation achieve a limited amount of cross-frontal exchange [14, 20]. Therefore, the permeability of, or the flux of substances across, the barrier region is of particular interest. Yet this is where certain conceptual and practical challenges arise.

First, the tracer flux across the barrier region is often difficult to evaluate because the barrier itself becomes highly active and deformed during the exchange. The barrier is a dynamic entity, constantly changing its shape and location. For example, the edge of the polar vortex is easily displaced from the pole, and its shape becomes elongated during the Rossby-wave breaking event. Spatially fixed (Eulerian) coordinates cannot trace the deformation of a barrier, and it is easy to confuse "reversible" transport associated with temporary displacement of the barrier with irreversible exchange of mass through it. It is the latter, not the former, that influences chemical characteristics of the fluid masses. To extract irreversible transport in the Eulerian coordinates, it is necessary to remove reversible transport by time averaging, but doing so obscures the definition of the barrier itself because its location is unsteady.

Second and perhaps more fundamental, the *maximal* tracer gradients observed in the barrier region appear to contradict the very notion of a barrier: a transport barrier implies a minimum in the flux, but if the flux is diffusive in nature, shouldn't it be *maximal* where the gradients are maximal?

Extant diagnostic methods have various shortcomings in addressing the above and other difficulties. The popular eddy diffusivity parameterization often performs poorly when the required scale separation is violated [21]: the width of the barrier region (where the tracer gradients are most concentrated) is comparable to, or even smaller than, the size of the driving eddy. This means that advective transport through the barrier region cannot be prescribed by local statistics of random eddy motion, clearly at odds with the eddy diffusivity idea [22, 23]. Another way to put it is that eddy diffusivity works only in nearly homogeneous environments. It is no surprise if it fails to identify a localized transport barrier.

The modern dynamical systems approach [24] focuses on the rate of divergence of neighboring particle trajectories as a mixing diagnostic. The method has been used widely, partly because the underlying theory is well developed for at least 2D, time-periodic problems (see [25] and references therein). Even for aperiodic problems, asymptotic diagnostics such as the Lyapunov exponents have been redefined for a finite-time interval and used, with some success, to identify regions of transport barriers [26, 27, 28]. Still, the trajectory-based approach requires, by definition, an analysis over a period of time and this can cause problems. One must make judicious choices of initial conditions and length of integration, which depend on the nature of the problem. When the flow is highly variable in time, it is possible that trajectories quickly move out of the barrier region and start to diverge, thereby failing to identify the barrier [29]. Even when the Lyapunov exponents robustly identify a barrier, their relationship to mass flux remains unclear. Given these, a new diagnostic that quantifies instantaneous and irreversible mass fluxes across the barrier region is desired.

1.4 Goals

This chapter advocates an alternative diagnostic formalism that uses a field of passive tracer as a coordinate of transport. A similar approach had been practiced among the atmospheric scientists working on stratospheric dynamics [30, 31], but it was developed only recently into a more viable mixing diagnostic [32, 33, 34, 35, 36, 37, 38, 39, 40].

In this approach, we demarcate a mass of fluid by the isosurfaces of a tracer field that obeys the advection–diffusion equation and ask how much mass is being exchanged through these surfaces. This essentially reduces the problem to one of 1D mass transport in the tracer coordinate, an intermediate model between (1) and (2). Using tracer as a coordinate has at least two advantages: (i) the geometrical complexity and the location of the barrier are absorbed in the coordinate so they do not hinder analysis, and (ii) instantaneous, irreversible transport can be extracted. The tracer isosurface is a material surface in the absence of diffusion, so the mass transport in the tracer coordinate arises solely from diffusion (therefore it is irreversible by definition). Yet its magnitude depends on the geometry of tracer and hence affected by the flow: where stirring is strong, the tracer geometry becomes complex and mass exchange is enhanced.

We will outline the formalism and demonstrate the utility of the diagnostic in the next section. We will also make connections with the *probability density function* (PDF) of tracer concentration that appears commonly in the statistical mechanics literature [8, 41, 42]. The key concept of effective diffusivity is introduced, which will then be used to quantify inhomogeneity in mixing. Section 3 deals with a strategy for partitioning mass fluxes into two opposing directions. This helps quantify asymmetry in wave breaking. In Sect. 4, we will revisit the Eulerian eddy diffusivity and point out its relationship to the current diagnostic.

2 Effective Diffusivity

2.1 Area-Mean Formalism

We start our analysis by associating the elements of fluid with values of tracer q^* governed by (1). For the sake of simplicity, we shall follow Nakamura [32] and consider advection–diffusion in 2D. We ask how much area (mass) the fluid occupies in regions where $q^* \leq q$ at a given time t, where q is a chosen value of q^*. (The following argument would be analogous for $q^* \geq q$. For 3D problems, the area should be replaced by volume; [39].) This area is

$$A(q,t) \equiv \int_{q^* \leq q} \mathrm{d}S. \tag{3}$$

Defined this way, A is a monotonically increasing function of q, from $A = 0$ for $q^* = q_{min}$ to $A = A_0$ for $q^* = q_{max}$, where q_{min} and q_{max} are the minimum and maximum values of q^*, respectively, and A_0 is the area of the entire domain (Fig. 2). The area A varies as a function of q and t as [32]

$$\frac{\partial}{\partial t} A(q,t) = -\frac{\partial}{\partial q} \int_{q^* \leq q} D\nabla^2 q^* \mathrm{d}S, \tag{4}$$

where the partial derivatives are defined with respect to the (q,t) space (see Appendix 1 for the derivation of (4)). The rhs is the mass flux converging into the region where $q^* \leq q$. This mass flux arises solely from diffusion because advection by nondivergent flow conserves area. In other words, in the absence of diffusion ($D = 0$), the tracer isosurface becomes a material surface. Notice that the location and geometry of the contours of q^* are arbitrary in (3) and (4). The contours can be broken, in which case contributions from all islands are summed up.

The rhs of (4) can be further transformed, using the divergence theorem and the identity

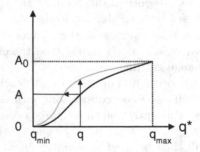

Fig. 2. Schematic of area–tracer relationship. In this example, the value of tracer contour increases with the enclosed area. Notice that an increase in the area for a fixed tracer value leads to a decrease in the tracer value for a fixed area. See text for details

$$\oint_{q^*=q} \frac{(\cdot)}{|\nabla q^*|} dl = \frac{\partial}{\partial q} \int_{q^* \le q} (\cdot) dS, \tag{5}$$

into

$$\frac{\partial}{\partial t} A(q,t) = -\frac{\partial}{\partial q} \left(\langle D|\nabla q^*|^2 \rangle_q \frac{\partial A}{\partial q} \right), \tag{6}$$

where the subscripted angular bracket denotes the average of a field variable on the tracer contour $q^* = q$, defined as

$$\langle \cdot \rangle_q \equiv \frac{\partial}{\partial A} \int_{q^* \le q} (\cdot) dS$$

$$\approx \left(\int_{q^* \le q+\delta q} (\cdot) dS - \int_{q^* \le q} (\cdot) dS \right) \Big/ (A(q+\delta q, t) - A(q,t)). \tag{7}$$

The last expression is a finite-difference approximation and useful for numerically evaluating this average. Since, the average is taken over the area between adjacent contours, instead of using line integral along the contour, it can be evaluated readily by box-counting methods.

2.2 Relationship to Tracer PDF

At this point, it is useful to introduce the *probability density function* (PDF) of tracer concentration. The quantity

$$p(q,t) \equiv \frac{1}{A_0} \frac{\partial A}{\partial q} \tag{8}$$

may be thought of as the PDF of q^*: it represents a fractional area between neighboring contours of q^*. If the domain were sampled randomly, probability of sampling a value between q and $q + \delta q$ would be $p \, \delta q$. Note that the probability enters solely through sampling; the flow does not have to be ergodic. Note also that $A(q,t)/A_0$ is the *cumulative density function* (CDF) of q^*, and that integration of (8) from $q_{\min}$ to $q_{\max}$ gives unity. By taking the derivative of (6) with respect to q, one can derive an equation for p:

$$\frac{\partial}{\partial t} p(q,t) = -\frac{\partial^2}{\partial q^2} \left(\langle D|\nabla q^*|^2 \rangle_q p \right). \tag{9}$$

Equation (9) is closely related to the PDF expressions reported in the statistical mechanics literature. For example, both Sinai–Yakhot [41] and Ching–Kraichnan [42] report a form of stationary PDF of the rms-normalized concentration. Their results can be derived from (6) and (9) (see Appendix 2).

Similar to the rhs of (2), $\langle D|\nabla q^*|^2 \rangle_q$ is the dissipation rate of tracer variance, but it is evaluated for each tracer value. For this reason, it is called *conditional dissipation rate* (CDR). If CDR is uniform, (9) is a diffusion equation with a negative diffusion coefficient. Thus, an initially gaussian p would quickly become a delta-function, corresponding to the homogenization of q^* toward its mean value. However, our interest lies in the cases in which CDR is *not* uniform.

2.3 Equivalent Length and Effective Diffusivity

For geophysical flows, it is more convenient to work with the inverted A–q relationship $q(A, t)$ than $A(q, t)$: the range of q diminishes as mixing proceeds, whereas the range of A is fixed and so it is better suited as a geophysical coordinate. This inversion is possible because the A–q relationship is one-to-one at a given time. The equation for q is derived using (6)

$$\frac{\partial}{\partial t} q(A, t) = -\frac{\partial q}{\partial A} \frac{\partial A}{\partial t} = \frac{\partial}{\partial A} \left(\langle D|\nabla q^*|^2 \rangle_q \frac{\partial A}{\partial q} \right). \tag{10}$$

The minus sign in the second expression is because an increase in A at a fixed q causes a decrease in q at a fixed A when A is an increasing function of q (Fig. 2). Equation (10) is further rewritten as [32]

$$\frac{\partial}{\partial t} q(A, t) = \frac{\partial}{\partial A} \left(DL_e^2 \frac{\partial q}{\partial A} \right), \tag{11}$$

$$L_e^2(A, t) \equiv \langle |\nabla q^*|^2 \rangle_q \Big/ (\partial q / \partial A)^2. \tag{12}$$

Equation (11) is a diffusion equation for q with respect to A, with DL_e^2 being the "diffusion coefficient." The quantity $-DL_e^2 \partial q / \partial A$ is the diffusive flux of q^* across its contour that encloses a fixed area A. Since the tracer contour is used as a coordinate, advective transport does not show up in (11) at all: it is absorbed in the motion of coordinate. Inhomogeneity in mixing is then understood in terms of inhomogeneity in the diffusion coefficient. Furthermore, the diffusion coefficient has a direct bearing on the geometry of the tracer: the variable L_e is closely related to the length of the tracer contour and termed *equivalent length* [32]. It can be shown [36], via the Cauchy inequality, that the geometrical length of the contour, L, is actually the lower bound of equivalent length, L_e:

$$L_e^2 = \left(\oint_{q^*=q} \frac{dl}{|\nabla q^*|} \right) \left(\oint_{q^*=q} |\nabla q^*| \, dl \right) \geq \left(\oint_{q^*=q} dl \right)^2 = L^2, \tag{13}$$

where (3), (5), and (7) were used to transform (12). This means that, where L_e is large, tracer contour is stretched and contorted, and this effectively enhances the cross-contour diffusivity. Where L_e is small, mixing is suppressed, suggesting the presence of a transport barrier. Since L_e is defined as a function of t and A, it is an instantaneous diagnostic of the structure of mixing.

The reason why the diffusive flux is proportional to the *square* of L_e (other than the dimensional consistency) is that the contour length has two effects. First, when a ring of fluid bounded by two adjacent contours is stretched without changing its area, the mean distance between the contours decreases and thus the tracer gradient increases in proportion to the contour length. This enhances diffusive flux *per unit contour length*. Second, the enhanced contour length provides more surface for diffusion to act upon.

The spatial coordinate in (11) is area, but it is more intuitive to work with a coordinate that has the dimension of length. Let

$$dr \equiv dA/L_0, \tag{14}$$

where L_0 (> 0) is the minimum possible value for L_e. The actual value of L_0 depends on the geometry of the domain. For example, for a closed tracer contour on an unbounded plane $L_0 = 2\sqrt{\pi A}$, since L_e is minimized when the contour forms a circle. Then, using r, (11) becomes

$$\frac{\partial}{\partial t}q(r,t) = \frac{\partial}{\partial r}\left(K_e\frac{\partial q}{\partial r}\right), \tag{15}$$

where

$$K_e(r,t) \equiv D\left(L_e(r,t)/L_0\right)^2 \tag{16}$$

is *effective diffusivity* [36, 37]. Effective diffusivity represents molecular diffusion magnified by the stretching of L_e relative to its minimum value. It reduces to molecular diffusivity when the tracer contour is unstretched. In this sense, the factor $(L_e(r,t)/L_0)^2$ is the *local Nusselt number*. Note that the diffusive flux $-K_e\partial q/\partial r$ is governed by both the tracer gradient and effective diffusivity: even when the tracer gradient is large, the flux can be minimal if effective diffusivity is locally very small. This solves the apparent paradox raised in the previous section: the tracer gradients can indeed be maximal at the barrier.

To compute effective diffusivity from the q^* field, first establish the A–q relationship by counting the grid boxes over the domain, using equally spaced bins between q_{min} and q_{max}. Then compute $|\nabla q^*|^2$ at every grid and evaluate (12) in the (q,t) space with the finite difference formula

$$L_e^2(q,t) \approx \left(\int_{q^*\leq q+\delta q}|\nabla q^*|^2\,dS - \int_{q^*\leq q}|\nabla q^*|^2\,dS\right)$$
$$\times (A(q+\delta q,t) - A(q,t))/(\delta q)^2. \tag{17}$$

Finally, map L_e^2 from (q,t) to (r,t) space.

Figure 3 shows potential temperature and associated effective diffusivity during a numerically simulated life cycle of Kelvin–Helmholtz instability. In this simulation, a small-amplitude, unstable normal-mode perturbation is allowed to grow on a stratified shear flow. Domain is periodic in x and bounded by rigid surfaces in z. Potential temperature is governed by a numerically

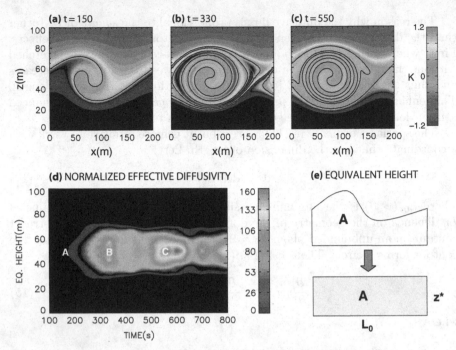

Fig. 3. (a)–(c) Life cycle of simulated 2D Kelvin–Helmholtz instability. Potential temperature (in K) in x–z plane is shown for three different stages. Potential temperature increases with height in general, but its contours roll up as the cat's eye grows. (d) Normalized effective diffusivity during the KH life cycle as a function of time and equivalent height. Letters correspond to the highlighted contours in the respective plates (a)–(c). See text for details. (e) Equivalent height (z^*) is the area underneath a potential temperature contour (A) divided by the channel length ($L_0 = 200$ m). For the details of simulation, see [32]

discretized version of (1) in the (x,z)-plane. As time goes on, a "cat's eye" forms in the center of the domain and potential temperature contours turn over (Figs. 3a–c). The corresponding effective diffusivity normalized by D ($\equiv L_e^2/L_0^2$) is plotted in Fig. 3d as a function of time and equivalent height. Here equivalent height is defined by (14), with A being the area between the chosen potential temperature contour and the lower boundary, and L_0 being the channel length in x (Fig. 3e). It is a measure of height because potential temperature generally increases with height.

Effective diffusivity is enhanced significantly inside the cat's eye but virtually unchanged outside. The letters in Fig. 3d correspond to the respective black contours in Figs. 3a–c. After saturated at $t \approx 330$ s, effective diffusivity inside the cat's eye fluctuates as contours of potential temperature are being dissipated and reoriented. Notice that the maximum effective diffusivity, more than two orders of magnitude greater than D, appears in the cat's eye at $t \approx 550$ s, after most potential temperature gradients are gone.

Figure 4 shows the vertical structures of N_2O mixing ratio (contours) and isentropic effective diffusivities (shades, in terms of $\ln(L_e^2/L_0^2)$) observed by the Upper Atmosphere Research Satellite as a function of equivalent latitude and potential temperature [34]. Equivalent latitude is the limiting latitude of a polar cap that encloses the same area A as enclosed by the contour of N_2O mixing ratio [31, 35]. It bears the sense of latitude because N_2O mixing ratio generally decreases with latitude (Fig. 1). The left frame shows the time average over the 7-day period ending 20 December 1992, for the Northern Hemisphere. It reveals minima in effective diffusivity at the edge of the stratospheric polar vortex and at the subtropical end of the "surf zone," where the horizontal gradients of N_2O mixing ratio are concentrated, consistent with Fig. 1. These barriers are robust in the middle to lower stratosphere, where they separate the more stirred polar vortex and surf zone. The right frame shows a late-winter cross-section for the Southern Hemisphere. The southern vortex is less perturbed and is characterized by a broader region of small

UARS CLAES N$_2$O AND EFFECTIVE DIFFUSIVITY

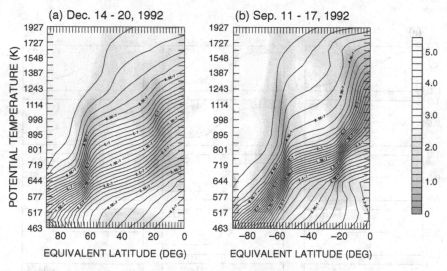

Fig. 4. (a)–(b) Equivalent latitude versus potential temperature plot of N_2O mixing ratio (solid contours, every 10 ppbv) and $\ln(L_e^2/L_0^2)$ (shades). To create these plots, analysis was done for each isentropic surface and then repeated for multiple levels. The vertical axis approximately spans the entire stratosphere. Dark shade indicates small effective diffusivity and therefore barriers to horizontal (isentropic) mixing. (a) 7-day average for 14–10 December 1992 for the Northern Hemisphere. (b) 7-day average for 11–17 September 1992 for the Southern Hemisphere. In both frames, the polar vortex is on the *left*. Notice the enhanced mixing in the surf zone and the polar vortex. Data at lowest levels may be contaminated by the Pinatubo aerosols that had been loaded in the previous year. Adapted from [34]

effective diffusivities. This insulates the vortex air chemically like a containment vessel, which sets the stage for ozone hole to develop in spring with little dilution. The barriers in the Southern Hemisphere reach higher altitudes, presumably due to a vertically more coherent westerly jet. Similar results have been obtained with offline advection-diffusion calculations driven by meteorological winds [36, 37, 38].

Tracer "edges," or concentrated gradients, can be created by a spatially varying effective diffusivity. Figure 5 shows three numerical solutions to (15) driven by prescribed K_e, starting from identical initial conditions. In all cases, K_e grows exponentially with time. The exponential growth is meant to mimic rapid stretching of tracer contours by fluid stirring. In Fig. 5a, the growth rate is maximal at the center of the domain, whereas in Fig. 5b, it is minimal at the center of the domain. In both cases, the spatial variation of growth rate is broad: it is a gaussian function of the normalized area A/A_0, with a characteristic width of 0.33. The locally maximal effective diffusivity creates two separating edges (Fig. 5a), whereas the locally minimal effective diffusivity creates an edge converging at the center (Fig. 5b). The former is analogous to the KH cat's eye (Fig. 3), whereas the latter is similar to the edge of the

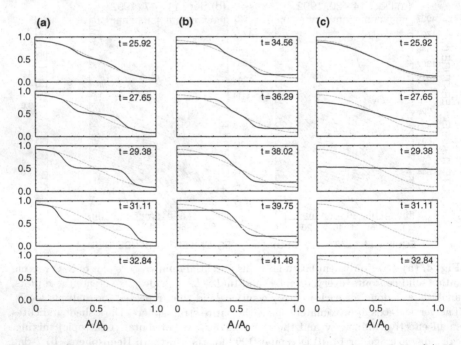

Fig. 5. Solutions for initial value problems using (15) with a prescribed K_e. (a) K_e grows exponentially and has a maximum near the center. (b) K_e grows exponentially and has a minimum near the center. (c) K_e grows exponentially and uniformly everywhere. The gray curves indicate the initial condition. Adapted from [32]

polar vortex (Fig. 4). Note that in both cases, the width of the edges is quite narrow, despite the broad spatial variation in K_e. In Fig. 5c, the growth rate (and K_e) is spatially uniform. No edge formation is visible in this case.

These are just a few examples of the utility of effective diffusivity. Unlike the Eulerian eddy diffusivity, it extracts true diffusive transport and is allowed to vary fully in space and time. Further relationships between effective diffusivity and the traditional Eulerian eddy diffusivity will be discussed in Sect. 4. Unlike the particle advection methods, effective diffusivity preserves a flux–gradient relationship. One need not solve (in principle) an initial-value problem to evaluate effective diffusivity. It depends solely on the instantaneous geometry of the tracer, which reflects the history of stirring in the recent past. Practical issues concerning the dependence of effective diffusivity on D will be touched upon in the concluding section.

3 Direction of Transport

Although effective diffusivity successfully quantifies spatio-temporal structure of mixing, it does not distinguish the direction of transport. Materials often cross the tracer contour in one preferred direction due to asymmetry in advection. For example, a patch of fluid demarcated by a tracer contour may be "pinched off" in the outward direction (Fig. 6a). In this case, as the filament of tracer is diffused, fluid inside the filament is exported to the outside. An opposite transport ensues when the pinching off occurs in the inward direction (Fig. 6b). More generally, transport proceeds in both directions simultaneously, though one is likely more pronounced than the other. However, the foregoing diagnostic measures only the *net* transport, unable to quantify the opposing fluxes separately.

The direction of transport is important because of the way it influences the embedded chemistry. Suppose a patch of air, chemically distinct from the surrounding air, expels its element (Fig. 6a). In this case, the patch loses mass

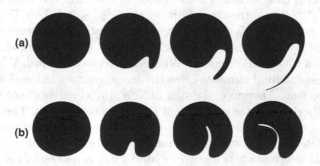

Fig. 6. Schematics of outward breaking (**a**) and inward breaking (**b**). See text for details

but its chemical composition remains intact. On the other hand, if the outside air is mixed into the patch, the patch's chemical composition will be affected (Fig. 6b). One of the reasons why the remnant of Antarctic ozone hole lingers through much of November (and sometimes into December) despite the (by-then) dynamically active vortex is that transport through the vortex boundary is primarily outward: the ozone-depleted air is detrained from the vortex, but the outside air does not get entrained easily [43].

It is relatively easy to detect asymmetry in mass transport. For example, when the transport is predominantly outward, the mass (volume) inside the tracer contour diminishes with time. However, to quantify mass fluxes in each direction separately proves more difficult. Traditionally, particles are advected numerically and the number of crossings is counted at a specified boundary over time [44, 45, 46]. Yet the flux so computed depends sensitively on the time that the particles spend between two crossings. Since many particles stay close to the boundary and spend very short time on either side of the boundary, fluxes calculated using particles with short residence time tend to be large [47]. In fact, it has been shown that the fluxes fail to converge at a vanishing residence time when the particle motion is random [48].

3.1 Partitioning of Fluxes

Here we address the flux Partitioning by extending the formalism of the previous section. Instead of transforming the rhs of (4) using the divergence theorem, we rewrite [40]

$$\frac{\partial}{\partial t} A(q,t) = -\frac{\partial}{\partial q} \int_{q^* \le q} D\nabla^2 q^* \mathrm{d}S = \dot{A}_{\mathrm{in}} - \dot{A}_{\mathrm{out}},$$

$$\dot{A}_{\mathrm{in}} = \frac{\partial}{\partial q} \int_{\substack{q^* \le q \\ \nabla^2 q^* \le 0}} D\left|\nabla^2 q^*\right| \mathrm{d}S, \quad \dot{A}_{\mathrm{out}} = \frac{\partial}{\partial q} \int_{\substack{q^* \le q \\ \nabla^2 q^* > 0}} D\left|\nabla^2 q^*\right| \mathrm{d}S. \quad (18)$$

The rhs of (4) is now partitioned into two contributions: one from the regions in which $\nabla^2 q^* > 0$, and the other from where $\nabla^2 q^* \le 0$. Where $\nabla^2 q^* > 0$, the rhs of (1) is positive. This means that the value of tracer increases following the motion of fluid. In other words, fluid elements cross the tracer contour in the upgradient (outward) direction. Whereas in regions in which $\nabla^2 q^* \le 0$, the reverse is true and fluid elements cross the tracer contour in the downgradient (inward) direction. Therefore, the sign of $\nabla^2 q^*$ (local curvature of tracer) can be used to discriminate the direction of transport (see Fig. 7). The two terms, $\dot{A}_{\mathrm{in}}$ and $\dot{A}_{\mathrm{out}}$, denote the inward and outward fluxes of mass across the tracer contour. Notice that they are both nonnegative. To evaluate this flux partition, no additional information is needed beyond what is necessary for computing effective diffusivity an instantaneous distribution of tracer is all it takes.

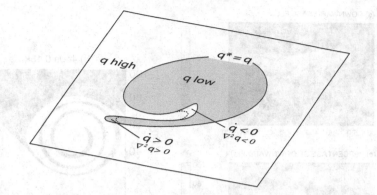

Fig. 7. Partitioning strategy for outward and inward mass fluxes. See text for details. Adapted from [40]

3.2 Examples

Let us apply this flux partition to the potential temperature field associated with the KH instability discussed earlier in Fig. 3. The result is shown in Fig. 8 as a function of time and equivalent height. Figure 8a shows the downward mass flux $\dot{A}_{in}$ (recall A is defined as the area between the lower boundary and the chosen contour of potential temperature). Due to the symmetry of the problem, the upward flux ($\dot{A}_{out}$) is a mirror image of Fig. 8a about the midlevel (not shown). In the early stage of life cycle, the downward mass flux emerges just above the midlevel. This corresponds to the growth of cat's eye through the entrainment of mass from above. After this stage, most of the downward mass flux occurs within the cat's eye. Figure 8b shows the geometry of the potential temperature contour at which the mass flux is maximal (denoted by letter A in Fig. 8a). The contour is characterized by an elongated filament. Since the filament coincides with the exrema in the Laplacian of potential temperature, it is susceptible to diffusion. As the filament is diffused, fluid elements are irreversibly transported down through the contour of potential temperature.

Figure 8c shows the percentage of the downward mass flux in the total mass exchange, $100\dot{A}_{in}/(\dot{A}_{in} + \dot{A}_{out})$. It is clear that the mass flux in the upper (lower) domain remains predominantly downward (upward). The anomalous values outside the cat's eye in the early stage may be due to minor breaking of gravity waves, but the fluxes themselves are so small that they have little dynamical significance. After about $t \approx 380$ s, the opposing fluxes in the cat's eye rapidly become comparable in size.

The same technique can be applied to other flows as well. A numerically synthesized tracer [40] the stratospheric winds quantifies isentropic mass fluxes associated with breaking Rossby waves [40]. By examining the equatorward and poleward fluxes, one can identify "episodes" of wave breaking that favor either equatorward or poleward transport. Figure 9 shows the tracer contours

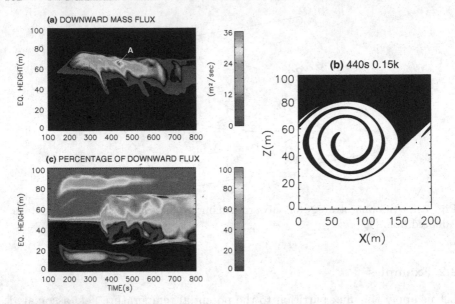

Fig. 8. (a) Downward mass flux through the contours of potential temperature during the life cycle of simulated KH instability (coordinates are the same as Fig. 3d). (b) Geometry of potential temperature contour corresponding to the letter A in (a). The area above the contour is shaded for the purpose of visualization. (c) Percentage of downward mass flux in the total mass exchange in the same coordinates as (a). See text for details. Adapted from [40]

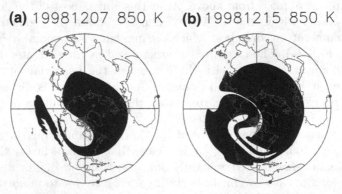

Fig. 9. (a) Geometry of a tracer contour on 850-K surface at which the equatorward mass flux was maximal during a Rossby-wave breaking event (7 December 1998). (b) Geometry of a tracer contour on 850-K surface at which the poleward mass flux was maximal during a poleward Rossby-wave breaking event (15 December 1998). The tracer was numerically synthesized by solving advection–diffusion problem on the isentropic surface using assimilated winds [38]. Contours in these plots were identified by first analyzing the poleward and equatorward mass fluxes and their asymmetry in the time-equivalent latitude space using (18). Adapted from [40]

on the 850-K ($\sim$ 32 km) surface during Rossby-wave breaking events, at which the directional asymmetry of mass fluxes was most pronounced. The geometry of the contours shown in Fig. 9 is very much akin to the schematics in Fig. 6. The detection of transport asymmetry has hitherto been largely subjective, relying on visual inspection of the geometry of material contours like Figs. 6 and 9 [16, 17, 49]. The present formalism provides a more objective measure of transport asymmetry. Furthermore, once a catalog of mass fluxes is computed, it allows one to identify mixing events with particular transport asymmetry far more economically and accurately than the traditional method.

4 Relationship to Eulerian Eddy Diffusivity

In Sect. 1 we touched on the difficulties with representing eddy advective transport using an Eulerian eddy diffusivity. The main concern was that the required scale separation between eddy and the mean field is not warranted in the atmosphere and ocean. As a result, the property of transport cannot be described in terms of random local statistics of eddy. Effective diffusivity introduced in Sect. 2 is free from this difficulty because it excludes advective transport altogether: it is absorbed in the motion of coordinate. The tracer contours are not restricted a priori to particular geographical locations, although they may have preferred locations due to specific geometry of the flow. Only *differential advection* (deformation) affects effective diffusivity through enhancement of equivalent length.

Therefore, Eulerian eddy diffusivity and effective diffusivity are fundamentally different, both qualitatively and quantitatively. However, as we will show below, a part of the traditional Eulerian eddy diffusivity can be attributed to instantaneous, irreversible mixing in a way similar to effective diffusivity. This new Eulerian diagnostic addresses one of the shortcomings of effective diffusivity: it cannot be evaluated for a geographical location fixed on the surface of the Earth because it is assigned to a moving contour of a tracer. We will demonstrate the utility of the new diagnostic by identifying the geographical locations of time-mean transport barriers.

4.1 Eulerian Mean Formalism

Let $\overline{(\,\cdot\,)}$ denote a low-pass filter, namely an Eulerian average ("mean") over a finite time or space (or both), and $(\,\cdot\,)'$ the departure from the mean ("eddy"). The definition of the mean is arbitrary as long as it commutes with both Eulerian time derivative and the gradient operator. By applying the average to (1),

$$\frac{\partial}{\partial t}\bar{q} + \nabla \cdot (\bar{\mathbf{v}}\,\bar{q}) = -\nabla \cdot (\overline{\mathbf{v}'\,q'}) + D\nabla^2\bar{q}. \tag{19}$$

Notice that the asterisks have been dropped. The eddy flux on the rhs is commonly cast into a flux–gradient relationship

$$\overline{\mathbf{v}' q'} = -\mathbf{K}\nabla\bar{q}, \tag{20}$$

where $\mathbf{K}$ is a second-order tensor. Upon substitution, one obtains

$$\frac{\partial}{\partial t}\bar{q} + \nabla\cdot(\bar{\mathbf{v}}\,\bar{q}) = \nabla\cdot(\mathbf{K}\nabla\bar{q}) + D\nabla^2\bar{q}. \tag{21}$$

If $\mathbf{K}$ is expressed in terms of the mean quantities, then (21) represents a flux closure. Instead, here we are interested in *diagnosing* $\mathbf{K}$ to quantify transport. Unfortunately, (20) does not determine $\mathbf{K}$ uniquely even if $\overline{\mathbf{v}' q'}$ and $\nabla\bar{q}$ are known: the number of elements of $\mathbf{K}$ (unknowns) is greater than the number of equations. However, there is a unique $\mathbf{K}$ that satisfies (20) and $\mathbf{K} = \mathbf{K}_S + \mathbf{K}_A$ such that [50] (see also [51, 52])

$$\mathbf{K}_S = K\mathbf{I}, \qquad K = -\frac{\overline{\mathbf{v}' q'}\cdot\nabla\bar{q}}{|\nabla\bar{q}|^2}, \tag{22}$$

$$\mathbf{K}_A = \begin{pmatrix} 0 & s_3 & -s_2 \\ -s_3 & 0 & s_1 \\ s_2 & -s_1 & 0 \end{pmatrix}, \qquad \begin{pmatrix} s_1 \\ s_2 \\ s_3 \end{pmatrix} = -\frac{\overline{\mathbf{v}' q'}\times\nabla\bar{q}}{|\nabla\bar{q}|^2} = \mathbf{s}, \tag{23}$$

where $\mathbf{I}$ is the identity matrix. ($\mathbf{K}_S$ and $\mathbf{K}_A$ are the symmetric and antisymmetric parts of $\mathbf{K}$. They are well conditioned everywhere except $\nabla\bar{q} = 0$.) The above corresponds to partitioning $\overline{\mathbf{v}' q'}$ into components normal and parallel to $\nabla\bar{q}$ locally. In other words, we sort the eddy transport according to the local orientation of the isosurfaces of $\bar{q}$. The significance of this partition is that $\mathbf{K}_S$ is isotropic, rendering the flux–gradient coefficient a scalar instead of a tensor. This becomes apparent after substituting (22) and (23) in (21):

$$\frac{\partial}{\partial t}\bar{q} + \nabla\cdot(\bar{\mathbf{v}}_e\,\bar{q}) = -\nabla\cdot((K+D)\nabla\bar{q}), \tag{24}$$

$$\bar{\mathbf{v}}_e = \bar{\mathbf{v}} + \nabla\times\mathbf{s}. \tag{25}$$

Notice that the antisymmetric part of $\mathbf{K}$ generates an additional transport velocity, which is now included in $\bar{\mathbf{v}}_e$ on the rhs of (25). Furthermore, using the eddy variance equation with (22), it is possible to show [50]

$$K + D \equiv K_k + K_m,$$

$$K_k = \left(\frac{\partial\overline{q'^2}}{\partial t} + \nabla\cdot\left(\bar{\mathbf{v}}\,\overline{q'^2} + \overline{\mathbf{v}'q'^2} - D\nabla\overline{q'^2}\right)\right)\Big/2\,|\nabla\bar{q}|^2, \tag{26}$$

$$K_m = DM_E, \qquad M_E = \overline{|\nabla q|^2}\Big/|\nabla\bar{q}|^2 = 1 + \overline{|\nabla q'|^2}\Big/|\nabla\bar{q}|^2. \tag{27}$$

The term K_k is related to the dispersion of tracer surfaces due to eddy advection, and in that sense analogous to the traditional eddy diffusivity [53]. Indeed, if one defines η' as the *normal* displacement of the q-surfaces from its mean position, that is,

$$q' = -\eta' \cdot \nabla \bar{q}; \qquad \eta' \parallel \nabla \bar{q}, \tag{28}$$

then in a small amplitude limit, (26) is approximately

$$K_k \approx \left(\frac{\partial}{\partial t} + \bar{\mathbf{v}} \cdot \nabla \right) \frac{\overline{\eta'^2}}{2}. \tag{29}$$

Thus, if $\overline{\eta'^2}$ increases following the mean velocity, K_k is positive. However, if the tracer surface is undergoing reversible undulation, positive values of K_k will be followed by negative values, so the time average of K_k will vanish. If the overbar already includes time mean and if the eddy statistics is stationary, the time derivative also vanishes. When eddy amplitude is large, the triple correlation term in the numerator of (26), $\nabla \cdot \overline{\mathbf{v}'q'^2}$, becomes important. To the extent that eddy statistics is homogeneous over the length scale of the mean field, this divergence is negligible. This is true when there is a separation of scales between eddy and mean [21]. However, in the presence of coherent structures in the flow, such scale separation easily breaks down. In that case, the divergence term can take either sign, so can K_k. The net advective transport can therefore be up or down the mean gradient. (K_k vanishes if the eddy statistics is stationary *and* homogeneous.)

In contrast, K_m is positive by definition. It arises from molecular diffusion, yet its magnitude is amplified by the factor M_E, which takes the minimum value of 1 when $q = \bar{q}$. In other words, M_E measures the enhancement of irreversible transport (molecular diffusion) by eddy stirring. This factor depends only on the geometry of the tracer. It measures the amount of small-scale details that is lost through averaging. (See Fig. 10 for the illustration of this point.) Therefore it is a measure of tracer "roughness" and hence analogous to the normalized squared equivalent length (Nusselt number) in (16). The difference is that M_E is defined locally in the Eulerian coordinate. We propose to use M_E (and K_m) to measure the structure of local eddy mixing.

4.2 Example

Figure 11 shows M_E calculated from potential vorticity (PV, *left*) and N$_2$O (*right*) simulated by GFDL's SKYHI general circulation model on the 320-K surface for the month of March (see [50] for details). This surface intersects with the tropopause and jet streams in the midlatitudes. (The jet streams correspond to the regions of packed streamfunction in the solid contours.) Although neither PV nor N$_2$O strictly obeys the advection–diffusion equation, both are mainly controlled by large-scale adiabatic flows at this altitude and hence have similar isentropic structures, at least in the extratropics. The two

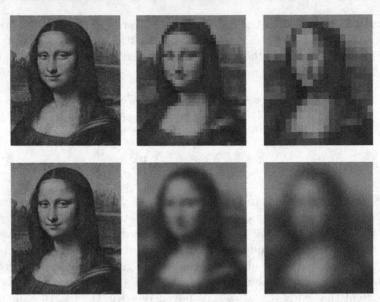

Fig. 10. Illustration of coarse-graining and diffusivity. Both "mosaic" (*top*) and "Gaussian blur" (*bottom*) filters remove the details of Mona Lisa. The loss of information is greatest where the fine-scale features abound (e.g., face) and less severe in the bland area (e.g., background). The "effective diffusivity" of the filters is therefore greater in the former region

rows in Fig. 11 correspond to different lengths of time averaging: 10 and 20 days. As the length of averaging doubles, more transient eddies are accounted for; this increases the overall magnitude of M_E. However, both PV and N_2O reveal very robust M_E minima along the axes of midlatitude jets and a weak barrier in the tropics. This illustrates very clearly that strong jets hinder mixing and act as transport barriers, and that mixing is stronger at the flanks of jets, where wave breaking associated with mobile baroclinic eddies takes place. Notice that the midlatitude barrier in the Northern Hemisphere shows more meandering and zonal localization than its counterpart in the Southern Hemisphere, reflecting the influence of stationary Rossby waves on the structure of the jet stream.

5 Summary and Discussion

We have outlined the theoretical underpinning and application of a diagnostic formalism for quantifying irreversible transport. Central to the idea is to characterize transport and mixing in terms of mass flux through isosurfaces of a passive tracer field, governed by advection and diffusion. Since the isosurface of the tracer is a material surface in the absence of diffusion, the mass flux arises solely from diffusion. However, the magnitude of the flux is enhanced

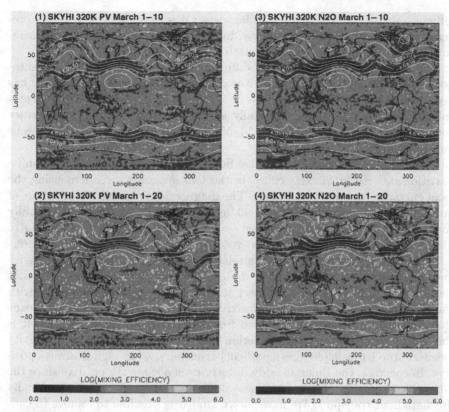

Fig. 11. *Left column*: In M_E computed from potential vorticity simulated by GFDL's SKYHI GCM on the 320-K surface. Solid contours are the time-mean streamfunction. *Right column*: same as the *left column* except it is computed from N_2O mixing ratio. *Top row*: 10-day average for 1–10 March. *Bottom row*: 20-day average for 1–20 March. Notice extremely small values along the axes of jet streams

by stirring (differential advection): the flux is large where the geometry of the tracer is complex and small where it is smooth.

By directly transforming the advection–diffusion equation, we have obtained a 1D transport equation that governs the mass distribution (PDF) of the tracer. Inverting the relationship for the tracer as a function of mass (CDF), this equation becomes a 1D diffusion equation with a diffusion coefficient that varies with space and time. The coefficient, termed effective diffusivity, arises from molecular diffusion but also reflects fluid dynamical stirring in the recent past [32]. Effective diffusivity is small where the tracer field is smooth (less stirred) and large where it is rough (highly stirred). A minimum in effective diffusivity is an indication of transport barrier. Since effective diffusivity is fully a function of time, it is an instantaneous measure of irreversible transport. Effective diffusivity addresses many of the difficulties with the extant mixing diagnostics described in Sect. 1. A comprehensive review of effective diffusivity, including comparison with other measures of mixing, is found in Shuckburgh and Haynes [39].

Since the structure of effective diffusivity depends only on the instantaneous geometry of the tracer field, it is readily calculable. The diagnostic has been successfully applied to a class of 2D flows (KH instability, the stratosphere, and ocean) using numerically synthesized tracers and actual trace constituents [32, 34, 35, 36, 37, 38, 54, 55, 56] to elucidate the inhomogeneous nature of mixing.

We have also shown that the mass flux through the tracer surfaces can be partitioned, with little extra effort, in the two opposing directions using the local curvature of the tracer surface. The diagnostic is capable, for example, of quantifying asymmetry in wave-breaking events [40]. The result is generally consistent with the visual impression of tracer geometry (e.g., shedding of filaments in one preferred direction) yet provides a more objective measure.

Finally, we have touched on the relationship between effective diffusivity and the traditional Eulerian eddy diffusivity. The two are different in that the former is diffusivity with respect to the tracer surfaces that move with the flow, whereas the latter is diffusivity about fixed locations. This means that advective transport is excluded from effective diffusivity, whereas it is the leading contributor to the Eulerian eddy diffusivity. Effective diffusivity extracts true irreversible (downgradient) transport, so its sign is always positive. In contrast, the Eulerian eddy diffusivity is not always positive since the advective excursion of tracer surfaces does not necessarily lead to a locally downgradient transport. We have shown that it is possible to remove the advective contribution from eddy diffusivity, and that the remainder represents local irreversible mixing. This last quantity is very much akin to effective diffusivity in that it captures molecular diffusion amplified by differential advection, except it is defined locally. This diagnostic is successfully used to identify the geographical locations of mixing barriers [50].

We shall conclude this chapter by addressing some of the frequently asked questions about the diagnostic method.

1. *When the tracer contour is multiply connected, why are all "islands" summed up together with the main body? When a blob is cut off from the main body of fluid mass, doesn't it constitute mass transport?* It is possible that a significant fraction of fluid mass is split from the main body with little diffusion. However, such cutoff blob stands a chance of being remerged with the main body. (Merger of vortices is common in geostrophic turbulence [57].) To exclude such "reversible" transport, we keep all identifiable islands lumped together. Only when the cutoff blobs and filaments lose their Lagrangian identity through diffusion, we count it as transport. This way the transport is guaranteed to be irreversible. It is also algorithmically challenging to take inventory of individual blobs.

2. *Effective diffusivity is calculated from the geometry of a tracer. Doesn't it mean it is a property of the tracer, and not of the flow?* In general, effective

diffusivity is indeed specific to the tracer. However, to the extent that tracers are passive and dictated by large-scale flows, they tend to form similar geometry and give rise to similar effective diffusivity. For example, Haynes and Shuckburgh [36] solve advection–diffusion problems for different initial conditions but with the same stratospheric winds and find that the effective diffusivities from all runs converge after a few months. This is related to the emergence of a "strange eigenmode," a complex pattern of tracer characterized by a stationary normalized PDF in the longtime limit [58, 59, 60, 61].

3. *It is claimed that effective diffusivity excludes advective transport. But in practice, due to coarse resolution of the tracer data, etc., one must parameterize processes at unresolved scales, including small-scale advection. Then molecular diffusion (D) must be replaced by a suitable subgrid eddy diffusion. How do you know if your choice of subgrid diffusion is "right?" How sensitive is effective diffusivity to the choice of subgrid diffusion?* If D is to be replaced by K_m in Sect. 3, advective transport is still kept out of effective diffusivity. See [50] for how K_m might be evaluated. If the primary role of diffusion is to remove the small-scale structure generated by large-scale stirring, then effective diffusivity is dictated by the large-scale process. It should be robust as long as subgrid diffusion is chosen so as to balance deformation (i.e., the Péclet number ~ 1) at the minimum resolved scale of the tracer. However, when stirring is weak (e.g., in the barrier region), effective diffusivity tends to increase with increasing numerical diffusion. The implication is that, when the resolution of the data/model is coarse and the subgrid diffusion is correspondingly large, it tends to overestimate effective diffusivity at the barriers, making them leakier than they actually are. See, for example, [34, 36, 50, 56].

4. *Most observed trace constituents in the atmosphere and ocean are also affected by processes other than advection and diffusion. What are the effects of the neglected processes on effective diffusivity?* This depends on the timescale of the processes involved. If stirring by large-scale motion is much faster than the other processes, effective diffusivity is primarily determined by transport and the neglected processes will have little effect. If the timescales of source/sink are much shorter than the transport timescale, then the properties of the flow will have less control on the effective diffusivity. The divergent component of the flow, neglected in (1), could also affect mixing [62]. In the case of the lower stratosphere, however, the difference is negligible at least in the extratropics [37].

5. *The concept of equivalent latitude works well when the tracer gradients are monotonic. But many stratospheric constituents have extrema in the tropics. How do you compute effective diffusivity there?* The effective diffusivity calculated from the real stratospheric constituents is indeed unreliable in the tropics where latitudinal gradients vanish. See [63] for details. One way to avoid this difficulty is to use potential vorticity (PV), which has monotonic pole-to-pole gradients. However, PV diagnosed from meteorological data may be susceptible to significant errors. A better approach is to numerically synthesize a PV-like tracer by solving advection–diffusion equation [35, 36, 37, 38].

6. *Can you use the diagnosed effective diffusivity to drive a 2D chemical-transport model?* Yes, as long as the model's coordinate is consistent with the effective diffusivity's (such as equivalent latitude). As we have seen, effective diffusivity can be significantly different from the Eulerian eddy diffusivity, so it should not be used to drive the Eulerian 2D model. The meridional circulation must also be specified in the same coordinate (this can be challenging).

7. *Is it possible to formulate a flux closure to predict effective diffusivity?* To predict effective diffusivity one must predict equivalent length, and to predict equivalent length one must predict the square of tracer gradient averaged on the tracer contour (CDR, see Sect. 2.2). Closure for the CDR is an active research topic in the turbulence community [64, 65, 66, 67, 68], but the theory is largely limited to homogeneous turbulence and very little has been done for flows with large eddy correlation length.

5.0.1 Acknowledgments

This work has been supported by NSF Grant ATM0230903. Stimulating discussions with the participants of the Aosta School 2004 helped shaping up this manuscript.

Appendix 1   Derivation of (4)

Consider the area of the region in which $q^* \leq q$. The change in this area from time t to $t + \delta t$ is

$$\delta \int_{q^* \leq q} \mathrm{d}S = \int_{q^*(t+\delta t) \leq q} \mathrm{d}S - \int_{q^*(t) \leq q} \mathrm{d}S$$

$$\approx -\oint_{q^* = q} \delta t \frac{\partial q^*/\partial t}{|\nabla q^*|} \mathrm{d}l = -\delta t \frac{\partial}{\partial q} \int_{q^* \leq q} \frac{\partial q^*}{\partial t} \mathrm{d}S. \qquad (30)$$

The last identity uses (5). By dividing both sides by δt, taking the limit $\delta t \to 0$ and substituting (1),

$$\frac{\partial}{\partial t} A(q,t) = -\frac{\partial}{\partial q} \int_{q^* \leq q} \left(D \nabla^2 q^* - \nabla \cdot (\mathbf{v} q^*) \right) \mathrm{d}S$$

$$= -\frac{\partial}{\partial q} \int_{q^* \leq q} D \nabla^2 q^* \mathrm{d}S + \frac{\partial}{\partial q} \int_{q^* \leq q} \nabla \cdot (\mathbf{v} q^*) \mathrm{d}S. \qquad (31)$$

However, using the divergence theorem,

$$\int_{q^*\leq q} \nabla \cdot (\mathbf{v}q^*)\, \mathrm{d}S = q \oint_{q^*=q} \frac{\mathbf{v}\cdot\nabla q^*}{|\nabla q^*|}\, \mathrm{d}l = q \int_{q^*\leq q} \nabla \cdot \mathbf{v}\, \mathrm{d}S = 0. \qquad (32)$$

Hence

$$\frac{\partial}{\partial t}A(q,t) = -\frac{\partial}{\partial q}\int_{q^*\leq q} D\nabla^2 q^*\mathrm{d}S. \qquad (33)$$

Appendix 2 Relationship to Sinai–Yakhot [41] and Ching–Kraichnan [42]

Consider a tracer field normalized by its standard deviation:

$$\theta^* = (q^* - \langle q^*\rangle)\Big/\sqrt{\langle (q^* - \langle q^*\rangle)^2\rangle}, \qquad (34)$$

where the angular bracket denotes the global average. Substituting in (1) and using (2),

$$\frac{\partial}{\partial t}\theta^* + \nabla \cdot (\mathbf{v}\,\theta^*) = D\nabla^2\theta^* + \langle D\,|\nabla\theta^*|^2\rangle\,\theta^*. \qquad (35)$$

Applying the same procedure used to derive (4),

$$\frac{\partial}{\partial t}A(\theta,t) = -\frac{\partial}{\partial\theta}\int_{\theta^*\leq\theta} D\nabla^2\theta^*\mathrm{d}S - \langle D\,|\nabla\theta^*|^2\rangle\frac{\partial}{\partial\theta}\int_{\theta^*\leq\theta}\theta^*\,\mathrm{d}S$$
$$= -\frac{\partial A}{\partial\theta}\left(\langle D\nabla^2\theta^*\rangle_\theta + \langle D\,|\nabla\theta^*|^2\rangle\,\theta\right), \qquad (36)$$

where (7) is used to derive the last expression. In a steady state, vanishing of the lhs leads to

$$\langle\nabla^2\theta^*\rangle_\theta = -\langle|\nabla\theta^*|^2\rangle\,\theta, \qquad (37)$$

Hence, diffusion of θ^* conditioned on θ is proportional to θ. Taking the θ derivative of (36) and using (6) and (8),

$$\frac{\partial}{\partial t}p(\theta,t) = -\frac{\partial}{\partial\theta}\left(\frac{\partial}{\partial\theta}\left(\langle D\,|\nabla\theta^*|^2\rangle_\theta\,p\right) + \langle D\,|\nabla\theta^*|^2\rangle\,p\theta\right). \qquad (38)$$

This is analogous to (9). Assuming a steady state,

$$\frac{\partial}{\partial\theta}\left(\langle\,|\nabla\theta^*|^2\rangle_\theta\,p\right) + \langle|\nabla\theta^*|^2\rangle p\theta = 0. \qquad (39)$$

The solution to the above Partial differential equation (PDE) is

$$p(\theta) = \frac{C}{\left\langle |\nabla\theta^*|^2 \right\rangle_\theta} \exp\left(-\int_0^\theta \frac{\left\langle |\nabla\theta^*|^2 \right\rangle}{\left\langle |\nabla\theta^*|^2 \right\rangle_{\theta'}} \theta' \, d\theta' \right), \tag{40}$$

where C is a constant. This expression corresponds to the result of Sinai and Yakhot [41]. Substituting (37) in (40),

$$p(\theta) = \frac{C}{\left\langle |\nabla\theta^*|^2 \right\rangle_\theta} \exp\left(\int_0^\theta \frac{\left\langle \nabla^2\theta^* \right\rangle_{\theta'}}{\left\langle |\nabla\theta^*|^2 \right\rangle_{\theta'}} \, d\theta' \right). \tag{41}$$

This last expression is the one derived by Ching and Kraichnan [42].

References

1. A. E. Roache, J. B. Kumer, R. M. Nightingale et al: J. Geophys. Res. **101**, 9679 (1996)
2. J. D. Mahlman, H. Levy and W. J. Moxim: J. Geophys. Res. **91**, 2687 (1986)
3. J. R. Holton: J. Atmos. Sci. **43**, 1238 (1986)
4. P. H. Haynes: Ann. Rev. Fluid Mech. **37**, 263 (2005)
5. M. E. McIntyre and T. N. Palmer: Nature **305**, 593 (1983)
6. R. A. Plumb: J. Meteor. Soc. Jpn **80**, 793 (2002)
7. S. E. Strahan and J. D. Mahlman: J. Geophys. Res. **99**, D10305 (1994)
8. L. C. Sparling: Rev. Geophys. **38**, 417 (2000)
9. J. L. Neu, L. C. Sparling and R. A. Plumb: J. Geophys. Res. **108**, Art. No. D4482 (2003)
10. D. W. Tarasick, V. E. Fioletov, D. I. Wardle et al: J. Geophys. Res. **110**, Art. No. D02304 (2005)
11. C. Gerbig, J. C. Lin, S. C. Wofsy et al: J. Geophys. Res. **108**, Art. No. D4756 (2003)
12. J. C. Lin, C. Gerbig, S. C. Wofsy et al: J. Geophys. Res. **109**, Art. No. D15304 (2005)
13. A. S. Bower, H. T. Rossby and J. L. Lilibridge: J. Phys. Oceanogr. **15**, 24 (1985)
14. S. E. Schollaert, T. Rossby and J. A. Yoder: Deep-sea Res. Part II-Top. Stud. Oceanogr. **51**, 173 (2004)
15. M. N. Juckes and M. E. McIntyre: Nature **328**, 590 (1987)
16. D. W. Waugh, R. A. Plumb, R. J. Atkinson et al: J. Geophys. Res. **99**, D1071 (1994)
17. R. A. Plumb, D. W. Waugh, R. J. Atkinson et al: J. Geophys. Res. **99**, D1089 (1994)
18. M. A. Shapiro: J. Atmos. Sci. **37**, 994 (1980)
19. M. Sprenger, M. C. Maspoli and H. Wernli: J. Geophys. Res. **108**, Art. No. D8525 (2003)
20. S. Dutkiewicz, L. Rothstein and T. Rossby: J. Geophys. Res.—Oceans **106**, 26917 (2001)
21. S. Corrsin: Limitations of gradient transport models in random walks and in turbulence. In: Advances in Geophysics, Vol. 18a, pp. 25–60. Academic Press, New York (1974)

22. M. E. McIntyre: Atmospheric dynamics: some fundamentals, with observational implications. In: The Use of EOS for Studies of Atmospheric Physics, pp. 313–386. North-Holland, Amsterdam (1982)
23. A. H. Sobel: J. Atmos. Sci. **56**, 2571 (1999)
24. R. T. Pierrehumbert: Phys. Fluids **3A**, 1250 (1991)
25. S. Wiggins: Ann. Rev. Fluid Mech. **37**, 295 (2005)
26. R. T. Pierrehumbert and H. Yang: J. Atmos. Sci. **50**, 2462 (1993)
27. G. Boffetta, G. Lacorata, G. Radaelli et al: Physica D **159**, 58 (2001)
28. B. Joseph and B. Legras: J. Atmos. Sci. **59**, 1198 (2002)
29. M. Bithell and L. J. Gray: Gophys. Res. Lett. **24**, 2721 (1997)
30. M. E. McIntyre: Phil. Trans. R. Soc. **A296**, 129 (1980)
31. N. Butchart and E. E. Remsberg: J. Atmos. Sci. **43**, 1319 (1986)
32. N. Nakamura: J. Atmos. Sci. **53**, 1524 (1996)
33. K. B. Winters and E. A. d'Asaro: J. Fluid Mech. **317**, 179 (1996)
34. N. Nakamura and J. Ma: J. Geophys. Res. **102**, D25721 (1997)
35. D. R. Allen and N. Nakamura: J. Atmos. Sci. **60**, 278 (2003)
36. P. H. Haynes and E. Shuckburgh: J. Geophys. Res. **105**, D22777 (2000)
37. P. H. Haynes and E. Shuckburgh: J. Geophys. Res. **105**, D22795 (2000)
38. D. R. Allen and N. Nakamura: J. Geophys. Res. **106**, D7917 (2001)
39. E. Shuckburgh and P. E. Haynes: Phys. Fluids **15**, 3342 (2003)
40. N. Nakamura: J. Atmos. Sci. **61**, 2735 (2004)
41. Y. G. Sinai and V. Yakhot: Phys. Rev. Lett. **63**, 1962 (1989)
42. E. S. C. Ching and R. H. Kraichnan: J. Stat. Phys. **93**, 787 (1998)
43. H. A. Michelsen, C. R. Webster, G. L. Manney et al: J. Geophys. Res. **104**, D26419 (1999)
44. A. H. Sobel, R. A. Plumb and D. W. Waugh: J. Atmos. Sci. **54**, 2241 (1997)
45. A. Gettleman and A. H. Sobel: J. Atmos. Sci. **57**, 3 (2000)
46. A. Stohl, H. Wernli, P. James et al : Bull. Am. Meteorol. Soc. **84**, 1565 (2003)
47. H. Wernli and M. Bourqui: J. Geophys. Res. **107**, 10.1029/2001JD000812 (2002)
48. T. M. Hall and M. Holtzer: Geophys. Res. Lett. **30**, Art. No. 1222 (2003)
49. M. Nakamura and R. A. Plumb: J. Atmos. Sci. **51**, 2031 (1994)
50. N. Nakamura: J. Atmos. Sci. **58**, 3685 (2001)
51. R. J. Greatbatch: J. Phys. Oceanogr. **31**, 2797 (2001)
52. A. S. Medvedev and R. J. Greatbatch: J. Geophys. Res. **109**, D07104 (2004)
53. R. A. Plumb: J. Atmos. Sci. **36**, 1699 (1979)
54. D. R. Allen, J. L. Stanford, M. A. Lopez-Valverde et al.: J. Atmos. Sci. **56**, 563 (1999)
55. H. E. Deese, L. J. Pratt and K. R. Helfrich: J. Phys. Oceanogr. **32**, 1870 (2002)
56. J. Marshall, E. Shuckburgh, H. Jones et al: J. Phys. Oceanogr. **36**, 1806 (2006)
57. I. Yasuda and G. R. Flierl: Dyn. Atmos. Ocean **26**, 159 (1997)
58. R. T. Pierrehumbert: Chaos, solitons. Fractals **4**, 1091 (1994)
59. J. Sukhatme and R. T. Pierrehumbert: Phys. Rev. **E66**, Art. No. 056302 (2002)
60. W. J. Liu and G. Haller: Physica D **188**, 1 (2004)
61. D. R. Fereday and P. H. Haynes: Phys. Fluids **16**, 4359 (2004)
62. J. V. Lukovich and T. G. Shepherd: J. Atmos. Sci. **62**, 3933 (2005)
63. E. Shuckburgh, W. Norton, A. Iwi et al: J. Geophys. Res. **106**, 14327 (2001)
64. R. H. Kraichnan: Bull. Am. Phys. Soc. **34**, 2298 (1989)
65. E. E. O'Brien and A. Sahay: Phys. Fluids **4**, 1773 (1992)

66. S. S. Girimaji: Phys. Fluids **4**, 2875 (1992)
67. Y. Kimura and R. H. Kraichnan: Phys. Fluids **5**, 2264 (1993)
68. G. W. He and R. Rubinstein: J. Turbulence **4**, doi:10.1088/1468-5248/4/1/029 (2003)

Lagrangian Statistics from Oceanic and Atmospheric Observations

J. H. LaCasce

Department for Geosciences, Norwegian Meteorological Institute, University
of Oslo, Oslo, Norway
j.h.lacasce@geo.uio.no

Abstract. We review statistical analyses made with Lagrangian data from the
atmosphere and ocean. The focus is on the types of measures used and on how
the results reflect the underlying dynamics. First we discuss how the most common measures come about and how they are related to one another. The measures
can be subdivided into those concerning single particles and those pertaining to
groups of particles. Single particle analysis is more typical with oceanic data. The
most widely-used such analysis involves binning velocities geographically to estimate
characteristics of the Eulerian flow, such as the mean velocities and the diffusivities.
Single particle statistics have also been used to study Rossby wave propagation, the
sensitivity to bottom topography and eddy heat fluxes. The dispersion of particle
pairs has been studied more in the atmosphere, although examples in the oceanic
literature have also appeared recently. Pair dispersion at sub-deformation scales is
similar in the two systems, with particle separations growing exponentially in time.
The larger scale behavior varies, possibly reflecting details of the large scale shear
flow. Analyses involving three or more particles are fairly rare but have been used
to measure divergence and vorticity, as well as turbulent dispersion.

1 Introduction

In a seminal paper in physical oceanography, Stommel and Arons [1] predicted
the structure of the abyssal (deep) circulation in the ocean. Their model exhibited a sluggish poleward interior flow linked to dense water formation sites
by boundary currents on the western side of the basin. If correct, the model
implied that a "Deep Western Boundary Current" (DWBC) should lie under
the Gulf Stream in the North Atlantic. This had never been observed, and
there was much interest subsequently in finding it. The difficulty was that
measuring currents at 3000 m at that time was a very difficult task.

Stommel believed that the current could be observed remotely if seeded
with neutrally buoyant, passive drifters, as one might do in a laboratory experiment. The English scientist, John Swallow, designed and built such a drifter.
The idea was that the drifter would sink to the desired depth and then follow

J. H. LaCasce: *Lagrangian Statistics from Oceanic and Atmospheric Observations*, Lect.
Notes Phys. **744**, 165–218 (2008)
DOI 10.1007/978-3-540-75215-8_8   © Springer-Verlag Berlin Heidelberg 2008

the current there. He made the device by hollowing out aluminum scaffolding tubes in baths of caustic soda, to obtain the correct compressibility.[1] He (with Stommel and Val Worthington from Woods Hole) deployed these "floats" off the east coast of North America and monitored their progress from shipboard with an acoustic receiver. The success of such an effort perhaps seems remote, but they located the current and confirmed, spectacularly, the theory [2].

Swallow made another important discovery in a subsequent experiment, the Aries expedition. The idea was to test for the sluggish interior flow predicted by Stommel and Arons. Again multiple floats were deployed from the ship and tracked acoustically. The expectation was that the floats would drift slowly, allowing them to be tracked for months (the oceanographers believed they could make calls to port and return to resume tracking). But instead, the floats shot away from the ship at speeds of order 10 cm/s and were soon lost [3]. The observers also found that pairs of floats deployed only 10 km apart quickly went in different directions. They deduced an active deep eddy field with relatively small spatial scales, something which was completely unanticipated. Word of this spread quickly, initiating what became a major effort in the following decades to understand the oceanic eddy field.

Lagrangian observations have been used profitably to study remote regions of both the atmosphere and ocean. In the following, we examine various analyses made using that data. The intent is not to present an exhaustive survey of Lagrangian experiments; such surveys can be found elsewhere [4]. The point rather is to consider in more detail statistical analyses based on data from the atmosphere and ocean, to illustrate the range of techniques used, as well as the results obtained.

1.1 Instruments

Before discussing those analyses, it will be useful to consider the instruments used. Most Lagrangian observations in the atmosphere have been made using constant level balloons. These rise to a preset pressure level and drift along, tracked by satellite. The heyday of large balloon experiments was the 1970s. Position errors at that time were on the order of 5 km [5].

In the ocean, two classes of instruments are used: one for tracking currents at the surface and the other at depth. The former, referred to as "drifters", are comprised of a transmitter and (but not always) a subsurface "drogue". The drogue usually resembles a large kite or sock and causes the transmitter to drift with the currents at the depth of the drogue, generally 5–50 m below the surface. The surface transmitter is monitored by satellite. Drifters tracked with the ARGO satellite system have positional errors on the order of 1 km. More recent models can be tracked by GPS and cellular phones, and these can offer 100 m accuracy and 10 min tracking [6].

---

[1] An entertaining history of the development of oceanic floats was written by John Gould of the Southampton Oceanographic Center.

The subsurface instruments are called "floats".[2] Floats sink to a depth determined by the float's compressibility and then follow currents there, like the balloons which rise to a designated pressure level. More recent designs track constant density surfaces instead, because fluid parcels tend to follow such "isopycnal" surfaces.

Because they are below the surface, floats cannot be tracked by satellite. Early floats (like those used by Swallow and the later Sound Fixing and Ranging or "SOFAR" floats [7]) were like large organ pipes, emitting low frequency sound pulses which were monitored by a network of microphones; the floats' positions were then determined by triangulation. Later, the inverse system (RAFOS, or "SOFAR" spelled backward [8]) was developed, with subsurface sound sources and much smaller floats carrying a microphone. These floats yield positions with an accuracy of roughly 1 km, from once to several times a day.

A recent addition is the ALACE float. Some 3000 of these floats have been deployed in the ARGO program and are currently drifting around the world. These drift at a constant depth and rise to the surface periodically to be located by satellite. They are not tracked at depth and so do not require subsurface sound sources (which are expensive and limit the sampling region). But because these floats drift for days or even weeks below surface without being tracked, the temporal resolution of the positions is much lower. So these floats will not be of use in the subsequent discussions.

An additional method of (Lagrangian) observation is to release a passive tracer and monitor its spread.[3] A typical example is of smoke spreading in the atmospheric boundary layer. While lacking the temporal resolution of a continuously tracked float, tracer evolution can provide information about the stirring, i.e. the change in the distribution of a cloud about its center of mass. We will touch only briefly on tracer release experiments, but focus instead on continuously tracked particles.

1.2 Analysis

Following Swallow's experiments, Lagrangian measurements have been used fruitfully for descriptive studies. This has been particularly true in the ocean, where direct sampling is labor intensive and costly. Floats have been deployed in eddies, such as Gulf Stream rings [9] and Meddies (the eddies formed by the outflow from the Mediterranean); [10], as well as in Rossby waves [11]. The trajectories yield information about both the paths and structures of these features. Floats and drifters have also been used to infer the structure

---

[2] As B. Warren remarked to me once, "drifters" float and "floats" sink.

[3] A colorful, unintentional example is when a container ship sank in the North Pacific in 1990, releasing some 61,000 Nike sneakers. These were swept eastward and many landed along the west coast of Canada and the USA. The information was used by C. Ebbesmeyer and colleagues to deduce surface drift patterns.

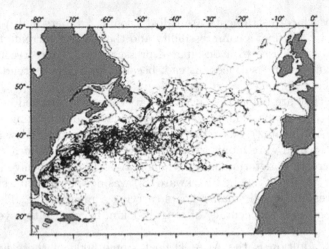

Fig. 1. Trajectories of surface drifters in the North Atlantic. From Richardson [12], with permission

of the Gulf Stream [12] (Fig. 1), the Norwegian-Atlantic Current [13], and the (infamously inhospitable) Antarctic Circumpolar Current [14].

While mean currents (the Jet Stream and Gulf Stream are examples) play a major role in the circulation of the atmosphere and ocean, both systems exhibit significant time variability. So two particles deployed at the same location at different times, or two particles deployed simultaneously at slightly different locations, generally follow very different paths. As recognized early on by turbulence researchers [15], such uncertainty necessitates a *statistical* description; the motion of a single particle is not as important as the probability of a given path, as inferred from multiple realizations.

Lagrangian statistics concerns the averages of positions, velocities, and related quantities over many realizations. It is useful to subdivide the statistics into those concerning single particles and those requiring two or more particles. Both single and multiple particle statistics are required for a full description of tracer evolution.

To see why, consider a group of particles, for instance, the four shown in Fig. 2. These constitute a simple "cloud" of tracer. Of interest is how the cloud moves, and how it spreads out. The movement can be quantified by the motion of the center of mass, the first moment of the particle displacements. In the x-direction, this is

$$M_x(t) = \frac{1}{N} \sum_{i=1}^{N} (x_i(t) - x_i(0)) .$$
(1)

The mean displacement is thus a single-particle measure because it derives from the average of individual displacements.

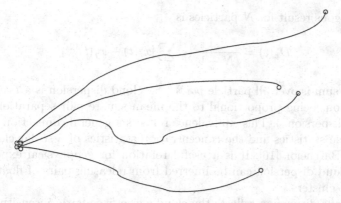

Fig. 2. Four hypothetical particles drifting and separating from one another

The spread of the cloud about its center of mass can be measured by the variance of the particle displacements, the second-order moment.[4] This is

$$D_x(t) = \frac{1}{N-1} \sum_{i=1}^{N} (x_i(t) - x_i(0) - M_x(t))^2 \, . \tag{2}$$

The variance is usually referred to as the "dispersion". We can rewrite the dispersion by expanding the RHS in (2). For instance, for three particles (and substituting x_i for the displacement from the initial position), we have

$$\left(x_1 - \frac{x_1 + x_2 + x_3}{3}\right)^2 + \left(x_2 - \frac{x_1 + x_2 + x_3}{3}\right)^2 + \left(x_3 - \frac{x_1 + x_2 + x_3}{3}\right)^2 =$$

$$\left(\frac{x_1 - x_2}{3} + \frac{x_1 - x_3}{3}\right)^2 + \left(\frac{x_2 - x_1}{3} + \frac{x_2 - x_3}{3}\right)^2 + \left(\frac{x_3 - x_1}{3} + \frac{x_3 - x_2}{3}\right)^2 =$$

$$\frac{1}{9} \left[2(x_1 - x_2)^2 + 2(x_1 - x_3)^2 + 2(x_2 - x_3)^2 + 2x_1^2 + 2x_2^2 + 2x_3^2 - 2x_1 x_2\right.$$

$$\left. -2x_1 x_3 - 2x_2 x_3\right] = \frac{1}{3}[(x_1 - x_2)^2 + (x_1 - x_3)^2 + (x_2 - x_3)^2].$$

---

[4] We divide by $N - 1$ to be consistent with the standard definition of the variance (one degree of freedom is lost determining the mean). Frequently, N is used instead.

The analogous result for N particles is

$$D_x(t) = \frac{1}{(N-1)N} \sum_{i \neq j} [x_i(t) - x_j(t)]^2 \,, \tag{3}$$

where the sum is over all particle *pairs*.[5] So cloud dispersion is a *two*-particle phenomenon, being proportional to the mean square pair separation (called "relative dispersion"). This equivalence reflects a general connection between two particle statistics and the concentration statistics of a scalar cloud, first noted by Batchelor [16]. It is a useful relation for geophysical experiments because cloud dispersion can be inferred from releasing pairs of floats rather than large clusters.

While the dispersion reflects the cloud's size, it is fairly insensitive to the cloud's distribution in space. Consider the two examples shown in Fig. 3. The *upper left panel* shows a group of particles undergoing essentially a random walk (generated by a stochastic advection scheme; Sect. 2.2), while the *upper right panel* shows particles advected by a 2D turbulent flow (Sect. 3.2). The cloud on the *left* is spreading out uniformly, but the one on the *right* is actually being drawn out into filaments. The dispersion in these two cases is similar, but the distributions are obviously different.

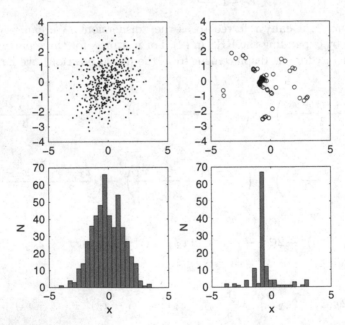

Fig. 3. Two examples of particle advection. The 484 particles in the *upper left panel* have been advected by a first-order stochastic routine (Sect. 2.2), while the 121 particles in the *upper right panel* move in a 2D turbulent flow (Sect. 3.2). The lower panels show histograms of the corresponding x-displacements

---

[5] Note we do not count duplicate pairs twice in this definition.

A way to distinguish them is with the probability density function (PDF) of the displacements. The PDF, a normalized histogram, is of fundamental importance statistically because all the moments (mean, dispersion, etc.) can be derived from it (Sects. 2, 3.1). Moreover its shape reflects how the cloud is dispersing. Binning the x-displacements from the center of mass for the cloud at *left* in Fig. 3 yields a nearly Gaussian histogram (*lower left panel*). The histogram for the turbulent flow on the other hand (*lower right panel*) has a peak near the origin and extended "tails", reflecting that most of the particles are near the origin but that a few have been advected far away.

The dispersion reflects the width of the PDF and, as noted, this is similar for the two cases shown in Fig. 3. Where they differ is in the higher order moments. A commonly used one is the *kurtosis*, the fourth-order moment:

$$\mathrm{ku}(x) \equiv \frac{\sum_i (x_i - M_x)^4}{(\sum_i (x_i - M_x)^2)^2} . \tag{4}$$

It is traditional to normalize the kurtosis by the squared second-order moment. The kurtosis has a value of 3 for a Gaussian distribution. In the random walk example in Fig. 3, the kurtosis is 2.97, whereas in the turbulence case, it is 6.62. The larger value reflects the extended tails.

Many of the measures discussed hereafter are variants on these basic quantities: the PDF and the moments (mean, dispersion, and kurtosis), either for displacements or velocities. One can, in addition, define corresponding quantities for either single particles or pairs of particles. We begin with single particles and continue with multiple particles thereafter.

2 Single-Particle Statistics

Single-particle statistics are the most frequently examined in geophysical Lagrangian studies. The following originates with Taylor's seminal work on diffusion by continuous movements [17]. A lucid summary, in the turbulence context, is given by [18]; a more recent treatise, which discusses the application to oceanic data and treats the problem of inhomogeneity, is given by [19]. We also note a seminal work in which many of these measures were applied to oceanic floats for the first time [20].

Consider a single fluid parcel. If this parcel is initially at $\mathbf{x} = \mathbf{x_0}$ at $t = t_0$, the probability that it arrives at $\mathbf{x}$ at time t can be expressed by a single-particle displacement PDF, $Q(x, t|x_0, t_0)$. If we have a group of parcels at various locations, the PDF permits us to predict where those parcels are likely to be subsequently:

$$P(x, t) = \int P(x_0, t_0)\, Q(x, t|x_0, t_0)\, dx_0. \tag{5}$$

Thus Q effectively maps the original positions to positions at the later time. If the flow is statistically *homogeneous* (invariant to changes in location), then Q is only a function of the displacement:

$$Q(x, t | x_0, t_0) = Q(x - x_0, t) \equiv Q(X, t). \tag{6}$$

If the flow is also *stationary* (invariant to changes in time), we simply have

$$Q(X, t) \equiv Q(X). \tag{7}$$

All the statistical moments (mean, variance, etc.) can be derived from the PDF. The first moment is the mean displacement. For homogeneous flows, this is

$$\overline{X}(t) = \int X Q(X, t) \, dX. \tag{8}$$

The second moment is the single-particle (or "absolute") dispersion:

$$\overline{X^2}(t) = \int X^2 Q(X, t) \, dX. \tag{9}$$

The absolute dispersion is *not* the same as the variance of a group of particles about their center of mass. Rather, it is the variance of the displacements relative to their starting positions, generally a very different quantity. With the group of particles shown in Fig. 2, the absolute dispersion would reflect both the spread about the center of mass *and* the drift from the cluster's starting location. So the absolute dispersion is affected by a mean flow (it is not Galilean invariant).

The time derivative of the single-particle dispersion is the "absolute diffusivity":

$$\kappa \equiv \frac{1}{2} \frac{d}{dt} \overline{X^2} = \overline{Xu} = \int_{t_0}^{t} \overline{u(t')u(t)} dt'. \tag{10}$$

The diffusivity is thus the integral of the velocity autocorrelation. If the flow is stationary, the velocity variance is constant, so that

$$\kappa = \int_{t_0}^{t} \overline{u(t')u(t)} dt' = \nu^2 \int_{t_0}^{t} R(t') dt', \tag{11}$$

where ν is the RMS particle velocity, and R is the normalized velocity correlation. The dispersion can also be written in terms of R:

$$\overline{X^2}(t) = 2\nu^2 \int_{t_0}^{t} (t - t') R(t') dt'. \tag{12}$$

With (12), one can make deductions about the dispersion under fairly general conditions. At early times, $R(t') \approx 1$ (as follows from a Taylor expansion in time). Then the dispersion grows quadratically in time:

$$\lim_{t\to 0} \overline{X^2}(t) \propto \nu^2 t^2 . \tag{13}$$

At long times, we have

$$\lim_{t\to\infty} \overline{X^2}(t) \approx 2\nu^2 t \int_{t_0}^{\infty} R(t')\mathrm{d}t' - 2\nu^2 \int_{t_0}^{\infty} t' R(t')\mathrm{d}t' . \tag{14}$$

If the integrals in (14) converge, the dispersion grows *linearly* in time and the diffusivity is constant. As recognized by [21], the system is then statistically equivalent to a diffusive one. The eddy stirring can then be represented as a diffusive process with the diffusivity determined as above. We emphasize though that this is only true in a statistical sense; individual events could vary greatly.

The linear dependence can fail under certain conditions, for instance if there is a long-time correlation in the velocity field, or if the spread of particles is restricted, as in an enclosed basin [22]. However, one can often adjust for this (Sect. 2.1). Between the initial and final asymptotic limits, one may observe *anomalous* dispersion, or dispersion with a power law dependence different than t^1 or t^2 (an excellent discussion is given by [23]). Anomalous dispersion has been observed in experiments [24] but has been difficult to resolve in geophysical flows.

Other quantities can also be derived from the velocity correlation. The integral of the normalized autocorrelation, the "integral time", $T_\mathrm{L} \equiv \int_0^{\infty} R(t')\mathrm{d}t'$, is an estimate of the time scale over which the Lagrangian velocities are correlated. This is a basic indicator of Lagrangian predictability. The Fourier transform of the autocorrelation is the Lagrangian frequency spectrum:

$$L(\omega) = 2 \int_0^{\infty} R(t) \cos(2\pi\omega t) \, \mathrm{d}t , \tag{15}$$

an equivalence first noted by Taylor [25]. The Lagrangian time scale can be shown to be one-half the value of the spectrum at the zero frequency, i.e. $T_\mathrm{L} = L(0)/2$. This implies the diffusivity is determined by the *lowest* frequency motion [26].

2.1 Advection–Diffusion

Applying these measures to data requires some modifications. The averaging in the previous expressions assumes an ensemble of nearly identical realizations, for example of particles deployed repeatedly at the same location. But particles in geophysical experiments are often deployed at different locations. If the flow is homogeneous, ergodicity permits using such particles as an ensemble. But most geophysical flows are not homogeneous, and one must modify the averaging to account for this.

Such issues have been considered in depth by R. Davis in a series of articles [19, 27, 28]. His methodology is widely used in the analysis of oceanic data. It derives from an assumption, a separation of scales, i.e. there is a slowly varying "mean" flow and a faster varying "eddy" or "residual" field, both of which are assumed to vary in space:

$$U(x, y, z) \equiv < u(x, y, z) >; \quad u'(x, y, z, t) \equiv u(x, y, z, t) - U(x, y, z) .$$

Here, the brackets represent regional averages, as discussed below. With these quantities, one can write a transport equation for a passive tracer, C:

$$\frac{\partial}{\partial t} C + U \cdot \nabla C = -\nabla \cdot < u'C' > \equiv \nabla \cdot (\kappa \nabla C) , \tag{16}$$

if C represents an average (in some defined sense) concentration and C' the departures from that average. The Reynolds transport of the perturbations is often parameterizing as a diffusive process, as shown above. This assumes of course that diffusivity exists. But if so, such a transport equation is potentially very useful, for instance in diagnosing the spread of pollutants in a sampled region.[6] All the terms in (16) can be calculated from single-particle statistics. We consider the mean first.

2.1.1 Mean Flow

The advective–diffusive formalism requires a separation in time scales, as noted. This is equivalent to saying that there is a gap in the velocity frequency spectrum. All indications are that such a gap does not exist, either in the atmosphere or ocean [29, 30]. We make the assumption nevertheless. But the question then is, what time scale should be used for determining the mean (the slow time field)? A practical choice is simply the length of the float experiment [19], which might be a year or two. The residual velocities then pertain to the shorter time scales.

To account for the inhomogeneity of the field, the mean is calculated by averaging velocities of particles passing through selected geographical bins. The implicit assumption is that the statistics are stationary (so while we anticipate variations in space, we neglect them in time). The geographical means apply to the flow at the vertical position, z_0, of the particles in the experiment.

In the atmosphere, Morel and Desbois [31] used the trajectories of balloons released during the EOLE experiment (Sect. 3.3) to map the mean atmospheric circulation in the southern hemisphere. The results indicated a zonal mean flow, with typical velocities of 10–30 cm/s. The mean had a standing wave pattern, with dominant wavenumbers 1 and 4. The latter perhaps reflects low frequency motion, like that due to quasi-stationary planetary waves, which has been subsumed into the mean.

---

[6] An alternate version, preferred by Davis, includes a second-order time derivative for the tracer field. This allows for wave propagation, and thus avoids the issue of tracer signals propagating infinitely fast, as in the diffusion equation (Davis, personal communication).

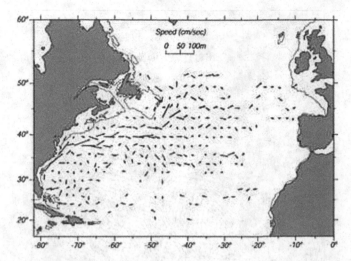

Fig. 4. Mean velocities obtained from averaging the velocities of surface drifters shown in Fig. 1. From Richardson [12]

In the ocean, regional averaging has been used widely. Averaging the surface drifter displacements shown in Fig. 1, Richardson [12] derived the mean velocities shown in Fig. 4. This clearly shows the Gulf Stream separating from the North American coast near the state of North Carolina and then flowing eastward and splitting. One branch then proceeds north off Newfoundland and another turns south.[7]

Another example, from drifter data in the Nordic Seas, is shown in Fig. 5. This captures the northern branch of the Gulf Stream (the "North Atlantic Current") as it crosses the basin, enters the Nordic Seas and proceeds toward the Arctic Ocean [34]. It also shows the Greenland Current rounding the southern tip of Greenland and proceeding south, along the Canadian coast. The large density of drifter data available here permits such a startlingly detailed picture.

While results like those in Fig. 5 would appear to justify the means (no pun intended), there are pitfalls with the binning method. First, the significance of the mean in a given bin varies with the number of particles which have been through. In most published analyses, the means are only plotted for bins with more than some minimum number of velocity realizations (e.g. 10 float days), but this is a subjective criterion.[8] Second, one must choose the size of the bins. Just as there is no spectral gap in the velocity frequency spectrum, neither is there one in the wavenumber spectrum. But by choosing a bin size,

---

[7] A more recent mapping, using an extensive set of drifter data from the North Atlantic during the 1990s, is given by [32].

[8] An alternate approach would be to show only those means which are significantly different than zero at a given level of confidence.

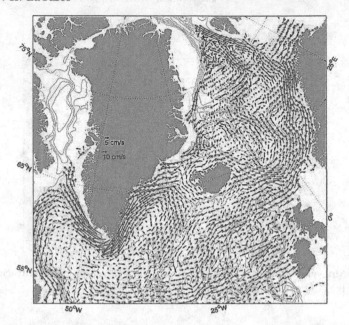

Fig. 5. Mean velocities derived from surface drifters in the North Atlantic and Nordic Seas. From Jakobsen et al. [33], with permission

one essentially picks the horizontal scale of the mean flow. Choosing too large bins yields an overly smooth mean while using too small bins subsumes eddy-like features. The Gulf Stream spans a large region, but it also has a relatively narrow core and worse, it *meanders*. Such aspects are difficult to capture with bin-averaging.

Third, uneven data coverage can result in non-smooth means. Several remedies for this have been explored, such as deriving the mean via a variational calculation with the binned velocities as input [14, 35]. Another approach is to fit the binned velocities with splines [36]. Both techniques yield smoother means. This in turn affects the residual velocities, and thus the diffusivity estimates (an example is given below).

But problems aside, the averaging of drifter and float data has been an invaluable tool, particularly in oceanography.

2.1.2 Diffusivity

As with the means, there are practical difficulties associated with calculating diffusivities. For instance, particles often visit regions with different characteristics during their lifetime. As such, it is not sensible to integrate the autocorrelation in (10) to $t = \infty$. And because the mean is not truly stationary, some fraction of the low frequency variance will remain in the residual velocity; this can produce long-term velocity correlations, hindering the convergence of the integral. As before, practical choices are required.

Often the autocorrelation integral is evaluated only up to a certain time, for instance a typical time that particles stay in a region or bin. This might be several times the integral time scale, T_L.[9]

An alternate approach is to calculate the diffusivity directly from the residual velocity and the residual displacement (the displacement minus that due to the mean velocity; [19]):

$$\kappa(\mathbf{x}, t) = - < u'_j(t_0|\mathbf{x}, t_0)d'_k(-t|\mathbf{x}, t_0) > , \qquad (17)$$

where

$$u'(t_0) = u(t_0) - U(\mathbf{x}), \quad d'(t) = d(t) - D(\mathbf{x}, t) .$$

The notation indicates that the quantities are calculated for a particle which lies at position $\mathbf{x}$ at $t = t_0$. One uses the time series for the velocity after t_0 but for the displacement *backward* in time from t_0.[10]

Zhurbas and Oh [38] used this method to map the diffusivity for the surface Atlantic (Fig. 6). Again, one is struck by the broad extent of the data coverage. The Gulf Stream is a region of heightened dispersion. So too are the Caribbean, the Equatorial region (where there are strong zonal currents and countercurrents), the region where the Antarctic Circumpolar Current flows northward along the South American coast and the region south of Africa (where the Agulhas current retroflects). Large diffusivities usually reflect large variances, and these regions are indeed the most energetic in the Atlantic.

There are several additional points with regards to Davis' method. First, the residual velocities should have a distribution which is not too different from a Gaussian. This has been tested in several locations and found to be approximately true (Sect. 2.3). Second, the diffusivity estimate can be biased by non-uniform float coverage; floats on average drift from regions with high float densities to those with low densities, and this can yield a false impression of a diffusivity gradient. Davis discusses correcting for this [19]. In addition, the determination of the mean affects the diffusivities, as suggested above. Using more advanced methods, like spline interpolation, can improve the convergence of the integral of the velocity autocorrelation. An example of this is shown in Fig. 7.

---

[9] Taken in reverse, one could *define* the bin size as the RMS distance particles spread from a central point over several integral times. Such an approach would help avoid subsuming eddies into the mean.

[10] This is conceptually similar to using reverse diffusion for deducing concentration moments, [37].

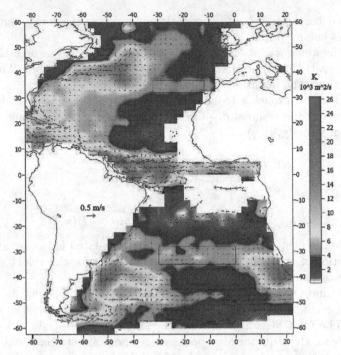

Fig. 6. Contours of diffusivity derived from surface drifters in the North Atlantic, from [38]. Superimposed are the mean velocity vectors, also deduced from the drifters. With permission

2.1.3 Diffusivity Scaling

The advective–diffusive formalism is a way of representing oceanic transport and dispersion of a passive tracer, such a spilled oil. However, to use it, one must have *already sampled* the affected region, and this is not always feasible. It would be advantageous if one could infer the means and/or diffusivities independently, for example from satellite measurements. To this end, some have sought simple parameterizations for the diffusivity.

For example, some studies have indicated the diffusivity scales with eddy kinetic energy. But the exact dependence is not always consistent. Some find that diffusivity is proportional to the RMS velocity, ν [39, 40], while others find a dependence on the kinetic energy, ν^2 [41, 42].

Recently, Lumpkin et al. [43] examined the dependence on ν systematically using drifter-derived velocities from the North Atlantic. They found that the dependence varied strongly with region, with a quadratic dependence in some places and a linear in others. They concluded that no single relation exists.

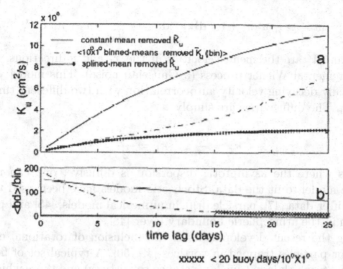

xxxxx < 20 buoy days/10°X1°

Fig. 7. The diffusivity deduced from drifters in the Pacific using three different mean fields: a constant one, one obtained from averaging in $10° \times 1°$ rectangles and one derived from spline-fitting. The latter method produces the best convergence. From [36], with permission

2.2 Stochastic Models

The Lagrangian equivalent of a diffusive system is one in which particles execute a random walk [17]. "Stochastic" routines for particle advection have been used in both the atmosphere and ocean to simulate the spread of tracer. In such models, it is a group of particles rather than a continuous tracer which is advected forward in time. Discussions of stochastic models are given by [37, 44, 45].

The most basic such model is the random or "drunkard's" walk. In this, it is the particle's position which is the "noised variable", i.e. the variable to which the stochastic perturbation is added. In higher order stochastic models, it is the velocity or the acceleration which is perturbed. In the random walk, the velocity autocorrelation is a delta function because each step is uncorrelated with the previous one. With the first-order model (in which the velocity is noised), the autocorrelation is a well-defined decaying function, and it is the autocorrelation of the acceleration which is a delta function.[11]

Consider the first-order model, which is given by a Langevin equation:

$$dx = (u + \overline{U})\, dt, \quad dy = (v + \overline{V})\, dt\,,$$

---

[11] This is an appropriate choice for simulating turbulence with very large Reynolds number [46].

$$\mathrm{d}u = -\frac{1}{T_x}u\,\mathrm{d}t + n_x\,\mathrm{d}W, \quad \mathrm{d}v = -\frac{1}{T_y}v\,\mathrm{d}t + n_y\,\mathrm{d}W, \tag{18}$$

where $\overline{U}$ and $\overline{V}$ are the mean velocities in the x- and y-directions, and $\mathrm{d}W$ is the incremental Wiener process (a Gaussian noise). This model yields an exponentially decaying velocity autocorrelation with two different time scales T_x and T_y. The diffusivities are simply

$$\kappa_x = \frac{n_x^2}{2}, \quad \kappa_y = \frac{n_y^2}{2}. \tag{19}$$

In regions where the asymptotic dispersion is diffusive, one can tune the stochastic model to fit the data. Stochastic models have been used to model surface drifter data [47], particle drift in numerical models, [45] and pollutant dispersion in the atmospheric boundary layer [48].

Among the recent developments is the inclusion of rotational effects to account for particles trapped in vortices [49, 50]. A typical set of float data includes floats which loop (due for instance to vortices) and those which don't. One can simulate the looping by including a rotational component in the stochastic equations:

$$\mathrm{d}u = -\frac{1}{T_x}u\,\mathrm{d}t + \Omega v\,\mathrm{d}t + n_x\,\mathrm{d}W, \quad \mathrm{d}v = -\frac{1}{T_y}v\,\mathrm{d}t - \Omega u\,\mathrm{d}t + n_y\,\mathrm{d}W. \tag{20}$$

The result is that the autocorrelations oscillate:

$$R_x = \mathrm{e}^{-t/T_x}\cos(\Omega t), \quad R_y = \mathrm{e}^{-t/T_y}\cos(\Omega t). \tag{21}$$

The rotational frequency, Ω, depends on the rate of swirling of the particles and can be determined from the data. Using such a model, Veneziani et al. [51] duplicated autocorrelations derived from floats at 700 m in the western North Atlantic. The model is very promising for simulating oceanic tracer dispersal.

Stochastic models must be applied cautiously however. A tracer which is well mixed initially should not develop mean gradients under these routines, the "well-mixed" criterion of Thomson [52], and this must be accounted for. Stochastic models are in addition usually non-unique except in the case of simple flows. The proper application of stochastic models for geophysical flows is thus an ongoing area of research. A comprehensive review is given by Sawford [37].

2.3 PDFs

As noted, the displacement moments derive from the PDF of the displacements, $Q(X, t)$. Closely related is the PDF of the residual velocities, $P(u', t)$. The advective–diffusive formalism of Davis [19] and the stochastic models assume that $Q(X, t)$ and $P(u', t)$ are approximately Gaussian. The first to check this assumption were evidently Swenson and Niiler [53], using drifter data from the California Current. Bracco et al. [54] did the same, using subsurface float data from the North Atlantic.

Let's consider how they did this. The problem again is that the eddy field is inhomogeneous and one must correct for this in averaging. The authors thus demeaned the velocities in geographical bins, to generate residual velocities, and then normalized the residuals by dividing by the local standard deviation. Then the normalized residuals were recombined to generate a PDF for a chosen region.

An example, from the shallow Northwest Atlantic, is shown in Fig. 8. The PDFs for both the zonal and meridional velocities deviate significantly[12] from a Gaussian. This deviation is most noticeable in the "wings" of the distribution, which are actually nearer to an exponential distribution (which would have straight wings in these plots) than a Gaussian.

The extended wings reflect an excess of *extreme* events, or velocities which are several times the standard deviation. These occur over the whole region (*lower panel*) and are often associated with coherent advection, e.g. a rapid translation or swirling motion which lasts a few days. This is where the Gulf Stream lies, and the energetic events are probably linked to the Stream and/or its eddies. Indeed, very similar velocity PDFs are found with fields of vortices and in numerical simulations of 2D turbulence [54, 56, 57, 58]. However Bracco et al. found similar PDFs in the deep western Atlantic (below the core of the stream) and *also* in the eastern Atlantic, a region of much weaker variability. Similar weakly non-Gaussian PDFs have been found from drifters data from the Adriatic Sea [59]. Only near the equator are the PDFs not significantly different from Gaussian [54].

In principle, the velocity PDF derived from Lagrangian data should be the same as that derived from Eulerian data [60]. The equivalence in the western North Atlantic was demonstrated by LaCasce [61], who examined a large number of velocity records from subsurface current meter moorings. These produced PDFs statistically identical (as determined by the K–S test) to those obtained with the floats (Fig. 9).

Because extreme events are rare, one requires long time series (or a large set of shorter series) to capture the wing deviations. The deviations from Gaussianity however are relatively minor; the kurtoses (Sect. 3.1) are typically not larger than 4.0. This implies that the distributions are probably near enough to normal to satisfy the requirements of Davis [19].

2.4 Alternate Coordinates

Up until now we have considered velocities and displacements in Cartesian coordinates, i.e. in the zonal and meridional directions. In *isotropic* flows,

---

[12] One can test the significance of deviations from a given PDF using a "goodness-of-fit test", such as the Kolmogorov–Smirnov (K–S) test [55].

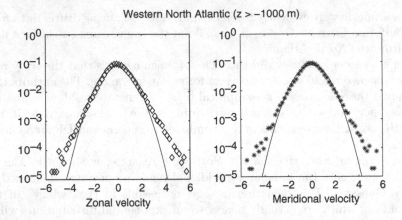

Western North Atlantic (z > −1000 m)

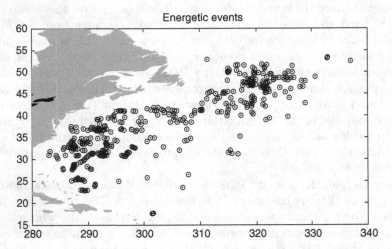

Fig. 8. Velocity PDFs from subsurface float data. The *upper panels* show the zonal and meridional velocity distributions, and the *lower panel* shows the locations of energetic events. The latter are spread over the region covered by the floats and often indicate multiday occurrences. From [54], with permission

the choice of coordinates is irrelevant, but it is important otherwise. The atmosphere is anisotropic, primarily due to the beta-effect (the latitudinal variation of the Coriolis parameter) and the Jet Stream; both favor zonal dispersion over meridional. Bottom topography in the ocean can also influence dispersion, but the direction favored generally varies with location.

Why would topography affect particle motion? Consider a barotropic fluid, under the shallow water equations. Taking the curl of the linearized momentum equations and invoking the continuity relation, one obtains the shallow water vorticity equation, which can be written as

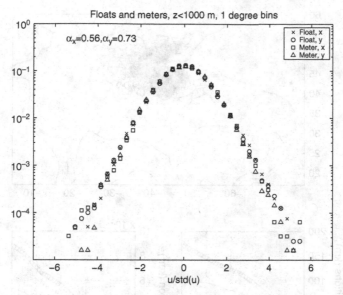

Fig. 9. The velocity PDFs derived from float data, normalized using $1°$ bins, and from current meter data, both from the western North Atlantic. Both distributions deviate from normal distributions in the wings, indicating excess energetic events. The numbers at *upper left* are the probabilities that the distributions are the same, from the Kolmogorov–Smirnov test, and indicate the null hypothesis (that they are different) cannot be rejected. From [61]

$$\frac{\partial}{\partial t} \nabla \times \bar{\mathbf{u}} + J\left(\psi, \frac{f}{H}\right) = \nabla \times \left(\frac{\tau_{\mathrm{w}}}{\rho H}\right) - \nabla \times \left(\frac{\tau_{\mathrm{B}}}{\rho H}\right), \qquad (22)$$

where $\bar{\mathbf{u}}$ is the depth-averaged velocity, ψ is the transport streamfunction, $H(x, y)$ is the water depth, and τ_{w} and τ_{B} are the surface wind and bottom stresses. Without forcing and dissipation, the vorticity changes only when there is motion across contours of constant f/H. As such, f/H yields a "restoring force" and hence can support waves. With a flat bottom, these are Rossby waves; over a sloping bottom, they are topographic waves; and with both effects present, the waves are a hybrid between the two [62].

If the ocean were barotropic and unforced, we would expect to see greater dispersion along f/H than across [63]. But how does one test for such anisotropy, given that topography varies spatially? One way is to project particle displacements onto and across the f/H contours and calculate the dispersion from the projected displacements [64]. An example, using floats from the North Atlantic Current (NAC) experiment in the shallow northwest Atlantic [65] is shown in Fig. 10.[13] The trajectories are shown in the

---

[13] These are subsurface floats designed to follow isopycnals. As such, the float depths vary, but are generally less than 800 m.

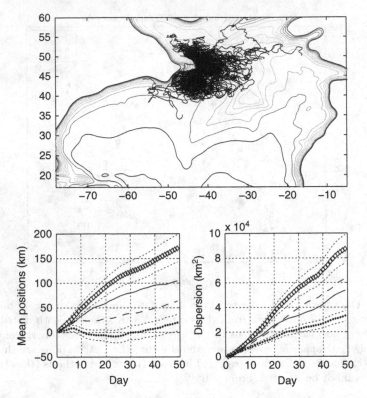

Fig. 10. The trajectories of floats from the western North Atlantic, superimposed on contours of f/H. The *lower panels* show the mean displacements and dispersion relative to latitude (*solid*) and longitude (*dashed*), and along (*diamonds*) and across (*dots*) f/H. The latter indicates a preferential tendency for translation and spreading along f/H. From [64]

upper panel, superimposed on f/H. Shown in the *lower panels* are the mean displacements and the dispersions as functions of time. In zonal/meridional coordinates, the dispersion is isotropic within the errors, and the mean displacements indicate a drift to the northwest. But the dispersion is significantly anisotropic with respect to f/H, with greater spreading occurring along the contours. The mean drift is also aligned with f/H.

We infer that the floats are steered by f/H, a fact not apparent from the statistics in $x - y$ coordinates. That topography affects floats at such shallow depths (order 100–200 m) is remarkable; indeed, the floats are constrained in their lateral spreading by the mid-Atlantic ridge, despite that the latter lies over 1000 m beneath the floats!

One way to test for topographic steering is to generate a set of synthetic trajectories which aren't steered, and make the same calculations. We do this by using the stochastic model given by (18) with the mean velocities, U and

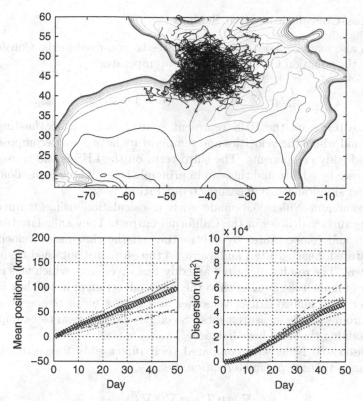

Fig. 11. Particle trajectories generated using a first-order stochastic model, with identical mean and variances as the floats, shown in Fig. 10. While the particles appear to cover roughly the same area, the displacement statistics reveal no sensitivity to the underlying f/H field. From [64]

V, and the integral time scales, T_x and T_y, calculated from the float data. We then project the displacements onto f/H and obtain the statistics shown in Fig. 11.

The stochastic trajectories themselves resemble those of the actual floats, spreading laterally to roughly the same extent. But the statistics are very different; now the dispersion is isotropic both with respect to latitude lines *and* to f/H. We conclude the stochastic particles do not "know" about the topography.

A similar approach can be used with other fixed fields. O'Dwyer et al. [66] compared float displacements to maps of in situ potential vorticity (derived from the density measurements) and found evidence for steering (suggesting that the PV is correlated with barotropic f/H).

2.5 Cross-Correlations with Scalars

One can also examine how dispersion relates to non-fixed fields. Consider for example the equation for the evolution of temperature:

$$\frac{\partial}{\partial t}T + \nabla \cdot (\mathbf{U}T) + \nabla \cdot (\mathbf{u}'T') = S + M \,, \tag{23}$$

where the terms on the RHS represent sources (e.g. surface heating) and mixing, and where the velocities and temperature have been decomposed into mean and eddy components. The third term on the LHS is the transport of temperature by eddies, and this can in principle be evaluated using float data, if the float simultaneously measures temperature.

Swenson and Niiler [53] made such a calculation using temperature-recording surface drifters in the California current. They calculated residual velocities and temperatures by subtracting off the time-mean velocity and temperature for each drifter, and averaged the estimates obtained from different drifters. The results suggested an eddy flux divergence which was roughly an order of magnitude smaller than the mean advection of heat (they inferred the latter from binned drifter velocities and from a mean temperature field derived from satellite measurements). So eddies did not seem to be important in maintaining the local heat balance.

Swenson and Niiler also compared their direct eddy flux estimate with that from a diffusive parameterization, i.e.:

$$\nabla \cdot (\mathbf{u}'T') \approx \nabla(\kappa \nabla \overline{T}) \,.$$

For this, they used a diffusivity, calculated from the drifter data and the satellite-derived mean temperature field. The two estimates agreed within the errors, a remarkable result which supported both the eddy divergence calculation and the diffusive parameterization.

Gille [35] made a similar calculation, using ALACE float data from the Southern Ocean. As noted previously, the ALACE float is not tracked continuously, but Gille was able to obtain flux estimates nevertheless. The results, which apply to the level of the floats (900 m depth), were consistent with the previous calculations using current meter data and also hydrography (density measurements). However, Gille found a poor correlation between the directly calculated flux and the mean temperature gradient, at odds with a simple diffusive parameterization. So more work is probably required to determine the applicability of such parameterizations.

2.6 Spectra

As noted in Sect. 2, the Fourier transform of the velocity autocorrelation is the Lagrangian frequency spectrum. What type of spectra should one typically expect from data? An exponentially decaying autocorrelation (a typical result from drifter data in many regions), yields

$$T(\omega) = 2 \int_0^\infty \exp(-t/T_L) \cos(2\pi\omega t)\, \mathrm{d}t = \frac{2T_L^{-1}}{T_L^{-2} + 4\pi^2\omega^2}. \qquad (24)$$

This exhibits an ω^{-2} decay at high frequencies and a white spectrum at low frequencies. The transition frequency is determined by the integral time, $\omega = (2\pi T_L)^{-1}$. The white spectrum at low frequencies occurs because the integral of the autocorrelation converges (a red spectrum would imply that no diffusivity exists [67]).

Examples of spectra calculated from oceanic data include [20, 68], from float data in the western North Atlantic, and [37, 67, 69], from surface drifter data in the Atlantic and Pacific. These calculations generally suggest a red spectrum at low frequencies, implying persistent low-frequency variability. This possibly reflects that the records aren't long enough or that the mean flow has not be captured properly (Sect. 2.1.1). The higher frequency behavior differs with the depth of observation. Near the surface, an ω^{-2} dependence is often observed, but the spectra at depth are steeper.

A possible explanation for the steeper spectra at depth can be inferred from the stochastic models (Sect. 2.2). The generic first-order model (with velocity as the noised variable) exhibits an ω^{-2} spectrum at high frequencies. The second-order model (with the acceleration as the noised variable) exhibits an ω^{-4} dependence at high frequencies. So the first-order model may apply better at the surface and the second-order model below the surface. If so, we would infer that surface drifters experience more rapid changes in acceleration, due to perturbations like the wind, than do subsurface floats.

2.7 Euler-Lagrange

Lastly, we consider the connection between single-particle statistics and the corresponding Eulerian quantities. If the Lagrangian integral time is 5 days, what can we say about the Eulerian integral time? Quantifying this connection is important for using Lagrangian measurements in (Eulerian) model parameterizations, but the subject has received only sporadic attention. Some recent results are nevertheless encouraging.

Corrsin [70] suggested a way to connect Eulerian and Lagrangian statistics. His idea, commonly referred to as "Corrsin's conjecture" is as follows. In the Eulerian frame, velocity correlations decay in both space and time, as illustrated graphically in Fig. 12. So the velocities at a single location will become decorrelated after a period of time (the Eulerian integral time). At the same time, two observers separated by more than a certain distance (the integral scale) will see uncorrelated velocities. A Lagrangian observer, by drifting, experiences both the temporal and spatial decorrelation simultaneously. So the integral time measured by the Lagrangian observer will generally be *shorter* than that measured by a fixed observer.

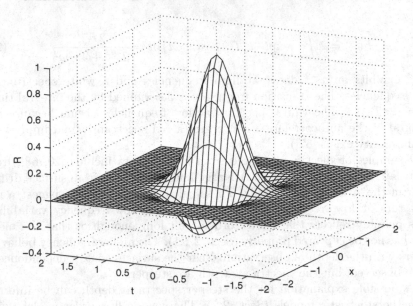

Fig. 12. A hypothetical Eulerian velocity autocorrelation in space and time. A Lagrangian observer drifting from its starting location experiences both the spatial and temporal decay in the field

Corrsin's conjecture states that the Lagrangian autocorrelation can be derived from the Eulerian spatial–temporal autocorrelation, if one knows the PDF of particle displacements:

$$R_{\mathrm{L}}(t) = \int \int R_{\mathrm{E}11}(\mathbf{r}, t)\, Q(\mathbf{r}, t)\, d\mathbf{r} \,, \tag{25}$$

where $Q(\mathbf{r}, \mathbf{t})$ is the displacement PDF and $R_{\mathrm{E}11}$ is the longitudinal Eulerian correlation (that related to the velocities parallel to the line connecting the two observation points). This makes sense because the integral over the displacement PDF reflects how far the particles wander from their starting positions and thus how much the spatial decorrelation affects the Lagrangian result.

Davis examined Corrsin's conjecture in the oceanic context in [26, 27] in developing his framework tracer transport. Middleton [71] modified the conjecture for application to geophysical data by assuming certain forms for the Eulerian energy spectrum. Both authors obtained analytical results by assuming that the displacement PDF was stationary and Gaussian[14]:

$$Q = \exp(-k^2 r^2/2) \,. \tag{26}$$

---

[14] As we have seen, the velocity PDF is weakly non-Gaussian, implying that the displacement PDF is similarly non-Gaussian. But the deviations from Gaussianity are not large.

Then it is possible to connect the Eulerian and Lagrangian integral times. The result depends on the ratio, denoted α, of the Eulerian integral time to the advective time, $T_a \equiv L/\nu$ (where L is a typical length scale and ν the RMS velocity). Middleton showed that

$$T_L/T_E \approx \frac{q}{(q^2 + \alpha^2)^{1/2}}, \quad q = (\pi/8)^{1/2} . \tag{27}$$

If $\alpha \ll 1$, the time scales are approximately the same. This would occur if the Eulerian field decorrelated much faster in time than in space. If $\alpha \gg 1$, the particle moves rapidly from its starting position and the spatial decorrelation dominates. Interestingly, Middleton found that relation (27) was relatively insensitive to the shape of the Eulerian spectrum.

An assessment of the applicability of (27) to in situ data has not yet been made (and would be difficult, given the need for extensive and concurrent Lagrangian and Eulerian data). But Lumpkin et al. [43] tested the relation using data from a numerical ocean model. They found (27) applies remarkably well, over a range of locations and of values of α (Fig. 13). Further confirmation with in situ data is desirable, but Lumpkin et al.'s results are very promising.

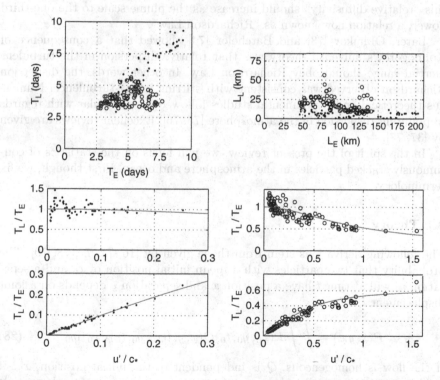

Fig. 13. The ratio of Eulerian and Lagrangian timescales from particles seeded in a numerical model of the North Atlantic. The solid curves are the prediction from Middleton [71]. From [43], with permission

3 Multiple Particles

As discussed in Sect. 2, the absolute dispersion, under fairly general conditions, exhibits a quadratic time dependence initially and a linear time dependence at late times.[15] One may also observe anomalous dispersion, between these two limits.

Relative dispersion, the mean square distance between pairs of particles, has the same early and late asymptotic behavior as absolute dispersion. It can also exhibit a range of different non-diffusive behavior at intermediate times. Which type of growth occurs depends on the character of the Eulerian flow, so relative dispersion is a more sensitive indicator of the Eulerian flow than absolute dispersion. In certain cases, relative dispersion can even be used to deduce the shape of the Eulerian energy spectrum. So relative dispersion has long been of interest to the turbulence community.

The seminal early work on relative dispersion was that of Richardson [72] who studied smoke plumes spreading from factory stacks. Richardson realized the rate of cloud dispersion *increased* with the size of the cloud, implying an effective diffusivity which was scale dependent. From observations, he deduced this "relative diffusivity" should increase as the plume scale to the one-third power, a relation now known as "Richardson Law".

Later, Obhukov [73] and Batchelor [74] showed that a consequence of Kolmogorov's universal theory was that relative dispersion in the turbulent inertial range should obey Richardson Law. In other words, the dispersion Richardson observed was consistent with stirring in 3D turbulence. Numerous theoretical and experimental studies followed, in particular with regards to pollutant dispersion in the atmosphere [75, 76]. Excellent reviews are given by [37, 77].

In the spirit of the present review, we will focus on the statistics of continuously tracked particles in the atmosphere and ocean. First though, we fix terminology.

3.1 Theory

The following derivations stem from those given by [16, 74, 77, 78, 79]. The probability that two particles with a mean initial position of x_0 and a separation y_0 will at time t have a position x and separation y depends on a joint displacement PDF:

$$P(x, y, t) = \int \int P(x_0, y_0, t_0) Q(x, y, t | x_0, y_0, t_0) \, dx_0 \, dy_0. \qquad (28)$$

If the flow is homogeneous, Q is independent of the initial position, x_0. If we integrate Q over all pair separations, y_0, we obtain the single-particle

---

[15] The latter fails when there is a long-time correlation in the velocity field, or if the spread of particles is restricted, as in an enclosed basin, [22].

displacement probability (Sect. 2). If we integrate instead over all initial positions, we obtain a PDF of pair separations:

$$q(y, t|y_0, t_0) = \int Q(x, y, t|x_0, y_0, t_0) \, dx_0 , \qquad (29)$$

which Richardson [72] called the "distance-neighbor function". Note if Q is independent of the initial position, then q and Q are equivalent. With the separation PDF, we can evaluate the probability of observing a given separation:

$$p(y, t) = \int p(y_0, t_0) \, q(y, t|y_0, t_0) \, dy_0 . \qquad (30)$$

We can use this to define moments, as we did for single particles. For example, the relative dispersion is

$$\overline{y^2}(t) = \int y^2 \, p(y, t) \, dy . \qquad (31)$$

Just as we defined the absolute diffusivity for single particles, we can define a relative diffusivity for pairs:

$$K \equiv \frac{1}{2}\frac{d}{dt}\overline{y^2} = \overline{yv} = \overline{y_0 v} + \int_{t_0}^{t} \overline{v(t)\, v(t')} \, dt' , \qquad (32)$$

where v is the pair separation velocity and the overbar again indicates an ensemble average. The relative diffusivity thus derives from the two particle velocity cross-correlation. However, there is an additional term which represents the correlation between the pairs' initial positions and their separation velocities. If the floats are deployed randomly, this term in principle should be zero. In practice, it tends to be small, but it is only vanishingly so when one has a large number of pairs [79]. The integral of the cross-correlation dominates only after this correlation, the "memory" of the initial state, has been lost.

We can rewrite the integral of the relative velocity correlation thus:

$$\int_0^t \overline{v(t)\, v(t')} \, dt' = \int_0^t \overline{(u_i(t) - u_j(t))\,(u_i(t') - u_j(t'))} \, dt' = 2\int_0^t \overline{u_i(t)\, u_i(t')} \, dt' -$$

$$2\int_0^t \overline{u_i(t)\, u_j(t')} \, dt' = 2\kappa(t) - 2\int_0^t \overline{u_i(t)\, u_j(t')} \, dt', \qquad (33)$$

where κ is the absolute diffusivity, defined in (10). So the relative diffusivity is less than twice the absolute diffusivity, so long as the particle velocities are correlated. As the particles drift further and further apart, the relative diffusivity asymptotes to twice the absolute diffusivity.

It is possible to make deductions about the relative dispersion under certain conditions. For this it is useful to use the square of the separation velocities (the "second-order Lagrangian structure function"):

$$\overline{v^2}(t) = 2\nu^2 - 2\overline{u_i u_j} \,, \tag{34}$$

where ν is the RMS single-particle velocity. If the two particles are initially very close, the velocity difference is approximately constant (as in a Taylor expansion), and the separation distance (as well as the relative diffusivity) grows linearly in time. When the particle separations are large enough, usually at scales greater than the size of the dominant eddies, the individual velocities are uncorrelated and the structure function is just twice the RMS single-particle velocity. So relative dispersion is like absolute dispersion at small and large scales.

At intermediate scales, the pair velocities are correlated, and in a way which depends on the flow. Consider a stationary, homogeneous, 2D flow. One can write [78]:

$$\overline{v^2} = \overline{(u(x+y,t) - u(x,t))^2}$$

$$= 2 \int_0^\infty E(k)[1 - J_0(ky)]\, \mathrm{d}k \,. \tag{35}$$

The Lagrangian and Eulerian velocity differences are only equivalent if the flow is homogeneous [27].[16] At the larger (intermediate) scales, we have

$$1 - J_0(ky) \approx \frac{1}{4}k^2 y^2, \quad ky \ll 1 \,, \tag{36}$$

and at the smaller (intermediate) scales,

$$1 - J_0(ky) \approx 1 + O(ky)^{-1/2}, \quad ky \gg 1 \,. \tag{37}$$

If we assume the Eulerian spectrum has a power law dependence, $E(k) \propto k^{-\alpha}$, as we did in Sect. 2.7, then we could write

$$\overline{v^2} \approx 2 \int_0^{1/y} k^{-\alpha} \left(\frac{1}{4}k^2 y^2 \right) \mathrm{d}k + 2 \int_{1/y}^\infty k^{-\alpha}\, \mathrm{d}k, \tag{38}$$

or

$$\overline{v^2} = \frac{1}{2}y^2 \frac{1}{3-\alpha} k^{3-\alpha}\big|_0^{1/y} + \frac{2}{1-\alpha} k^{1-\alpha}\big|_{1/y}^\infty \,. \tag{39}$$

The first term diverges if $\alpha \geq 3$ (steep spectra), while the second diverges if $\alpha \leq 1$. Consider the intermediate case, where $1 < \alpha < 3$; then

$$\overline{v^2} \propto y^{\alpha-1} \,. \tag{40}$$

The corresponding diffusivity can be shown to be

---

[16] If the flow isn't homogeneous, the energy spectrum isn't a useful concept anyway.

$$K = \frac{1}{2}\frac{d}{dt}\overline{y^2} \propto y^{(\alpha+1)/2} . \tag{41}$$

This is termed "local dispersion" because the motion of pairs with a separation L is dominated by eddies of the same scale. Richardson's Law occurs here, as the spectrum has the inertial range scaling of $\alpha = \frac{5}{3}$. So we have

$$K \propto y^{4/3} . \tag{42}$$

For steep spectra ($\alpha \geq 3$), the relative dispersion is "non-local" because the stirring is dominated by the largest eddies. Then

$$\overline{v^2} \approx \frac{1}{2}y^2 \int k^2 E(k)\,dk = c_1 \Omega y^2 , \tag{43}$$

where c_1 is a constant and Ω is the total enstrophy (the integrated square vorticity). The corresponding diffusivity is

$$K = \frac{1}{2}\frac{d}{dt}\overline{y^2} = c_2 T^{-1}\overline{y^2} . \tag{44}$$

The time scale, T, is proportional to the mean rate of strain. Relation (44) implies an *exponential growth* of pair separations. This was evidently first deduced by Batchelor [80], who considered pair dispersion in the 3D turbulent dissipation range.[17] Note that exponential growth occurs with all spectra *steeper* than $\alpha = 3$. So observing exponential growth does not imply a single spectrum [78, 79].

As with single-particle statistics, the displacement PDF plays a central role with multiple particle measures. Richardson [72] considered the PDF of pair separations, as noted, and proposed it should obey a Fokker–Planck equation. Thus if one knows the PDF initially, one can predict its subsequent evolution [84]. Bennett [78] showed the kurtosis of pair separations is *constant* for local dispersion, with a value which depends on the spectral slope, α. He also showed that ku(y) grows *exponentially* under non-local dynamics. So exponential stretching is accompanied by increasingly non-Gaussian PDFs. The PDF in the *lower right panel* of Fig. 3 is an example of a strongly non-Gaussian distribution occurring due to turbulent stirring.

3.2 2D Turbulence

As a specific example, consider the case of isotropic, homogeneous turbulence in two dimensions. Two-dimensional turbulence is complimentary to 3D turbulence and has been a useful test bed for Eulerian and Lagrangian theory [85, 86, 87]. Unlike with 3D turbulence, where energy is transported via nonlinear interactions from the large scales to the small, dissipative scales, energy in 2D turbulence moves from small to large scales. This "inverse cascade"

---

[17] Exponential growth of pair separations is also obtained in strongly shearing regions near hyperbolic points [81, 82, 83].

thereby shifts energy away from the dissipative scales, requiring a large scale dissipation mechanism (such as bottom drag). At the same time, enstrophy, the squared vorticity, is transferred downscale. So if energy is injected at a single scale (for example at the internal deformation radius due to baroclinic instability [88]), there will be two different inertial ranges (Fig. 14).

In the energy cascade range, the spectrum has the same slope as in the 3D inertial range, $\alpha = 5/3$, so the Richardson-type scalings apply:

$$\overline{y^2} \propto \epsilon t^3, \quad K \propto \epsilon^{1/3} y^{4/3}, \quad ku(y) = \text{const.}, \tag{45}$$

where ϵ (with units of L^2/T^3) is the energy dissipation rate, assumed constant. The enstrophy cascade range is steeper, with $\alpha = 3$ [89], so the dispersion is weakly non-local and pair separations grow exponentially [90]:

$$\overline{y^2} \propto \exp(c_3 \eta^{1/2} t), \quad K \propto y^2, \quad ku(y) \propto \exp(c_4 \eta^{1/2} t), \tag{46}$$

where η (with units of $1/T^2$) is the enstrophy dissipation rate, also assumed constant. Thus, a pair of particles with an initial separation smaller than the injection scale would experience exponential growth until the separation reached the injection scale, after which the square separation would grow cubically in time, up to the scale of the largest eddies.

3.2.1 FSLEs

When computing relative dispersion, one averages squared separations between available pairs at fixed times. Thus one averages pairs with different

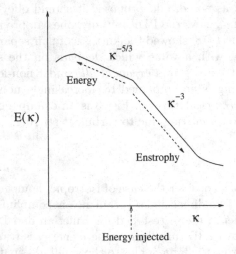

Fig. 14. The energy spectrum for 2D turbulence driven by a source at an intermediate scale. The enstrophy (squared vorticity) cascades to small scales where it is dissipated and the energy to large scales. The enstrophy range has a κ^{-3} dependence, and the energy range a $\kappa^{-5/3}$ dependence

separations. Under local dispersion (as in the Richardson regime), such averaging could possibly blur the temporal dependence.[18] An alternate approach is to change the dependent variable and average *times* at fixed distances. This is the idea behind the "Finite Scale Lyapunov Exponent". In this, one selects a set of distances, increasing multiplicatively, i.e., [22, 92]

$$d_n = rd_{n-1} = r^n d_0 \, .$$

Then one records the time required for individual pairs to grow from one distance to the next. These "exit times" are then averaged. Of particular interest is the mean inverse time scale, which converges to the largest Lyapunov exponent (minimum e-folding time):

$$\lambda_S(n) = \log(\alpha) < \frac{1}{T_n} > \, . \tag{47}$$

With exponential growth, the FSLE is constant, reflecting a constant Lyapunov exponent (or a constant e-folding time). With local dispersion, in which the diffusivity $K \propto y^{(\alpha+1)/2}$, the mean inverse exit time scales as $D^{2/(3-\alpha)}$, from dimensional considerations.

Results from numerical simulations [93] suggest using distance as the independent variable is superior to using time in a Richardson regime. The FSLE also uses *all* available pairs, not just those pairs deployed together, and this can greatly increase the degrees of freedom. However, the danger with the FSLE is that it ignores the dependence on the initial separation, a crucial point with in situ data. Generally, exponential growth will not be observed before the "memory" of the pair's initial velocities is lost [79], so if the pairs are too far apart initially, the FSLE can be degraded [91]. We illustrate this below.

3.2.2 Chaotic Advection

All of the measures discussed thus far ignore flow inhomogeneity. Pairs in regions of intense dispersion are averaged together with those in weakly dispersive regions. "Dynamical systems theory" is concerned with treating such inhomogeneity. One differentiates between "elliptic" and "hyperbolic" regions, with pair separations growing algebraically in the former and exponentially in the latter [94]. The theory is also concerned with the stable and unstable *manifolds*, the time dependent material curves which evolve from saddle points in the flow [81, 82, 83]. Identifying manifolds requires detailed estimates of the flow and/or high densities of in situ particles. So the dynamical systems ideas have been applied mostly to model data [95]. Two exceptions are Lozier et al. [96] and Kuznetsov et al. [94], who applied the methodology in relation to floats in the Gulf Stream and surface drifters in the Gulf of

---

[18] With non-local dispersion, this is not a problem because all the pairs respond to large-scale stirring [91].

Mexico, respectively. In future, such analyses will become more common (a promising application, for example, uses surface radar to diagnose the Eulerian field [98]). We will not treat this type of analysis, but a lucid account is given by [99].

3.3 Atmosphere

Now we turn to the observations. During the 1970s, two large experiments were undertaken to study dispersion in the southern hemisphere stratosphere. These were the EOLE experiment, with 483 constant level balloons at 200 mb, and the TWERLE experiment, with 393 constant level balloons at 150 mb. In both cases, the balloons were launched in pairs or clusters, specifically to measure the relative dispersion. The relative dispersion in these experiments was described in two seminal papers, by Morel and Larcheveque [100] and by Er-El and Peskin [101].

Pair statistics in general demand larger numbers of realizations for satisfactory convergence than do single-particle statistics. So Morel and Larcheveque increased their numbers of pairs by using balloons which happened to drift near one another at some time after deployment. This procedure is potentially problematic because the separations of these *chance pairs* are more likely to be correlated with their separation velocities, and this can potentially affect statistical convergence (Sect. 3.1). But Morel and Larcheveque found that the statistics from chance pairs were nevertheless identical to those from deployed pairs.

The relative dispersion for the EOLE pairs is plotted in Fig. 15. The growth is exponential over roughly the first 6 days, with an e-folding time scale of 1.35 days, up to scales of about 1000 km. The growth at larger scales is consistent with a linear increase. If the atmosphere were a 2D turbulent fluid, we would infer an enstrophy cascade below an energy-containing eddy scale of 1000 km and diffusive spreading at larger scales. However, as stated, exponential growth does not necessarily imply an enstrophy cascade; any spectrum steeper than κ^{-3} will also cause exponential growth.

Er-El and Peskin [101] obtained similar statistics with the TWERLE balloons. The relative dispersion (Fig. 16) exhibited exponential growth below 1000 km, during the first week after deployment (or initialization for the chance pairs, which they also used). Er-El and Peskin suggested that the growth at large scales was possibly consistent with a D^3 dependence (Fig. 17). If so, this could reflect a Richardson regime at large scales and perhaps an inverse energy cascade (the scales are too large for 3D turbulence). However, the results were fairly noisy and other dependences could not be ruled out.

There were other points of interest. Morel and Larcheveque gauged isotropy by plotting the ratio of the RMS zonal pair separations to the RMS meridional separations (Fig. 18). Below 1000 km, in the exponential growth range, the pairs spread out equally in both directions. But the spreading was preferentially zonal at larger scales, perhaps due to the mean circulation (which

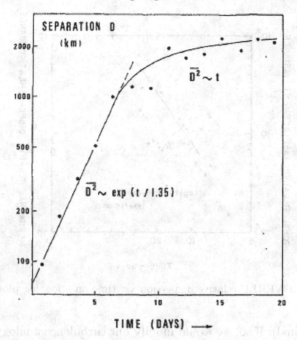

Fig. 15. Dispersion vs. time for the EOLE balloon pairs. From [100]

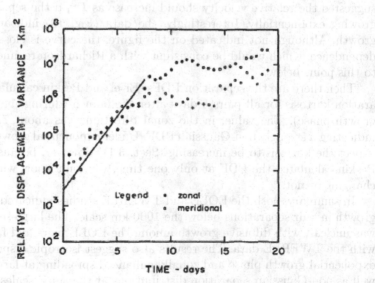

Fig. 16. Relative dispersion vs. time for the TWERLE balloons. From [101]

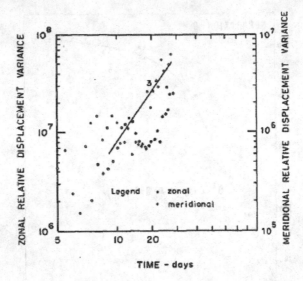

Fig. 17. The TWERLE relative dispersion vs. time on a log–log plot. From [101]

is primarily zonal). If so, we would modify our turbulence analogy to include, in addition to·1000 km eddies, a large scale zonal flow.

Morel and Larcheveque also examined the dependence of the mean square relative velocity (Sect. 3.1) on pair separation. From scaling arguments, they suggested the relative velocity should increase as D^2, if the separations were growing exponentially. Interestingly, the data (Fig. 19) indicated a slower growth. Although not indicated on the figure, the curve is closer to a $D^{4/3}$ dependence, which would be consistent with a Richardson regime. We return to this point below.

Then there are the separation PDFs. Er-el and Peskin calculated the separation kurtosis for all pairs 5 days after the launch (during the exponential growth phase). The value, in the zonal direction, was about 7.5 (Fig. 20), indicating a strongly non-Gaussian PDF. Under exponential growth, we would expect the kurtosis to be increasing (Sect. 3.1). However, because Er-El and Peskin calculated the PDF at only one time, we don't know whether it was changing or not.

In summary, both the EOLE and TWERLE studies indicated exponential growth in pair separations below the 1000 km scale. The large-scale behavior was unclear, with diffusive growth among the EOLE pairs and faster growth with the TWERLE data. The results also suggest isotropic dispersion in the exponential growth phase and zonally enhanced spreading at larger scales, as well as non-Gaussian separation distributions at the small scales.

Interestingly, the balloon results are not quite consistent with independent Eulerian analyses. Gage and Nastrom [29] calculated velocity spectra from a large set of aircraft measurements. They found a κ^{-3} inertial range at scales of 500–2000 km and a $\kappa^{-5/3}$ range, at *smaller* scales (see also [102]). So this

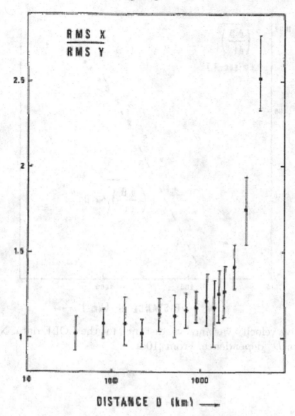

Fig. 18. The ratio of zonal to meridional relative velocity variance as a function of distance for the EOLE data. The small scales are approximately isotropic and the large scales zonally anisotropic. From [100]

is opposite to the situation sketched in Fig. 14. Later work suggested that the large-scale range behaves as an enstrophy cascade range, but that the smaller range has a flux of energy toward smaller rather than larger scales [103].

Recently, Lacorata et al. [104] reexamined the EOLE data set, using the FSLE measure described earlier. As noted, the FSLE is complimentary to relative dispersion and potentially superior under local dispersion. Their results (Fig. 21) indicate a power law rather than an exponential growth at small scales. The slope here is $D^{-2/3}$, consistent with a Richardson scaling. Second, there is a transition just below 1000 km to a different regime where the dependence is D^{-2}, consistent with diffusion. Third, the FSLE based on the total dispersion falls off more slowly than that based on meridional displacement, indicating zonal anisotropy (Fig. 22).

The FSLE is thus consistent with Morel and Larcheveque's relative dispersion at large scales but is inconsistent with exponential growth at small scales. Lacorata et al. suggested that there was a brief period of exponential

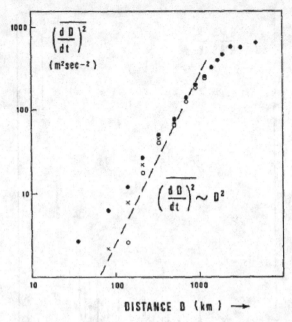

Fig. 19. Relative velocity variance vs. distance for the EOLE data. Note, the data do not support a D^2 dependence. From [100]

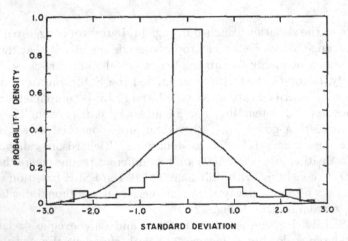

Fig. 20. The PDF of relative zonal displacements 5 days after deployment from the TWERLE balloons. From [101]

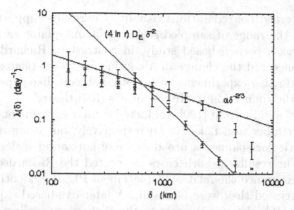

Fig. 21. The FSLE from the EOLE data. The two curves represent the full displacements and the meridional ones (the latter decay faster). From [104]

growth prior to the Richardson regime, with a very rapid e-folding time (0.4 days). But this applied only to scales less than 100 km.

3.4 Ocean

The origins of relative dispersion experiments in the ocean are colorful. Intrigued by Richardson's work, Henry Stommel visited the scientist in England, and the two subsequently conducted a pair dispersion experiment at the surface of Loch Long in Scotland. For this they used pairs of parsnip pieces[19] and

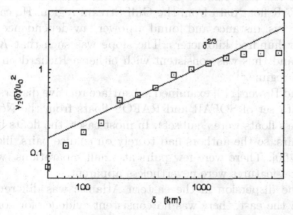

Fig. 22. Relative velocities as function of distance for the EOLE balloons. From [104]

---

[19] As noted by the authors, parsnips are easily visible and float just below the surface, reducing wind drag. In a further note, they lamented the necessity of observing from a pier because of interference from the support posts. "A suspension bridge would have been an ideal platform", they suggested.

monitored the growth of separations visually. The results supported Richardson's law over the range of sampled scales [105]. As quaint as it sounds, it nevertheless was a *particle*-based study, in contrast to Richardson's earlier work which concerned the change in a continuous tracer (smoke). Stommel [106] describes further experiments (using other objects, like paper cards) and also discusses the connection to Kolmogorov's [107] theory.

Okubo [108] and Sullivan [109] conducted similar experiments, at the surface of the North Sea and Lake Huron respectively, but using dye. Okubo's surveys in particular spanned a broad range of horizontal scales (from 10 m to 100 km). The results in both cases supported the Richardson scaling.[20] Sullivan [109] examined the relative displacement PDFs, evidently the first to do so, and suggested they were Gaussian. A later dye-based experiment, by Anikiev et al. [110], also lent support to the Richardson scaling at the ocean surface.

The first to analyze pairs of continuously tracked surface drifters was Kirwan et al. [111], in the North Pacific. The primary result of their analysis was an apparent transition from relative to absolute dispersion at the 50–100 km scale. However, they did not examine the dependence of the diffusivity on distance at smaller scales. Davis [112] did so, using drifter data from the California Current region. However, he found no consistent distance dependence and concluded that the dispersion had different characteristics in different regions. Davis also calculated separation PDFs. He found these were non-Gaussian soon after deployment and then became Gaussian (at larger separations). This suggests a shift from correlated to uncorrelated pair velocities.

The first to calculate relative dispersion with subsurface floats was Price [41] using SOFAR float data from the Gulf Stream region. He calculated relative diffusivity vs. distance and found a power law dependence on scales of less than a few hundred kilometers. The slope was such that $K \propto y^n$ with $4/3 \leq n \leq 2$, and thus was consistent with either a Richardson or an exponential growth regime.[21]

LaCasce and Bower [113] examined subsurface relative dispersion by using a historical data set of SOFAR and RAFOS floats from the North Atlantic (of which Price's floats were a subset). In most cases, the floats had not been deployed in pairs, so the authors had to rely on chance pairs, like Morel and Larcheveque [100]. There were few pairs at small separations (with $y_0 < 10$ km), but several features were nevertheless apparent.

For one, the dispersion in the eastern Atlantic was different than that in the west. In the east, there was no consistent evidence for correlated pair velocities at any of the sampled scales. Evidently, the energy-containing eddies

---

[20] Because Okubo's results pertain to such large scales, it is unlikely that 3D turbulence was responsible for the observed dependence, as noted by Bennett [77].

[21] The number of float pairs was fairly small and Price was uncomfortable with asserting a particular dependence (Price, personal communication).

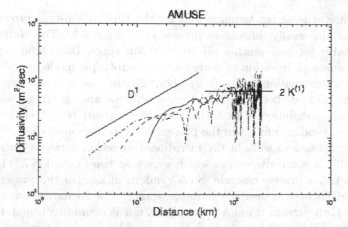

Fig. 23. The relative diffusivity vs. distance for a float experiment in the eastern North Atlantic. The D^1 dependence is consistent with the relative dispersion growing quadratically in time. The diffusivity asymptotes to twice the absolute diffusivity at scales larger than about 50 km. From [113]

were comparable to the smallest resolved scales, so that the relative dispersion was like absolute dispersion, with a quadratic growth initially and diffusive growth later on (Fig. 23). The transition to diffusion occurred at about 50 km (as with Kirwan et al.'s Pacific drifters).

In contrast, the pair velocities in the western Atlantic were clearly correlated at scales smaller than about 200 km. Furthermore, the relative dispersion indicated a Richardson-type dependence (Fig. 24). The behavior at scales larger than 200 km was consistent with diffusion.

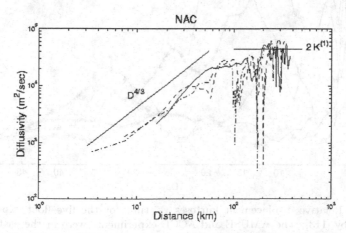

Fig. 24. The relative diffusivity vs. distance for a float experiment in the western North Atlantic. The $D^{4/3}$ dependence is consistent with the Richardson law and proceeds up to about 200 km. From [113]

There are at least two explanations for the Richardson-type growth here. One is that this really reflects an inverse energy cascade. The Gulf Stream is an unstable jet and pinches off 100–200 km rings. Baroclinic instability typically causes an injection of energy to the barotropic mode at the scale of the internal deformation radius (here about 20 km). So it is possible there is a barotropic cascade between the deformation scale and the ring scale.

A second possibility is that the statistics instead reflect shear dispersion, that is, random mixing in the presence of a background shear. Particles undergoing a random walk in the meridional direction across a linear zonal shear exhibit a zonal dispersion which grows as time cubed [77, 114, 115], exactly as in an inverse cascade. So a random mixing in the presence of a background shear can produce the appearance of a Richardson regime. Of course the Gulf Stream is not a linear shear, nor is it unidirectional. But it is sheared nevertheless and might thus be responsible for the perceived growth.

LaCasce and Bower also examined the displacement PDFs and tracked how they changed in time. Shown in Fig. 25 is the displacement kurtosis plotted against time for the five different data sets examined. The two sets from the

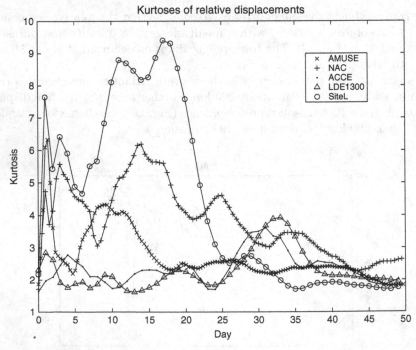

Fig. 25. Relative displacement kurtoses vs. time for the five float experiments examined by [113]. The AMUSE and ACCE experiments were in the eastern and central North Atlantic, while the NAC, LDE, and SiteL experiments were in the west. The latter three exhibit non-Gaussian kurtoses during the first 20 days. From [113]

eastern Atlantic have kurtoses near three for the entire period, consistent with Gaussian distributions; the three western sets in contrast exhibit a rapid growth in kurtosis followed by a period (from 10 to 20 days) in which the kurtosis is elevated and hovering around a fixed value. At late times, the kurtoses decrease back toward three. The elevated kurtoses are consistent with correlated pair velocities; that they are approximately constant during this time is consistent with local dispersion, as noted earlier [78].

More recently, Ollitrault et al. [116] examined pair dispersion using different floats, deployed in the middle North Atlantic. Their results suggest a Richardson regime up to 300 km, in *both* the western and eastern Atlantic. This is consistent with LaCasce and Bower's results in the west, but not in the east; the reason for the difference is unknown. They also found some support for exponential growth below the deformation radius in the eastern basin. Their separation PDFs closely resembled those of Davis, indicating non-Gaussian distributions at 10 km but Gaussian ones at 30 km. The latter would seem to imply uncorrelated pair velocities and so, perhaps, shear dispersion. But the authors argue this is unlikely, because the dispersion is isotropic.

In the aforementioned studies, the dispersion below the deformation radius was unresolved or marginally resolved. But these scales were resolved by LaCasce and Ohlmann [91] using drifters in the Gulf of Mexico. The SCULP program [117] involved over 700 drifters, many of which were deployed near one another; the result was 140 pairs with $r_0 \leq 1$ km. Because the deformation radius is somewhat larger in the Gulf of Mexico (roughly 45 km), this afforded a glimpse into the sub-deformation scale dispersion.

The SCULP relative dispersion is shown in Fig. 26. There is exponential growth up to roughly 50 km, or over the first 10 days of the pair lifetimes, with an e-folding time of roughly 2 days. The dispersion during this time moreover was isotropic and comparable in the sampled regions (i.e. it was approximately homogeneous). The dispersion at late times was consistent with a power law growth, i.e. $D^2 \propto t^n$. The exponent, $n \approx 2.2$, was less than for a Richardson regime, but suggests local dispersion.

If the latter is true, then the temporal averaging involved with relative dispersion may be problematic during the late phase. To check this, the authors calculated the FSLE (Fig. 27). This also clearly indicates two regimes, one where the FSLE is constant with distance, at scales less than 10 km, and a power law regime at larger scales. The flat FSLE is consistent with exponential growth, and the e-folding time of about 3 days is comparable to the 2 day estimate based on relative dispersion. However, the estimates do not agree at the larger scales where the FSLE, with a $D^{2/3}$ dependence, indicates a Richardson regime.

In addition, the FSLE differs from relative dispersion in that it predicts a smaller transition (10 km) scale between the early and late phase. This evidently stems from the FSLE using pairs whose initial separation is nearer 10 km and who therefore have not yet lost the memory of their initial separation

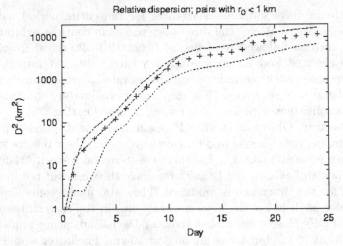

Fig. 26. The relative dispersion vs. time for surface drifters in the SCULP experiment in the Gulf of Mexico. An exponential growth period at early times is clearly seen. From [91]

velocity. By calculating the FSLE using *only* the pairs used for the relative dispersion calculation, they obtained the same transition scale (50 km).

As with the western Atlantic floats, the power law at late times here could be caused by either an energy cascade or by a mean shear. The mean square

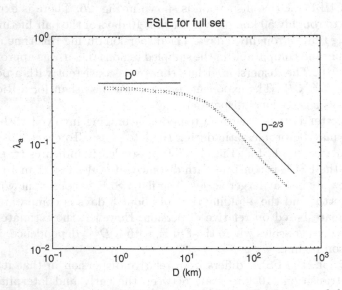

Fig. 27. The FSLE from the SCULP surface drifters. The *flat part* of the curve is consistent with exponential growth in pair separations, and the power law dependence consistent with a cubic temporal dependence. From [91]

velocities suggested the pair velocities were correlated only for the first 20–25 days whereas the power law growth persisted twice as long. As noted before, uncorrelated pair velocities would favor the shear interpretation. The relative dispersion never settled into a diffusive stage and this too would favor the shear interpretation. On the other hand, the displacement kurtoses were elevated out to 40 days. So the interpretation remains unclear.

An important point is that the ocean surface is actually *divergent* because fluid can upwell from or sink into the interior. Because surface drifters remain at the surface, they cannot track this vertical motion and therefore diverge from parcel motion. We examine divergence effects more in the following section.

3.5 Three or More Particles

Using larger groups of particles can shed more light on the character of the mixing. For instance, the *folding* of material lines can be detected with three particles [118]. Geophysical studies with three or more particles are relatively uncommon, but are of interest nevertheless.

As with two particles, we will focus on the intermediate scales, after the particles have lost the memory of their initial states but before they are uncorrelated. We subdivide this range into smaller and larger scales; these might apply to scales from 100 m up to the deformation radius, and from the deformation radius to several hundred kilometers, respectively.

3.5.1 Small Scales

Imagine we have three particles, confined to a 2D surface (e.g. the ocean surface). The area of the triangle formed by the three changes in response to flow divergence:

$$\frac{1}{A}\frac{dA}{dt} = \frac{\partial u}{\partial x} + \frac{\partial v}{\partial y} \tag{48}$$

So one can diagnose the divergence by monitoring the change in the area of a triplet of drifters.

This idea was explored by Molinari and Kirwan [119], with drifters from the western Caribbean. The authors also used triangles to calculate vorticity, stretching, and shearing deformations, by using a clever construction due to Saucier [120]. Saucier's method involves rotating the instantaneous velocity vectors of the constituent drifters. For instance, by replacing,

$$u \to v', \quad v \to u',$$

one obtains for the vorticity:

$$\frac{\partial v}{\partial x} - \frac{\partial u}{\partial y} = \frac{\partial u'}{\partial x} + \frac{\partial v'}{\partial y},$$

which equals

$$\frac{1}{A'}\frac{dA'}{dt}.$$

Here A' is the area enclosed by the cluster with the vertices formed by the rotated velocity vectors. Similar transformations can be used to obtain the shearing and stretching terms.

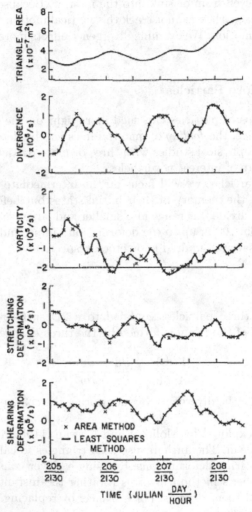

Fig. 28. An example of diagnostics calculated from a triangle of drifters by Molinari and Kirwan. The quantities are area, divergence, vorticity, stretching, and shearing deformations. The *solid lines* represent quantities derived using the least squares method and those derived by the area method by *x*s. With permission

One can also calculate the vorticity and other terms simply by differencing the particle velocities [119, 121]. The differences between individual velocities and that of the center of mass can be used to deduce $\partial u/\partial x$, $\partial v/\partial y$ and so on. The various estimates can then be combined in a least squares sense. The larger the number of particles, the better the results are; Okubo and Ebbesmeyer [121] suggested 6 as a reasonable lower bound, although Molinari and Kirwan found similar results using groups of only 3 and 4.

Molinari and Kirwan applied both methods to triplets of the Caribbean drifters. The two methods produced very similar results (Fig. 28), supporting the assumptions underlying both. The drifters must be fairly close together for this to work (think in terms of a Taylor expansion of the velocity about the center of mass), and indeed the RMS separation in Molinari and Kirwan's triangle was a few kilometers. The estimated quantities were also of reasonable magnitude (of order 10^{-5}s^{-1}). However the quantities tended to oscillate between positive and negative values on periods of days (Fig. 28). One might wonder if the estimates were adversely affected by noise (since drifter velocities are obtained by differencing positions, and the shear involves taking two differences). But Molinari and Kirwan used the vorticity and divergence estimates to check a Lagrangian vorticity balance:

$$\frac{\mathrm{d}}{\mathrm{d}t}(\zeta + f) + (\zeta + f)D = \text{resid.}. \tag{49}$$

Although the residuals were comparable to the two terms on the LHS, there was a clear indication that those two terms were balancing one another (Fig. 29).

Now if the surface flow were actually non-divergent, triangle areas would be conserved, from (48). So exponential growth in one direction would be accompanied by exponential contraction in the perpendicular direction. This was noted by Batchelor [80] in the context of dispersion in the presence of a constant strain, and by Garrett [122] in relation to 2D turbulence. Thus a triangle might see its base grow while its height collapsed (i.e. the triangle would be drawn out into a filament).

The SCULP drifters in the Gulf of Mexico were of sufficient density to yield a small number of "chance" triplets. One such is shown in Fig. 30. The evolution is somewhat difficult to see, but the triangle is drawn out early on. Later it grows, shifting back to a more equilateral shape.

LaCasce and Ohlmann tracked the evolution of about 30 triangles. The mean triangle base (defined as the longest leg) grew approximately exponentially in time, at a rate consistent with the mean pair dispersion (Fig. 31). However, rather than collapsing, the triangle height also grew. This implies the triangle areas were also growing, as with Molinari and Kirwan's triplets (Fig. 28).

The interpretation of this however is not straightforward. Even if the surface were non-divergent, one could not observe the mean height shrink below 1 km, the spatial resolution of the drifter positions. But the triangle height in

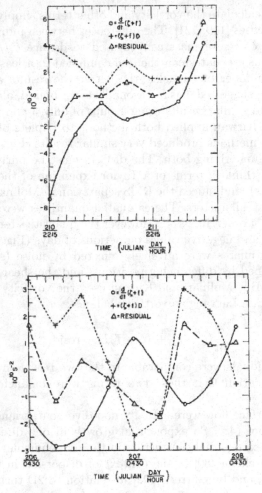

Fig. 29. The terms in (49) evaluated for a triangle of drifters. Note there is tendency for the tendency and divergence terms to balance, despite that the residuals are of comparable size. From [119], with permission

Fig. 31 is clearly growing. Divergence could be responsible for this, however the areas seem to be increasing monotonically, suggesting only positive divergences. Convergences would cause the areas to decrease.

There is however another possible explanation, that the surface flow is effectively non-divergent but that the drifters are suffering random displacements, due to wind forcing. Such perturbations could induce a random walk, in addition to the drifters normal motion. This would mean that in addition to the exponential stretching, the triangle legs would also grow diffusively. La-Casce and Ohlmann's results were not sufficiently well constrained to test this

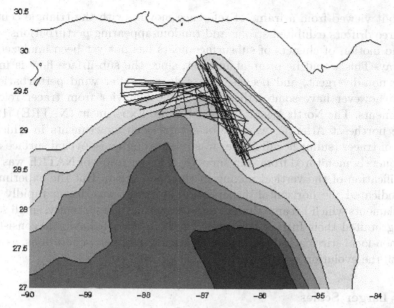

Fig. 30. A triplet of drifters in the SCULP experiment. The group is initially stretched out, then expands and takes on a more equilateral shape. From [91]

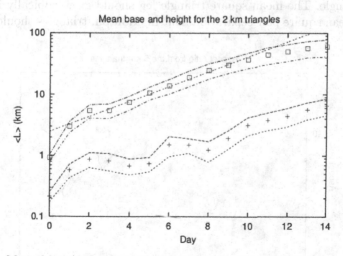

Fig. 31. Mean base (defined as the longest leg) and height of 32 triangles from the SCULP experiment. The base is growing exponentially, but the height is also increasing. From [91]

idea, but viewed from a frame of reference moving with the triangle center, the three drifters exhibited strong and random-appearing perturbations.

The motion of clusters of subsurface floats has not yet been analyzed in this way. This would be of great interest, since the subsurface flow is more nearly non-divergent, and because floats do not suffer wind perturbations. We do however have some indications of the behavior from tracer release experiments. The North Atlantic Tracer Release Experiment (NATRE) [123], in the northeast Atlantic, was one of several such experiments in which a patch of tracer (sulfur hexafluoride) was released on an isopycnal surface and subsequently monitored from ship surveys. A major result of NATRE was the quantification of the vertical mixing in the open ocean. But the experiment also indicated the horizontal dispersion. The tracer was drawn rapidly out into filaments which became thinner and thinner until, apparently, small scale mixing limited their further collapse [122, 124]. Such behavior is consistent with non-local stirring and the exponential growth of pair separations, as well as with the evolution envisioned by Batchelor [80].

3.5.2 Larger Scales

The SCULP pair results suggest a transition from non-local stirring below the deformation radius to local stirring at larger scales. What would happen to particle clusters in such a case? Cluster behavior under local dispersion has been studied recently, in the context of the Richardson regime [125, 126]. In this case, the triangle area is not conserved because fluid is mixed in and out of the triangle. The mean square triangle leg should grow cubically in time, like the mean square pair separation. But in addition, triangles should evolve

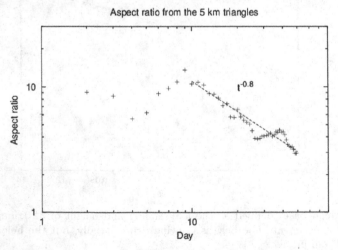

Fig. 32. The mean aspect ratio (defined as the base, the longest leg, divided by the height) for the 32 triplets of drifters in the SCULP experiment. From [91]

toward a more equilateral shape. The SCULP triangles are initially stretched out, due to the non-local dispersion. But if there is a Richardson cascade at larger scales, they should then shift back toward an equilateral shape.

The growth of the RMS leg among the SCULP triangles exhibited a similar power law growth to the pairs, with an exponent of $n \approx 2.2$. But more interestingly, the mean triangle aspect ratio (defined as the base divided by the height) systematically decreased during this period (Fig. 32). So the triangles were indeed shifting toward a more equilateral shape. The aspect ratio also exhibited an approximate power law dependence, with an exponent of roughly -1.

As noted earlier, some aspects of the late time relative dispersion resembled those in a Richardson regime while others pointed to shear dispersion. The change in aspect ratio is probably more consistent with a turbulent cascade. While diffusive mixing could also increase the aspect ratio of a strained-out triangle, it would likely do so less rapidly than observed. This too demands further study.

4 Summary and Conclusions

We have examined the statistics of single and multiple particles in the ocean. Single-particle studies dominate the literature, and among these, the favored analysis technique is that of Davis [19]. In this, Lagrangian velocities are binned geographically and averaged to produce regional estimates of the Eulerian mean velocity and the Eulerian lateral diffusivities. The method has been subsequently refined and used to map currents over large regions of the world ocean. Single-particle studies have also been used to detect Rossby wave propagation, to test the sensitivity of eddies to bottom topography and to determine the importance of eddy heat fluxes in maintaining lateral temperature gradients. Recent work suggests there is hope in relating Lagrangian statistics to Eulerian statistics, which would improve the use of Lagrangian information in numerical models.

Two-particle dispersion is fundamental to understanding actual Lagrangian dispersion (the spreading of a cloud of tracer). Pair dispersion calculations in both the atmosphere and ocean suggest that separations increase exponentially in time at scales less than the internal deformation radius, and that growth is approximately isotropic. Calculations with more than two drifters or floats are fairly rare, but clusters have been used to diagnose vorticity and divergence, and to compare with recent theoretical work on cluster dynamics in turbulent flows.

Numerous fundamental questions do however remain with regards to geophysical dispersion. Several different observations suggest that exponential stretching occurs at sub-deformation scales and if this is so, it would be interesting to discover why. One possibility is that energy injection at or near the deformation radius, due for instance to baroclinic instability, is creating an enstrophy cascade to smaller scales. If exponential growth is indeed common at these scales, it could open the way for sub-gridscale mixing parameterizations in numerical models.

Relative dispersion at scales larger than the deformation radius evidently varies with location. We have some indications of a Richardson regime in both the ocean and atmosphere. However, the dynamical reason for this growth is not settled, and could well be the product of shear dispersion due to the large scale flow. Studies with synthetic floats in ocean models could help elucidate this, in particular to distinguish the role of the lateral shear.

Then there is the question of divergence at the ocean surface. Surface drifter results are suggestive on one hand of non-divergent 2D turbulence and, on the other, of divergence having an order one role in the vorticity balance. It would be useful to sort this out, perhaps again with drifters in a numerical model (as these need not suffer random perturbations due to wind forcing). Cluster calculations using subsurface floats would also be useful, as divergence is much less important at depth.

In addition, there is the effect of low-frequency fluctuations in the Eulerian flow. Floats and drifters exhibit red frequency spectra, due most likely to variability in the general circulation. We usually think of particle dispersion in terms of a stationary mean and eddies, but the actual situation is one with a continuum of scales. What is the best way to represent oceanic advection and diffusion in such cases?

Of course, it would also be valuable to have more relative dispersion experiments. In most existing data sets, drifters and floats were deployed alone, necessitating using "chance pairs" to study relative dispersion. Systematic deployments of pairs and clusters of floats and/or drifters would improve greatly the statistical reliability of the results. Float deployments could also be made in conjunction with tracer releases, so that each data set could compliment the other. Experiments like this would greatly improve our perception of the time and scale dependencies of Lagrangian dispersion, and the utility of Lagrangian data.

Acknowledgments

Thanks to Russ Davis, Sarah Gille, Antonello Provenzale, and Maurizio Brocchini for detailed comments, and to the authors who kindly allowed to reproduce their figures.

References

1. H. M. Stommel and A. B. Arons: Deep-Sea Res. **6**, 140-154 (1960)
2. J. C. Swallow and L. V. Worthington Nature **179**, 1183–1184 (1957)
3. J. C. Swallow: Phil. Trans. Lond., A **270**, 451–460 (1971)

4. R. E. Davis: Ann. Rev. Fluid Mech. **23**, 43–64 (1990)
5. P. Jullian, W. Massman and N. Levanon : Bull. Am. Meteor. Soc. **58**, 936–948 (1977)
6. J. C. Ohlmann, P. F. White, A. L. Sybrandy and P. P. Niiler: GPS-cellular drifter technology for coastal ocean observing systems. J. Atmos. Ocean Tech. **22**, 1381–1388 (2005)
7. T. Rossby and D. Webb: Deep-Sea Res. **17**, 359–65 (1970)
8. T., Rossby D. Dorson and J. Fontaine: J. Atmos. Ocean Tech. **3**, 672–79 (1986)
9. R. E. Cheney and Richardson, P. L.: Deep-Sea Res. **23**, 143–155 (1976)
10. P. L. Richardson, D. Walsh, L. Armi, M. Schroder, J. F. Price: J. Phys. Oceanogr. **19**, 371–383 (1989)
11. J. F. Price and T. Rossby: J. Mar. Res. **40**, 543–558 (1982)
12. P. L. Richardson: J. Geophys. Res. **88** (C7), 4355–4367 (1983)
13. P. M. Poulain, A. Warn-Varnas and P. P. Niiler: J. Geophys. Res. **101**, 18237–18258 (1996)
14. R. E. Davis: J. Geophys. Res. **103**, 24619–24639 (1998)
15. G. K. Batchelor: The Theory of Homogeneous Turbulence. Cambridge University Press, Cambridge (1953)
16. G. K. Batchelor: Proc. Cambridge Philos. Soc. **48**, 345–362 (1952a)
17. G. I. Taylor: Proc. Lond. Math. Soc. **20**, 196–212 (1921)
18. G. K. Batchelor and A. A. Townsend: Surv mech. **23**, 352–398 (1953)
19. R. E. Davis: Deep-Sea Res. **38** Suppl. S531–S571 (1991)
20. H. J. Freeland, P. B. Rhines and T. Rossby : J. Mar. Res. **33**, 383–404 (1975)
21. A. Einstein: *Ann. Physik* **17**, 549–560 (1905)
22. V. Artale, G. Boffetta, A. Celani, M. Cencini and A. Vulpiani: Phys. Fluids **9**, 3162–3171 (1997)
23. W. R. Young: Lectures on stirring and mixing. Proceedings of the Woods Hole Summer Program in Geophysical Fluid Dynamics (1999, available at www-pord.ucsd.edu/ wryoung/)
24. T. H. Solomon, E. R. Weeks and H. L. Swinney: Phys. Rev. Lett. **71**(24), 3975–3978 (1993)
25. G. I. Taylor: Proc. R. Soc. A **64**, 476–490 (1938)
26. R. E. Davis: J. Fluid Mech. **114**, 1–26 (1982)
27. R. E. Davis: J. Mar. Res. **41**, 163–194 (1983)
28. R. E. Davis: J. Mar. Res. **45**, 635–666 (1987)
29. K. S. Gage and G. D. Nastrom: J. Atmos. Sci. **43** 729–740 (1986)
30. X. Zang and C. Wunsch: J. Phys. Oceanogr. **31**, 3073–3095 (2001)
31. P. Morel and M. Desbois: J. Atmos. Sci. **31**, 394–408 (1974)
32. D. M. Fratantoni: J. Geophys. Res. **106**, 22067–22093 (2001)
33. P. K. Jakobsen, M. H. Rikergaard, D. Quadfasel, T. Schmith and C. W. Hughes: J. Geophys. Res. **108**, doi:10.1029/2002JC001554 (2003)
34. C. Mauritzen: Deep-Sea Res. **43**, 769–806 (1996)
35. S. Gille: J. Phys. Oceanogr. **33**, 1167–1181 (2003)
36. S. Bauer, M. S. Swenson, A. Griffa, A. J. Mariano and K. Owens: J. Geophys. Res. **103** 30,855–30,871 (2002)
37. B. L. Sawford: Ann. Rev. Fluid Mech. **33**, 289–317 (2001)
38. V. Zhurbas and I. S. Oh: J. Geophys. Res. **109**, C05015 (2004)
39. W. Krauss and C. W. Böning: J. Mar. Res. **45**, 259–291 (1987)
40. K. H. Brink, R. C. Beardsley, P. P. Niiler, M. Abbott, A. Huyer, S. Ramp, T. Stanton and D. Stuart: J. Geophys. Res. **96** (C8), 14693–14706 (1991)

41. J. C. McWilliams, E. D. Brown, H. L. Bryden, C. C. Ebbesmeyer, B. A. Elliot, R. H. Heinmiller, B. Lien Hua, K. D. Leaman, E. J. Lindstrom, J. R. Luyten, S. E. McDowell, W. Breckner Owens, H. Perkins, J. F. Price, L. Regier, S. C. Riser, H. T. Rossby, T. B. Sanford, C. Y. Shen, B. A. Taft and J. C. van Leer: The local dynamics of eddies in the western North Atlantic. In Eddies in marine science, Springer, Berlin. 609 pp (1983)

42. P. M. Poulain, and P. P. Niiler: J. Phys. Oceanogr. **19**, 1588–1603 (1989)

43. R. Lumpkin, A.-M. Treguier and K. Speer: J. Phys. Oceanogr. **32**, 2425–2440 (2002)

44. A. Griffa, K. Owens, L. Piterbarg and B. Rozovskii : J. Mar. Res. **53**, 371–401(1995)

45. P. S. Berloff and J. C. McWilliams: J. Phys. Oceanogr. **32**, 797–830 (2002)

46. B. L. Sawford: Phys. Fluids A **3**, 1577–1586 (1991)

47. P. Falco, A. Griffa, P. Poulain and E. Zambianchi: J. Phys. Oceanogr. **30**, 2055–2071 (2000)

48. J. D. Wilson and B. L. Sawford, B. L.: Bound. Layer Meteor. **78**, 191–210 (1996)

49. M. S. Borgas, T. K. Fleisch and B. L. Sawford: J. Fluid Mech. **332**, 141–156 (1997)

50. B. L. Sawford: Bound. Layer Met. **93**, 411–424 (1999)

51. M. Veneziani, A. Griffa, A. M. Reynolds and A. J. Mariano: J. Phys. Oceanogr. **34**, 1884–1906 (2004)

52. D. J. Thomsom: J. Fluid Mech. **180**, 529–556 (1987)

53. M. S. Swenson and P. P. Niiler: J. Geophys. Res. **101**, 22,631–22,645 (1996)

54. A. Bracco, J. H. LaCasce and A. Provenzale: J. Phys. Oceanogr. **30**, 461–474 (2000)

55. R. B. D'Agostino and M. A. Stephens: Goodness-of-fit Techniques. Marcel-Dekker, New York (1986), 576 pp

56. J. Jimenez: J. Fluid Mech. **313**, 223–240 (1996)

57. J. B. Weiss, A. Provenzale and J. C. McWilliams: Phys. Fluids **10**, 1929–1941 (1998)

58. A. Bracco, J. H. LaCasce, C. Pasquero and A. Provenzale: Phys. Fluids **12**, 2478–2488 (2000)

59. A. Maurizi, A. Griffa, P. M. Poulain and F. Tampieri: J. Geophys. Res. **109** (C4), C0401010.1029/2003JC002119 (2004)

60. H. Tennekes and J. L. Lumley : A first course in turbulence 300 pp. MIT Press (1972), 300 pp.

61. J. H. LaCasce J. Phys. Oceanogr. **35**, 2327–2336 (2005)

62. P. B. Rhines: Geophys. Fluid Dyn. **1**, 273–302 (1970)

63. J. H. LaCasce and K. G. Speer: J. Mar. Res. **57**, 245–274 (1999)

64. J. H. LaCasce: J. Mar. Res. **58**, 61–95 (2000)

65. H. M. Zhang, M. D. Prater and T. Rossby: J. Geophys. Res. **106**, 13,817–13,836 (2001)

66. J., O'Dwyer, R. G. Williams, J. H. LaCasce and K. G. Speer: J. Phys. Oceanogr. **30**, 721–732 (2000)

67. A. Colin de Verdiere:J. Mar. Res. **41**, 375–398 (1983)

68. V. Rupolo, B. L. Hua, A. Provenzale and V. Artale: J. Phys. Oceanogr. **26**, 1591–1607 (1996)

69. R. Lumpkin and P. Flament: J. Mar. Sys. **29**, 141–155 (2001)

70. S. Corrsin: Advances in Geophysics, Vol. 6, Academic Press, New York 161–162 (1959)
71. J. Middleton: J. Mar. Res. **43**, 37–55 (1985)
72. L. F. Richardson: Proc. R. Soc. London, Ser. A **110**, 709–737 (1926)
73. A. M. Obhukov: Izv. Akad. Nauk. SSR, Ser. Geogr. Geofiz. **5**, 453–466 (1941)
74. G. K. Batchelor: Q. J. R. Meteorol. Soc. **76**, 133–146 (1950)
75. F. N. Frenkiel and I. Katz: *J. Meteor.* **13**, 388–394 (1956)
76. F. Gifford: J. Meteor. **14**, 410–414 (1957)
77. A. F. Bennett: Rev. Geophys. **25**(4), 799–822 (1987)
78. A. F. Bennett 1984: J. Atmos. Sci. **41**, 1881–1886 (1984)
79. A. Babiano, C. Basdevant, P. LeRoy and R. Sadourny: J. Fluid Mech. **214**, 535–557 (1990)
80. G. K. Batchelor: Proc. R. Soc., A **213**, 349–366 (1952b)
81. J. Guckenheimer and P. Holmes: Nonlinear oscillations, dynamical systems and bifurcations of vector Fields. Springer, Berlin (2002), 484 pp.
82. A. J. Lichtenberg and M. A. Lieberman: Regular and chaotic dynamics. Springer, Berlin (1992), 714 pp
83. J. M. Ottino: The kinematics of mixing. 378 pp. Cambridge University Press, Cambridge (1989)
84. T. S. Lundgren: J. Fluid Mech. **111**, 27–57 (1981)
85. R. H. Kraichnan and D. Montgomery : Rep. Prog. Phys **43**, 547–619 (1980)
86. P. B. Rhines: Ann. Rev. Fluid Mech. **11**, 401–41 (1979)
87. R. Salmon: Lectures on geophysical fluid dynamics 378 pp. Oxford University Press, Oxford. (1998)
88. R. Salmon: Geophys. Astrophys. Fluid Dyn. **10**, 25–52 (1980)
89. R. H. Kraichnan: Phys. Fluids **10**, 1417–1423 (1967)
90. J.-T. Lin: J. Atmos. Sci. **29**, 394–395 (1972)
91. J. H. LaCasce and C. Ohlmann: J. Mar. Res. **61**, 285–312 (2003)
92. E. Aurell, G. Boffetta, A. Crisianti, G. Paladin and A. Vulpiani: J. Phys. A: Math. Gen. **30**, 1–26 (1997)
93. G. Boffetta and A. Celani: Physica A **280**, 1–9 (2000)
94. D. Elhmaïdi, A. Provenzale and A. Babiano: J. Fluid Mech. **257**, 533–558 (1993)
95. A. M. Rogerson, P. D. Miller, L. J. Pratt and C. Jones: J. Phys. Oceanogr. **29**, 2635–2655 (1999)
96. M. S. Lozier, L. J. Pratt, A. M. Rogerson and P. D. Miller: J. Phys. Oceanogr. **27**, 2327–41 (1997)
97. L. Kuznetsov, M. Toner, A. D. Kirwan, Jr, C. K. R. T. Jones, L. H. Kantha and J. Choi: J. Mar. Res. **60**, 405–429 (2002)
98. F. Lekien, C. Coulliette and J. E. Marsden: 7th Experimental Chaos Conf., AIP. 162–168 (2003)
99. S. Wiggins: Ann. Rev. Fluid Mech. **37**, 295–328 (2005)
100. P. Morel and M. Larcheveque: J. Atmos. Sci. **31**, 2189–2196 (1974)
101. J. Er-el and R. Peskin: J. Atmos. Sci. **38**, 2264–2274 (1981)
102. E. Lindborg: J. Fluid Mech. **388**, 259–288 (1999)
103. J. Y. N. Cho and E. Lindborg: J. Geophys. Res. **106**, 10,233–10,232 (2001)
104. G. Lacorata, E. Aurell, B. Legras and A. Vulpiani: J. Atmos. Sci. **61**, 2936–2942 (2004)
105. L. F. Richardson and H. Stommel: J. Meteorology **5**(5), 38–40 (1948)

106. H. M. Stommel: J. Mar. Res. **8**, 199–225 (1949)
107. A. N. Kolmogorov: *Dokl. Akad. Nauk SSSR* **30**, 9–13 (1941)
108. A. Okubo: Deep-Sea Res. **18**, 789–802 (1971)
109. P. J. Sullivan: J. Fluid Mech. **47**, 601–607 (1971)
110. V. V. Anikiev, O. V. Zaytsev, T. V. Zaytseva and V. V. Yarosh: Izv. Atmos. Ocean Phys. **21**, 931–934 (1985)
111. A. D. Kirwan, G. J. McNally, E. Reyna and W. J. Merrell : J. Phys. Oceanogr. **8**, 937–945 (1978)
112. R. E. Davis: J. Geophys. Res. **90**, 4756–4772 (1985)
113. J. H. LaCasce and A. Bower: J. Mar. Res. **58**, 863–894 (2000)
114. K. F. Bowden: J. Fluid Mech. **21**, 83–95 (1965)
115. Riley and S. Corrsin: J. Geophys. Res. **79**, 1768–1771 (1974)
116. M. Ollitrault, C. Gabillet, A. Colin de Verdiere. J. Fluid Mech. **533**, 381–407
117. J. C. Ohlmann and P. P. Niiler: A two-dimensional response to a tropical storm on the Gulf of Mexico shelf. Prog. Oceanogr. **29**, 87–99 (2005)
118. J.-L. Thiffeault: Phys. Rev. Lett. **94**(8), 084502 (2005)
119. R. Molinari and A. D. Kirwan, Jr: J. Phys. Oceanogr. **5**, 483–491 (1975)
120. W. J. Saucier: Principles of meterorological analysis. University of Chicago Press, Chicago, IL, 438 pp. (1955)
121. A. Okubo and C. Ebbesmeyer: Deep-Sea Res. **23**, 349–352 (1976)
122. C. Garrett: Dyn. Atmos. Oceans **7**, 265–277 (1983)
123. J. R. Ledwell, A. J. Watson and C. S. Law: J. Geophys. Res. **103**, 21499–21529 (1998)
124. M. A. Sundermeyer and J. F. Price: J. Geophys. Res. **103**, 21,481–21,497 (1998)
125. A. Celani and M. Vergassola: Phys. Rev. Lett. **86**, 424–427 (2001)
126. G. Falkovich, K. Gawedzki and M. Vergassola : Rev. Mod. Phys. **73**(4), 913–975 (2001)

The Modulation of Biological Production by Oceanic Mesoscale Turbulence

Marina Lévy

LOCEAN/IPSL, CNRS, BC 100, 4 place Jussieu, 75252 Paris Cedex 05, France
marina@lodyc.jussieu.fr

Abstract. This chapter reviews the current state of knowledge on bio-physical interactions at mesoscale and at sub-mesoscale. It is focused on the mid-latitudes open ocean. From examples taken from my own studies or selected in the literature, I show how high-resolution process-oriented model studies have helped to improve our understanding. I follow a process oriented approach; I first discuss the role of mesoscale eddies in moderating the nutrient flux into the well-lit euphotic zone. Then I address the impact on biogeochemistry of transport occurring on a horizontal scale smaller than the scale of an eddy. I show that submesoscale processes modulate biogeochemical budgets in a number of ways, through intense upwelling of nutrients, subduction of phytoplankton, and horizontal stirring. Finally, I emphasize that mesoscale and submesoscale dynamics have a strong impact on productivity through their influence on the stratification of the surface of the ocean. These processes have in common that they concern the short-term, local effect of oceanic turbulence on biogeochemistry. Efforts are still needed before we can get a complete picture, which would also include the far-field long-term effect of the eddies.

1 Introduction

The photosynthesis of phytoplankton represents roughly half of the biological production on the planet. This Primary Production (PP) supports almost all marine life. It plays a key role in the global carbon cycle because phytoplankton growth, and subsequent death and sinking, transports vast quantities of carbon out of the surface layer where it can be sequestrated for long times [1, 2, 3]. Phytoplankton require nutrients for growth and reproduction. PP occurs in the sunlit surface layer of the ocean where photosynthesis can take place. In this well-lit euphotic layer[1] available nutrients are rapidly assimilated. Then, it is generally the supply of new nutrients from deeper water that limits productivity in the ocean. The dynamical mechanisms that control this supply occur over a large range of temporal and spatial scales. On the

---

[1] The euphotic layer is generally defined by the 1% light level.

M. Lévy: *The Modulation of Biological Production by Oceanic Mesoscale Turbulence*, Lect. Notes Phys. **744**, 219–261 (2008)
DOI 10.1007/978-3-540-75215-8_9 © Springer-Verlag Berlin Heidelberg 2008

planetary scale, and for timescales beyond the year, the transport of nutrients is controlled by the thermohaline and the wind-driven circulations. These circulations regulate the subsurface nutrient distribution [4]. On the seasonal timescale, the convective supply of nutrients is strongly modulated at mid and high latitudes [5]. At shorter timescales, vertical advection is controlled by the three-dimensional circulation associated with baroclinic eddies [6, 7, 8] and by the processes of frontogenesis and frontolysis [9, 10, 11, 12, 13], an effect that can be crucial in subtropical gyres, i.e. over large areas of the ocean [14]. The relative importance of these different processes for the PP is still under debate [15, 16, 17, 18]. There is however growing evidence that PP occurring both at the scale of eddies (the mesoscale) and at the scale of frontogenesis (the submesoscale) contributes significantly to the global budgets.

From the point of view of observations, direct PP measurements are sparse because they involve long incubations and heavy isotope techniques [19, 20]. However, satellite retrieved ocean color[2] [21] and continuous measurements of other related parameters such as fluorescence or carbon dioxide partial pressure (pCO2) make clear that ocean productivity is highly variable at the (sub)-mesoscale[3] [22, 23, 24, 25, 26]. Jenkins [27] has also suggested that the discrepancy between the large rates of export of organic matter estimated from biogeochemical budgets in the oligotrophic North Atlantic subtropical gyre compared with the lower rates of measured productivity could be explained by the undersampling of eddy induced PP.

PP is a process where dynamics, biogeochemistry, and radiation play equal roles and interact strongly. It is only since the early 1990s that models have been developed where all these processes are coupled together [28, 29]. To develop such models, the first natural step is to introduce in dynamical Ocean General Circulation Models (OGCM) conservative schemes for the transport of minor species, the second is to introduce into those models realistic biogeochemical schemes for the life cycle of phytoplankton. Mostly because of computational limitations, most of the OGCMs today use rather coarse horizontal grids ($1/2°$ to $2°$). Some preliminary modeling studies suggest that such a resolution can result in errors near 30% in the estimation of PP [15, 30, 31, 32]. Some simulations at even higher resolution show that incorrect representation of submesoscale frontogenesis can result in even larger errors (up to 50%, [33]).

---

[2] Due to the absorption properties of chlorophyll-a (the primary photosynthetic pigment), measures of solar light retrodiffusion by the ocean in the green and blue wavebands by optical sensors onboard earth viewing platforms provide an accurate estimate of the concentration of chlorophyll-a at the surface, the "color" of the sea.

[3] The terminology (sub)-mesoscale is used in this chapter to define the scale range including both the submesoscale and the mesoscale. Approximately, the mesoscale covers the range 20–100 km, the submesoscale covers the range 2–20 km, and thus the (sub)-mesoscale the range 2–100 km.

The growing evidence that the (sub)-mesoscale variability of PP is large is a challenge for the measurement networks and for the OGCMs used to study the climate: none of them resolve those scales. Therefore, there is an imperative need to understand the mechanisms by which the (sub)-mesoscale physical dynamics are reflected by the biological processes. Providing we can predict the (sub)-mesoscale dynamics from the larger scale dynamics, this should enable the development of parametrizations of the (sub)-mesoscale biophysical interactions for application to large-scale models. The ultimate goal is to better predict the evolution of the oceanic carbon cycle at climatic scales.

The purpose of this chapter is to give an overview of the current state of knowledge on the (sub)-mesoscale biophysical interactions. It complements previous review papers: Lewis [34] presented the discussions about the observed discrepancy between the rate of PP of organic matter and its export; Flierl and McGillicuddy [35] reviewed the impact of mesoscale and submesoscale physics on biological dynamics; Williams and Follows [4] focused on the transport processes regulating nutrient distribution in the global ocean; Martin [36] discussed the patchy distribution of phytoplankton at mesoscale and submesoscale from an observational and theoretical point of view.

The present review is focused on the mid-latitudes open ocean. Section 2 ("Generalities") introduces some basic ingredients of the oceanic biogeochemical cycles, of the turbulent motion in the ocean, and of the influence of the latter on the former. Section 3 ("Modelization of biophysical interactions") describes the models that are used to analyze these interactions. It includes a discussion of the transport equations for plankton and how the physical dynamics enter the equations. Then, we adopt a process-oriented approach, and review our current knowledge of the biogeochemical interactions with the mid-latitudes turbulent oceanic eddy field, jets, and the isolated vortices in the open ocean. Using few examples, we will show how high-resolution process-oriented model studies improve our understanding. Given that the primary source of variability of PP is due to variations in nutrient input, Sect. 4 ("Transport by mesoscale eddies") will discuss the role of mesoscale eddies in moderating the nutrient flux into the well-lit euphotic zone. In Sect. 5 ("Transport by submesoscale dynamics"), we discuss the impact on biogeochemistry of transport occurring on a horizontal scale smaller than the scale of an eddy. We will show that submesoscale processes modulate biogeochemical budgets in a number of ways, through intense upwelling of nutrients, subduction of phytoplankton, and horizontal stirring. In Sect. 6 ("Biophysical interactions through changes in stratification"), we show that mesoscale and submesoscale dynamics have a strong impact on productivity through their influence on the stratification of the surface of the ocean. We will conclude with some insight on the remaining way to go before a complete understanding of the impact of oceanic mesoscale turbulence on phytoplankton productivity and, more generally, on ocean biogeochemical cycles, is established.

2 Generalities

2.1 Interplay of Transport and Biology

2.1.1 Role of Phytoplankton in the Carbon Cycle

Phytoplankton play a major role in the oceanic carbon cycle and these floating, microscopic single-cell plants are the foundation of the marine food web. Like land plants, phytoplankton fix carbon through photosynthesis, making it available for higher trophic levels. Phytoplankton generally have limited or no swimming ability and are advected through the water by currents ("plankton" is actually derived from "πλανκτειν", "to wander"). Also as in terrestrial plants, the chlorophyll pigment in the phytoplankton absorbs light, which is used as an energy source to fuse water molecules and carbon dioxide into carbohydrates. The major environmental factors that influence phytoplankton growth are light [37] and inorganic nutrients [38]. When favorable conditions are encountered, phytoplankton can undergo rapid population growth usually referred to as "blooms" [39]. However, most of the time phytoplankton growth is either light limited (e.g. in winter at high latitudes) or nutrient limited (e.g. in the subtropical gyres). Because light attenuates dramatically with depth, phytoplankton growth is restricted to the euphotic layer. Limiting inorganic nutrients such as nitrogen and phosphorus are constantly removed from the surface waters by the growing phytoplankton. Dissolved inorganic carbon is also consumed by phytoplankton. It is exchanged at the sea surface and always plentiful in the surface layer, contrary to limiting nutrients (Fig. 1).

Most of the phytoplankton are consumed locally by zooplankton, so that the nutrients comprising their biomass are regenerated at the surface and made available for another round of production. The Regenerated Production (RP) is that portion of PP fueled by the limiting inorganic nutrients remineralized within the euphotic zone [40]. The Export Production (EP) is the fraction of PP that finds its way to the deep sea through the settling of dead cells and detritus, through zooplankton diel migrations, and sometimes by downwelling or mixing. Most of the EP is ultimately assimilated by bacteria, which regenerate it into inorganic forms (end product of respiration and excretion). At steady state and at large scale, the biotically mediated downward flux of organic matter is balanced by an upward return flux of inorganic nutrients that fuels the New Production (NP).

The collective action of this so-called biological pump [41] is to create a sharp vertical gradient of nutrients. Minimum nutrient concentrations are found at the surface due to photosynthetic consumption and maximum concentrations between 500 and 1000 m due to remineralization. This pump plays a central role in the global carbon cycle. It sequesters carbon away from the atmosphere in the deep sea.

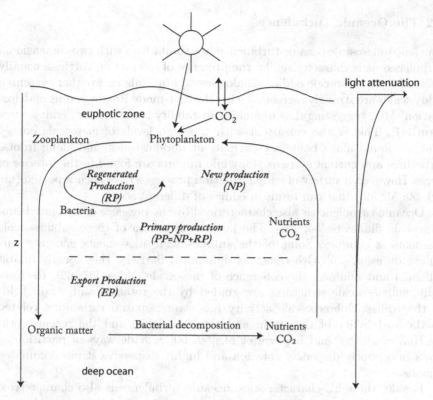

Fig. 1. Schematic representation of the biological pump in the ocean

2.1.2 Transport Modulation of Biotic Rates

Transport modulates NP, RP, and EP in various ways. Primarily, the flow transports the biological actors, i.e. phytoplankton, zooplankton, bacteria, dissolved and particulate organic matter. When this transport is directed out of the euphotic zone, it directly contributes to EP. Secondly, the flow transports inorganic nutrients, providing them to the euphotic layer through vertical diffusion or upwelling. Thirdly, the ocean physics can also modulate the rates of the biological processes. The most common manifestation is due to the vertical movements which displace phytoplankton within the light gradient and thereby affect phytoplankton growth rate [37]. Also, zooplankton growth rate depends on the encounter rate between zooplankton and its prey, which is partially controlled by the transport [42]. Note also that phytoplankton retroacts on transport: large concentrations of phytoplankton significantly alters the penetration of solar radiation, which in turn modulates density, and transport through the thermal-wind balance [43, 44]. This effect is often neglected in models.

2.2 The Oceanic Turbulence

The (sub)-mesoscale oceanic turbulence has similarities with two-dimensional turbulence. It is characterized by the presence of interacting vortices, usually referred to as mesoscale eddies (analogous to atmospheric weather systems). Eddy scales are strongly correlated with the first-mode Rossby radius of deformation[4] [46], suggesting that baroclinic instability is the primary eddy source term [47]. This is also consistent with the high levels of mesoscale energy that are found along boundary currents at mid-latitudes, and along frontal structures and current systems [48] while minima are found in the interior of gyres. However, a variety of eddy formation processes have been reported (i.e. [49, 50, 51, 52]) that can result in eddies of different scales.

Oceanic turbulence is also characterized by the presence of elongated sub-mesoscale filaments [54, 198]. The large scale action of these submesoscale filaments is complex. Some of the submesoscale movements generate turbulent diffusion [55]. Others act as dynamical barriers; they locally inhibit diffusion and reinforce the coherence of mesoscale eddies [57, 62]. Reciprocally, submesoscale structures are guided by the rotation and strain fields of the eddies. Submesoscale activity has strong spatial variability, related to the variabillity of the strain and rotation fields, and the recent works by Hua et al. [58] and Lapeyre et al. [59, 60] provide ways of partitioning flows into poorly dispersive rotation and highly dispersive strain dominated regions.

Besides these 2D-characteristics, oceanic turbulence is also characterized by its vertical structure. Baroclinic mid-ocean eddies often have vertical structure of the first baroclinic mode (cyclones characterized by doming isopycnals in their core and anticyclones by shoaling isopycnals). Actually, oceanic mesoscale eddies are better described by quasi-geotrophic dynamics than by 2D-turbulence. They are associated with vertical velocities (w) of the order of 1–10 m/d. The typical w distribution in the quasi-geostrophic (QG) approximation is a multipolar structure with alternate upwellings and downwellings along meandering fronts (with upwelling occurring downstream of the trough and downwelling occurring downstream of the ridge, [61] and around the eddies [7, 8, 9, 62]). These patterns, which extend down to the zero-crossing of the first baroclinic mode (i.e. approximately 1000m), are induced by the curvature and by eddy–eddy interactions.

Submesoscale structures are particularly intense, close to the surface (above 200 m approximately). They are strongly ageostrophic; their relative vorticity can be of the order of the planetary vorticity. They can be

---

[4] The first mode Rossby radius of deformation R_d is the length scale at which rotation effects become as important as buoyancy effects. R_d decreases from 300 km at the equator to a few kilometers at high latitudes, and ranges between 30 and 50 km at mid-latitudes [45]. Within the approximation of homogeneous buoyancy, $R_d = NH/f$, with N the buoyancy frequency, f the Coriolis parameter, and H the depth.

described by the dynamics contained in the Primitive Equations (PE). Associated with the strong submesoscale vorticity gradients, the vertical velocities are one order of magnitude stronger than the vertical velocities of QG dynamics [11, 12, 13, 63, 64]. These vertical velocities are characterized by dipolar structures astride the vorticity gradients [12, 33, 65]; see also Fig. 7. When the flow in the filament of vorticity accelerates, a secondary circulation develops across the filament leading to upwelling on the anticyclonic side and downwelling on the cyclonic side. When the flow decelerates, a secondary circulation of opposite sign is formed.

Numerical simulations [33] suggest that submesoscale vertical velocities are maximum at around 100 m depth where they take over mesoscale QG vertical velocities, which are maximum at around 1000 m depth.

2.3 Observed Variability of Phytoplankton

2.3.1 Basin Scale Variability of Phytoplankton

The production patterns at basin scale are strongly constrained by the wind-driven vertical circulation (Ekman pumping). The general features of this Ekman transport are upwelling in subpolar gyres, at the equator, and at eastern boundaries and downwelling in subtropical gyres. Sea-color satellite images reveal the signature of the Ekman transport on the distribution of phytoplankton [66]. Figure 2a shows a climatology of the surface chlorophyll distribution in the North Atlantic. Strong spatial inhomogeneities of the phytoplankton distribution are revealed, with maxima located in regions of ascendance (at the equator, in the subpolar gyre and at the coasts) and minima in regions of subsidence (in the center of the subtropical gyre). The maintenance of different levels of phytoplankton production by these dominant oceanographic features enables the division of the world's oceans in so-called biogeochemical provinces [68].

2.3.2 Seasonal Cycles of Phytoplankton Production

Seasonal variations of winds and solar radiation drive seasonal cycles of phytoplankton production. These variations are mediated through the mixed-layer (ML) seasonal cycle. For instance, from 30° N to 50° N in the northeast Atlantic, three different seasonal regimes can be distinguished, depending on the strength of winter mixing [67]. These regimes and their boundaries exhibit an intense variability from 1 year to the next, which are driven both by the synoptic and by the lower frequency variability in the atmosphere [67, 69, 70]. Nevertheless, general patterns can be drawn:

When the winter ML is deeper than twice the euphotic layer depth (Ze) (Fig. 2b), Sverdrup's [71] conditions are encountered: production is inhibited, cells being continuously mixed below the euphotic layer for periods greater

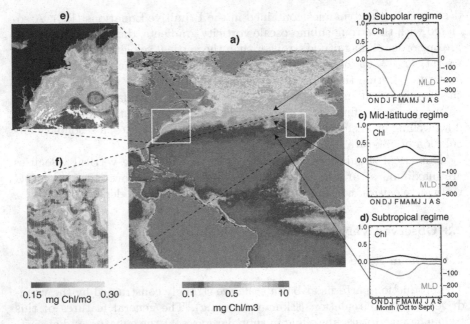

Fig. 2. (a) Climatology of sea-surface chlorophyll from space (Classic CZCS scene, from NASA Web site). (b)–(d) Typical seasonal cyclings of sea-surface chlorophyll versus mixed-layer depth in the northeast Atlantic [67]. The *grey line* shows the seasonal cycle of the mixed-layer depth, and the *black line* the seasonal cycle of the surface chlorophyll concentration. (e) and (f) High-resolution snapshot of sea-surface chlorophyll from space (e: classic CZCS scene, NASA Web site; f: Lehahn, personal communication). Locations of the cyclings and of the high-resolution images are indicated on the climatological map

than their doubling time. At the same time, this deep mixing efficiently supplies the surface with nutrient, thus enabling an intense spring bloom to occur as soon as the mixed-layer stratifies in spring. The rapid exhaustion of nutrients leads to an oligotrophic situation in summer. The deepening of the ML in fall leads to an entrainment bloom. This cycle is encountered in the subpolar gyre of the North Atlantic.

When the winter ML never exceeds Ze (Fig. 2d), nutrient limitation prevails. The seasonal cycling is characterized by a single weak entrainment bloom, that starts with the deepening of the ML, peaks when the ML is at its deepest, and ends with the exhaustion of nutrients [72].

When the winter ML is comprised between Ze and 2Ze (Fig. 2c), the seasonal cycling is characterized by a single bloom that lasts longer than any other bloom and corresponds to the merging of the subpolar spring bloom with the subpolar fall bloom. This bloom is initiated in fall by the deepening of the ML (as the entrainment bloom), and peaks after restratification (as the spring bloom) [67].

A general feature of these different cyclings is the alternation between periods of light limitation and periods of nutrient limitation. When light limitation prevails, PP is controlled by the ML depth. Phytoplankton concentration is maximum in the ML and decreases below. When nutrient limitation prevails, PP is controlled by the supply of nutrients. The distribution of phytoplankton is characterized by a subsurface maximum, located at the base of the nutricline [41]. As we will see, the impact of mesoscale turbulence on biological production depends on which situation prevails, and therefore varies both regionally and seasonally.

2.3.3 Observations of Mesoscale and Submesoscale Variability of Phytoplankton

The development of towed vehicles and the advent of sea-color remote sensing now permits us to observe the distribution of phytoplankton at high resolution. These observations have revealed considerable variability in the (sub)-mesoscale range [23, 26, 73, 74, 75, 76], as shown for example in the sea-color snapshots of Fig. 2e and f. In situ observations complement the sea-color view and enable the association of plankton variability with specific hydrologic structures such as fronts, meanders, eddies, and filaments.

In the 1980s, most of such observations were concerned with Gulf Stream "warm core rings" [77, 78, 79, 80]. In 1991, Falkowski et al. [81] reported an enhancement of production by a cyclonic eddy in the subtropical Pacific. Since then, the number of in situ observations at the mesoscale has kept increasing. Allen et al. [82] measured PP within and outside a cyclonic eddy, and found that photosynthetic rates near the edge and at the center of the eddy were approximately 50% higher, than outside the eddy. Mooring data in the Sargasso Sea [83, 84] provide evidence of waters rich in nutrients and chlorophyll within a cyclonic eddy (Fig. 3). Other striking correlations between the presence of eddies and the chlorophyll distribution have also been reported by Robinson et al. [85] during NABE (North Atlantic Bloom Experiment), by Aristegui et al. [86] around the Canaria Islands, by Letelier et al. [87] in the North Pacific subtropical gyre, by Moran et al. [88] in the Algerian Basin, by Barth et al. [89] in the Antartic polar front, and by Garcia et al. [90] in the Brazil–Malvinas Confluence region, among others.

Correlations between chlorophyll and dynamical features have also been observed at the submesoscale. For instance, Hitchcock et al. [91], during a series of transects across the Gulf Stream, identified a maximum of chlorophyll at the periphery of a warm core ring. Strass [92], using a towed, undulating vehicle in the open North Atlantic during summer revealed patches of high chlorophyll concentration of scales 10–20 km. These patches were located on the warm side of a temperature front. Perez et al. [93], during a summer oceanographic cruise in the Azores front region, located with good accuracy maximum chlorophyll concentrations on the South side of the front and at the border of an anticyclonic eddy. Other examples of submesoscale variability

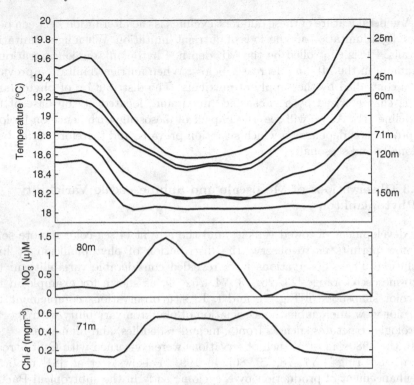

Fig. 3. Observations from Bermuda Testbed Mooring deployment 3 during the summer 1995. *Upper panel*, temperature records at various depths; *lower two panels*, nitrate concentration at 80 m, chlorophyll fluorescence at 71 m. All signals have been filtered via a 6-day moving average. From McGillicuddy et al. [83]

have been observed in the Gulf Stream [94], in the Almeria-Oran front [95] and in the Antarctic Circumpolar Current [96].

There is also some evidence of (sub)-mesoscale variability in zooplankton and in bacteria [97, 98, 99, 100, 101, 102], and in particle fluxes [103]. Watson et al. [25] provided the first evidence of the impact of (sub)-mesoscale dynamics on oceanic pCO_2, which has been confirmed by drifting buoys observations [24]. Section 4 will present theories that attempt to explain the impact of oceanic (sub)-mesoscale turbulence on the biological pump.

3 Modelization of Bio-physical Interactions

In this section, we present the basis for the modelization of the interaction between biogeochemical cycles and ocean dynamics.

3.1 Transport

The description is restricted to transport occurring on space scales much larger than the cell distribution scale. In this case, plankton are described as the continuous concentration (in space and time) of a constitutive element. Nitrogen provides a natural currency for biological quantities. A planktonic population, like, any tracer T expressed for instance in $\mathrm{mmole}N/\mathrm{m}^3$ is then assumed to obey the transport equation, which reads, in its Eulerian form:

$$\frac{\partial}{\partial t}T = -\boldsymbol{\nabla}\cdot(T\mathbf{v}) + B(T)\,, \tag{1}$$

where $\mathbf{v}$ is the velocity field and $B(T)$ is the budget between the biological sources minus the biological sinks for tracer T. Advection is written here in a flux form assuming that the ocean is incompressible. Diffusion is negligible at the scale at which plankton population can be described as a concentration and has therefore been disregarded. Because biological tracers are positive quantities, often close to zero, the numerical resolution of the transport equation requires the use of positive advection schemes [104, 105].

3.2 Biological Source/Sink Terms

The biogeochemical schemes describe the interactions between the various forms of plankton, organic and inorganic material. They vary in complexity from very simple models with one or two tracers (nitrate, or nitrate and phytoplankton), to much more complicated models with more than 20 tracers [106]. In these complex models, different species of phytoplankton are considered, as well as different limiting nutrients. A common trade-off in complexity is the use of models with 4–6 tracers (so-called NPZD models, i.e. [107]). Biogeochemical fluxes are exchanged by the tracers. These fluxes are empirical functions (parametrizations) of the biological variables, often non-linear, and sometimes of other environmental conditions such as light or temperature, derived from laboratory experiments [108]. The determination of the parameters used in these parameterizations is a difficult task and a large source of model error and uncertainties. Nowadays, inverse modeling is the most objective way of tuning parameters in biogeochemical models [109, 110, 111].

3.3 Reynolds Equation

The Reynolds equation is derived by applying to the transport equation (1) an operator $\overline{\bullet}$ defined as:

$$\overline{\bullet} = \frac{1}{\mathcal{V}}\int_V \bullet\, dv\,, \tag{2}$$

where $\mathcal{V}$ is an element of volume of scale $\mathcal{S}$. In numerical models, the scale $\mathcal{S}$ is set by the size of the grid and the averaging is done over a grid cell.

Any variable (such as T or $\mathbf{v}$) can then be decomposed into the mean $\overline{T}$, comprising the variability above the cut-off scale $\mathcal{S}$, and $T' = T - \overline{T}$, the variability below the cut-off scale, which verifies $\overline{T'} = 0$.

Thus averaging (1) with the linear operator (2), and then using, for simplicity, the notation T instead of $\overline{T}$ and $\mathbf{v}$ instead of $\overline{\mathbf{v}}$, gives the Reynolds equation:

$$\frac{\partial}{\partial t}T = -\boldsymbol{\nabla}\cdot(T\mathbf{v}) + B(T) - \boldsymbol{\nabla}\cdot\overline{(T'\mathbf{v}')} + \overline{B'(T)} \tag{3}$$

Compared to (1), the Reynold equation (3) has additional terms on the RHS (third and forth terms). These terms represent the impact that the scales below the cut-off (or sub-grid scales) have on the larger scales. The third term is the transport Reynolds term and represent the effects of motions on scales smaller than $\mathcal{S}$. The fourth term is the biological Reynolds term and represents the effects of the inhomogeneous distribution of the biological tracers within the grid.

To close equation (3), the effects of the sub-grid scales must be represented entirely in terms of large scale quantities. In the transport term, these effects appear as the divergence of turbulent fluxes (i.e. fluxes associated with the mean correlation of small-scale perturbations). To assume a turbulent closure is equivalent to chose a formulation (or parametrization) for these fluxes, usually called the sub-grid scale physics.

Although much progress has been made in the parameterization of turbulent fluxes for ocean models [112, 113, 114, 115, 116, 117, 118], there is still a long way to go before (sub)-mesoscale transport is correctly represented in coarse resolution or eddy-permitting models.

On the contrary, I am not aware of any reference regarding sub-grid scale biology. Practically, it is always omitted in models. Interestingly, an analogous problem concerns the chemical reactions in the stratosphere. For instance, Edouard et al. [119] show that ozone depletion in the Artic is sensitive to filament-scale inhomogeneities in the distribution of reactant species because of the non-linearities in the chemical rate laws. Their modeling study suggests that the effect is of the order of 40%. Another study by Vinuesa and Vila-Guerau de Arellano [120] shows that heterogeneous mixing in the atmospheric boundary layer can slow down the reaction rates of ozone formation and depletion. Similar behaviors may be expected for biological species in the ocean, since the interactions are often non-linear.

4 Transport by Mesoscale Eddies

4.1 Vertical Transport Associated with Coherent Mesoscale Eddies

The "eddy-pumping mechanism" [81, 83, 121, 122] rests on the fact that within a cyclonic eddy, isopycnals are deflected upward, pushing subsurface nitrate rich waters into the euphotic zone (Fig. 4). The same upward deflection applies

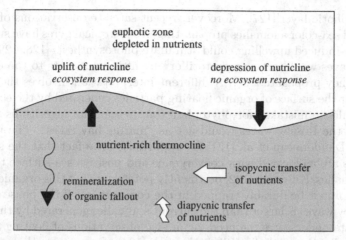

Fig. 4. Schematic figure depicting the ecosystem response to an uplift and depression of the nutricline. When nutrient-rich isopycnals are raised into the euphotic zone, there is biological production. Conversely, when the nutrient-rich isopycnals are pushed into the dark interior, there is no biological response. In order for the transient upwelling to persist, there needs to be a process maintaining the nutrient concentrations in the thermocline, which might be achieved by remineralization of organic fallout, diapycnal transfer or a lateral influx of nutrients from the time-mean or time-varying circulations. After Williams and Follows [4]

to anticyclonic mode water eddies [123]. Conversely, within an anticyclonic eddy the nutricline is depressed and so there should be no biological response [83]. These vertical displacement are though to occur during events of eddy intensification, for example through eddy–eddy interaction.

This 1D-vertical view on the scale of the eddy is based on observed distributions of nitrate or chlorophyll across cyclones which highlight the surfacing of nutrient-rich waters within cyclones and which suggest that the surfacing nutrients come from upwelling [74, 83, 84, 123], see also Fig. 3.

This idea led to various estimations to the contribution of eddies to the nitrate supply to the euphotic layer from satellite altimetry [124]. These estimates should be handled with care, since they strongly depend on the efficiency of pumping [125] and of its recurrence time.

4.2 Eddy Propagation

The propagation of eddies with doming isopycnals on the beta-plane (northwestward for cyclones and south-westward for anticyclones) can cause upwelling that uplift nutrients to the euphotic layer [35, 125, 126]. In the limit of linear propagation, this transport mechanism can be interpreted as a propagation flux; the passing of an eddy can be thought as a wavelike upward displacement of the isopycnals, resulting in an injection of nutrients in the euphotic layer [127]. Moreover, recent satellite observations of Rossby waves and sea-color anomalies propagating in subtropical gyres have suggested that wave-induced upwelling could stimulate photosynthesis [128, 129]. Thus, Rossby waves would act as a "rototiller" by lifting nutrients to the euphotic layer as they propagate [130]. A different interpretation involves the convergence near the surface of organic floating particles generated by the ecosystem which could be mistaken for phytoplankton by sea-color algorithms [131]. In this case, the Rossby waves would act as "marine hay rakes". The interpretation of Dandonneau et al. [131] is supported by the fact that the sea-color anomalies are co-located with convergence and positive sea-surface temperature anomalies. Killworth [132] pertinently points out that this organic detritic material cannot be durably trapped in the convergences if the phase speed of the Rossby waves is larger than the current anomalies generated by the waves. The debate underscores the need for in situ observations of floating material [133]. Finally, the study by Killworth et al. [134] also suggests that upwelling might not be the main mechanism responsible for the observed wave-like signal in sea color. Instead, their analysis shows that horizontal advection of surface chlorophyll against its background gradient accounts for most of the observed propagation in ocean color.

4.3 Horizontal Transport by Coherent Mesoscale Eddies

If the eddy exhibits a strongly non-linear behavior, trapping waters within it for long periods, it is a coherent feature. Its passing can be visualized as the translation of a solid obstacle which moves surrounding waters around it. Observations of eddies traveling for several months and over hundreds of kilometers, while maintaining the chemical characteristics of their source waters, have been reported [135, 136]. Provenzale [137] shows that this horizontal transport by coherent barotropic vortices is possible because they are highly impermeable to inward and outward particle fluxes.

Lévy's [138] numerical experiments illustrate such a case: a cyclone C and an anticyclone AC are formed by baroclinic instability of a density front (Fig. 5a and b), as schematized on Fig. 5g. The raised isopycnals of C are not the signature of upwelling as in the eddy-pumping mechanism. They result from the *horizontal* displacement of a water column into a warmer environment (Fig. 5g). This is all the more contrasting with the eddy-pumping mechanism than the formation of C is associated with a downward stretching of the water column. Indeed, as put forward by Williams and Follows [4], the conservation of potential vorticity (defined as $(\zeta + f)/h$, with ζ the relative vorticity, f the Coriolis parameter, and h the thickness of an isopycnic layer), implies that the increase of ζ must be balanced by an increase of h (which corresponds to the stretching).

Lévy [138] carried two experiments where they varied the initial nitrate distribution. Experiment 1 (Fig. 5c and d) describes an oligotrophic front; nitrate is distributed along isopycnal surfaces below the euphotic layer, and drops to

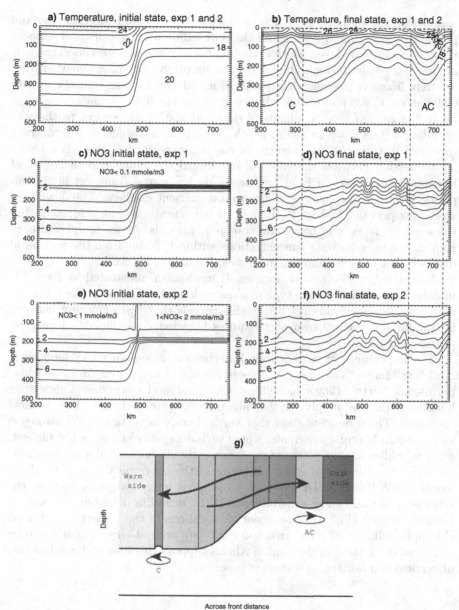

Fig. 5. (**a**) Initial temperature front in the experiments of Lévy et al. [138]. (**b**) Temperature front at the final state (after 1 month of simulation). The *dashed lines* delimit a cyclone C and an anticyclone AC that have been formed in the course of the simulation. (**c**) Initial nitrate concentration in experiment 1. (**d**) Nitrate concentration at the end of experiment 1. (**e**) Initial nitrate concentration in experiment 2. (**f**) Nitrate concentration at the end of experiment 2. (**g**) Schematic representation of the formation of cyclone C and of anticyclone AC during the experiments

zero within the euphotic layer. Experiment 2 (Fig. 5e and f) describes a more productive front. As in Exp. 1, nitrate is distributed along isopycnal surfaces below the euphotic layer. Within the euphotic layer, nitrate concentrations are low but not null. Due to biological consumption, they have departed from a purely linear regression with density. Fig. 5d and f shows that the nitrate distribution in AC and C depend very much on the initial situation.

In the case of Exp. 1, nitrate is depleted in C with respect to the surrounding waters down to 200 m depth. In Exp. 2, nitrate is on the contrary increased in C at the same depth. In the two experiments, there is actually a competition between the effect of downward stretching and the effect of horizontal transport. In Exp. 1, there is no horizontal transport in the euphotic layer because there is no horizontal nutrient gradient. The downward stretching prevails. In Exp. 2, there is an important nutrient gradient within the euphotic layer, and horizontal transport prevails. As for temperature, in Exp. 2, the raised nitrate concentrations within C result from the horizontal displacement of C. Similar arguments can be drawn for AC.

Interestingly, the (mostly horizontal) mechanism illustrated by Exp. 2 is in agreement with the same type of observations (i.e. Fig. 3) as the (vertical) eddy-pumping mechanism. Clearly, this reveals that distinguishing between horizontal and vertical transport requires knowledge on the eddy formation process and history.

The importance of the horizontal mechanism is supported by large-scale modeling studies. Oschlies [18] regional budgets computed from an eddy-permitting North Atlantic basin biogeochemical model experiment show that horizontal nitrate supply to the oligotrophic gyre is larger than vertical transport. These budgets show that vertical eddy advection is the strongest near western boundary currents, where turbulent eddy energy is the highest, whereas eddies supply nutrients predominantly via horizontal advection near the quieter southern and eastern margins of the subtropical gyre. The above works led Williams and Follows [4] to propose a generalized version of the scheme of McGillicuddy and Robinson [121], including horizontal as well as vertical transfer (Fig. 4). The recent observations of the impact of eddies on chlorophyll distribution by Aristegui et al. [86] around Gran Canaria and by Crawford et al. [139] in the Gulf of Alaska support this idea of a combination of vertical and horizontal transport processes.

5 Transport by Submesoscale Dynamics

5.1 Vertical Advection at the Submesoscale

5.1.1 Upwelling of Nutrients in Filaments

In order to explore vertical advection of nutrients at scales smaller than the eddies, Mahadevan and Archer [31] report the change of primary production

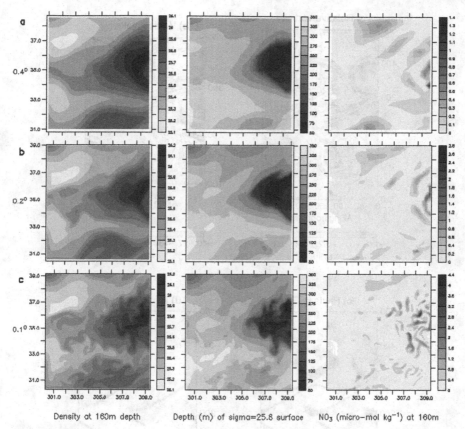

Density at 160m depth   Depth (m) of sigma=25.8 surface   NO₃ (micro–mol kg⁻¹) at 160m

Fig. 6. The density and nitrate distribution plotted at the base of the euphotic zone alongside the depth of a particular isopycnal. In each column, fields from (**a**) 0.4, (**b**) 0.2, and (**c**) 0.1 model resolution are shown. The pattern of new production is very similar to the nitrate distribution pattern. From Mahadevan and Archer [31]

induced by the change of horizontal resolution in a model (Fig. 6). They explore the range of resolution from 10 to 40 km in a model representing a limited area of the ocean where PP is limited by the availability of nutrients. They show a tremendous increase in PP (up to a factor three) in response to increasing model resolution. The increase is related to a better representation of the mesoscale range; increasing the model resolution results in more undulation of the isopycnal surfaces and in an increased length of the frontal zone (or isopycnal outcropping).

Lévy et al. [33] follow a similar approach and increase their model resolution up to 2 km in order to resolve the submesoscale features. They report a factor 2 change in PP when changing the resolution from 10 to 2 km. This increase is due to the resolution of intense vertical velocities, captured within filaments of strong vorticity gradients which surround the eddies or which

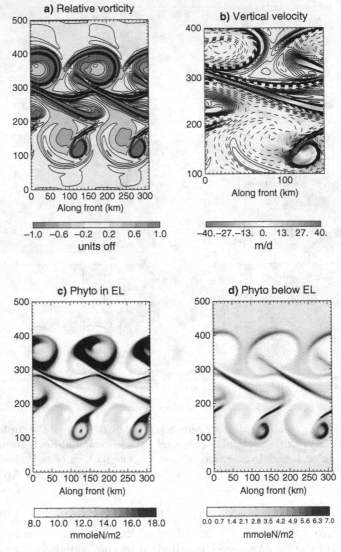

Fig. 7. (**a**) Relative vorticity at the surface, (**b**) vertical velocity at 100 m (zoom, in color, superposed on vorticity lines), (**c**) phytoplankton within the euphotic layer (0–120 m) and (**d**) export of phytoplankton below the euphotic layer (120–240 m), simulated with a primitive equation model with a horizontal resolution of 2 km. The initial state is an unstable baroclinic front and the fields shown are after 22 days of simulation. From Lévy et al. [33]

are ejected by the eddies (Fig. 7a and b). This strongly ageostrophic surface dynamics cannot be captured in the frame of the QG approximation, since it involves strong surface density gradients. This explains why it is not seen in modeling studies using QG models [121, 126, 140, 141, 142], nor in modeling studies based on primitive equation models but where the horizontal resolution is not sufficient enough to accurately resolve these gradients, and the associated submesoscale vertical transport [15, 18, 31, 32, 143].

As the submesoscale vertical velocities are in phase with the vorticity gradients, regions of nutrient input coincide with regions of elevated strain. This results in phytoplankton distribution being concentrated and elongated in filaments, either isolated filaments or filaments around eddies (Fig. 7c). First obtained in a simulation of decaying turbulence [33], this result has then been generalized to the situation of forced turbulence [144], Fig. 8.

A more careful examination of the fields in Figs. 7 and 8 reveals that phytoplankton mainly develop in filaments of negative vorticity within the euphotic layer. Moreover, the spectral slope of phytoplankton, zooplankton,

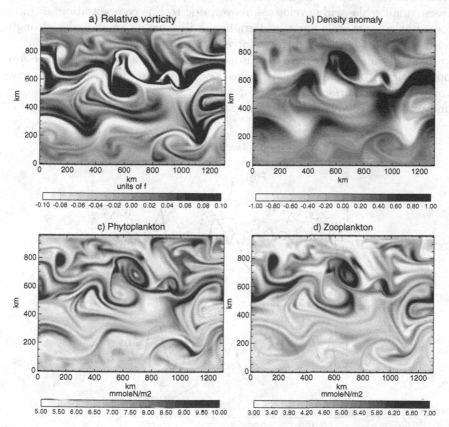

Fig. 8. Surface relative vorticity, density anomaly, phytoplankton and zooplankton simulated with a primitive equation model on the beta-plane with a horizontal resolution of 6 km. Fields shown are after 1600 days of simulation. The domain is periodic in longitude and turbulence is forced by restoring to a background density gradient. From Lévy and Klein [144]

and vorticity are close to -1.5 while the density spectrum is steeper with a slope close to -3 (Fig. 9).

These results [33, 144] can be explained with the rationalization of Klein et al. [145] regarding the density field: to prevent a thermal-wind imbalance, the physical system locally organizes the vertical and horizontal velocity fields such that submesoscale vertical and horizontal advection of density tend to compensate each other. This phase relationship between vertical and horizontal advection of density explains why very few small-scale features are present in the density field (slope close to -3). It is also consistent with motions being almost parallel to the isopycnals. This local compensation between horizontal and vertical advection holds for any tracer forced by the same large-scale vertical and horizontal gradients than density. However, it does not hold for a tracer forced by either the horizontal or the vertical gradient. In the last two cases, small scales will develop. Moreover, due to this compensation, a tracer forced by a large-scale horizontal gradient will have its small scales strongly anticorrelated to those of a tracer forced by a large-scale vertical gradient.

Vorticity and nitrate have their isopleths almost orthogonal (and therefore both inclined to the isopycnals), since potential vorticity (close to relative vorticity in the surface layers) is forced by a large-scale horizontal gradient and nitrate by a large-scale vertical gradient. This favors the development of

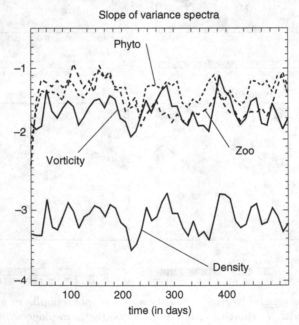

Fig. 9. Time evolution of the spectral slope of surface relative vorticity, density, phytoplankton, and zooplankton, in a primitive equation model of forced turbulence (same simulation as that shown on Fig. 8). Adapted from Lévy and Klein [144]

small scales for vorticity and for nitrate (slope close to −1.5). Moreover, small scales of vorticity should be strongly anticorrelated to those of nitrate, and ultimately to those of phytoplankton.

5.1.2 Subduction of Phytoplankton in Filaments

The simulations of Lévy et al. [33] and Lévy and Klein [144] also evidence that downwelling velocities associated with vorticity filaments are responsible for an export flux out of the euphotic zone. This export is located in filaments of positive vorticity (Fig. 7a and d), for the same reasons as developed above regarding nutrients and vorticity (the vertical phytoplankton gradient has a sign opposite to that of the vertical nutrient gradient). In situ observations with high-resolution towed vehicles confirm such patterns of localized submesoscale subduction [95, 96, 146]. Nevertheless, they are too sparse to provide a number for the magnitude of this export compared with more traditional form of export like the sedimentation of detritus [147] or the convective export of organic matter [148].

5.1.3 Net Impact of Submesoscale Structures on PP

In oligotrophic situations, phytoplankton undergoes two antagonistic effects of (sub)-mesoscale transport: production is favored through the inputs of nutrients within the euphotic layer, and is inhibited through the removal of phytoplankton cells from the euphotic layer. Phytoplankton production is favored in filaments of negative vorticity and is inhibited in filaments of positive vorticity. Its development in a turbulent field therefore requires that its growth is fast enough to balance the losses.

In order to get some further insight on this balance, we compare the experiment of Lévy et al. [33] (LKT experiment) with the experiment in Lévy and Klein [144] (LK experiment). Both experiments are run with the same biological model, the same model parameters and the same initial condition for nitrate (a nutricline located at 100 m depth). The experiments differ in their horizontal resolution (6 km in LK versus 2 km in LKT) and in their forcing (decaying turbulence in LKT, forced turbulence in LK). They also differ in the size of the domain, the duration of the experiment, and the width of the unstable front that forces turbulence. These differences all together result in vertical velocities one order of magnitude lower in LK (maxima around 10 m/d) compared with LKT (maxima around 100 m/d). One interesting contrasting result is that mesoscale turbulence increases PP by a factor 3 in LKT, but does not significantly change PP in LK. This indicates that the supply of nutrients prevails over the subduction of phytoplankton in the LKT experiment, while the two effects compensate in the LK experiment.

One possible explanation lays in the different order of magnitude of the vertical velocity field in the two experiments. A linear increase in upwelling velocities results in an exponential increase of the phytoplankton growth rate.

Indeed, upwelling displaces the phytoplankton subsurface maximum and the nitracline closer to the surface [33], and the phytoplankton growth rate increases exponentially with decreasing depths in response to the exponentially increasing light. On the other hand, phytoplankton decrease through subduction responds in a linear manner to a linear change in downwelling velocities because the expression of advection is linear in w. Hence, with increasing vertical velocities, and assuming a comparable range of change for upwelling and downwelling velocities, phytoplankton production should increase more rapidly than phytoplankton subduction. This view suggests that vertical velocities must be large enough to have a positive influence on the increase in phytoplankton, and the experiments suggest that they should be larger than 10 m/d. This view is in agreement with Smith et al. [141], who found that the net effect of eddies on the rates of primary production is small in their QG simulation (QG vertical velocities are below 10 m/d).

Let's now consider a water parcel below the euphotic layer, which is advected up to the euphotic layer and back below it. The parcel is initially loaded with nutrients. The parcel will induce a net flux of nutrients into the euphotic layer which will have an effect on phytoplankton only if the time it spends in the euphotic layer is long enough to enable nutrient uptake by phytoplankton. Otherwise, the parcel will pass through the euphotic layer with no net effect. Typically, phytoplankton growth rate is of the order of 1–2 days and the euphotic layer is 100 m depth. If the vertical velocity of the water parcel is less than 100 m/d, then its journey in and out of the euphotic layer will last more than 2 days: nutrients will be consumed during that period. Vertical velocities must therefore be less than 100 m/d to induce a net nutrient flux to the euphotic layer. Faster vertical velocities, for instance those associated with convective plumes (of the order of 1000 m/d) are likely not to induce a net transport : nutrients do not remain long enough in the euphotic layer to be consumed.

Vertical velocities associated with submesoscale activity, typically 10–100 m/d, fall in the range imposed by the two above constrains. Those associated with mesoscale activity (1–10 m/d) have a more marginal impact because the growth of phytoplankton that they induce is not strong enough to balance the loss of phytoplankton through subduction that they also induce.

In bloom situations, nutrients are plentiful in the euphotic layer. submesoscale transport mainly acts as a sink for phytoplankton and tends to decrease primary production.

5.2 Horizontal Transport at Submesoscale

5.2.1 Stirring

Another aspect concerns the ability of phytoplankton distribution to be stirred by mesoscale turbulence. This ability depends on the ratio between the tracer

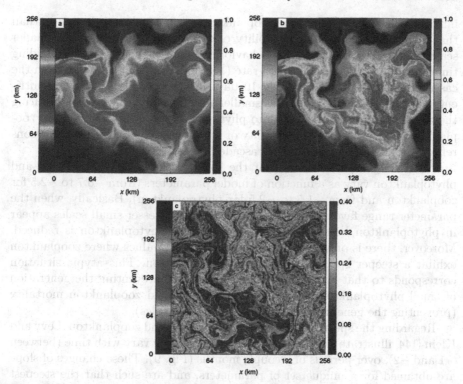

Fig. 10. Snapshots at the end of a high-resolution model run. (a) Carrying capacity (the equivalent for nutrients). (b) Phytoplankton. (c) Zooplankton. The *strip* at the *left* shows the zonally varying distributions the populations would have in the absence of advection while the *bar* on the *right* gives the values associated with the different colors. From Abraham [152]

decay rate and the advection timescale of the flow [149]. Phytoplankton decay rate is approximately 1 month (the mean time-lag between mid-latitude phytoplankton and zooplankton blooms). This is long compared to the transport time within a filament, typically 2–10 days. It is therefore reasonable to expect phytoplankton to be stirred by the flow.

Production of small-scale filaments of phytoplankton in the ocean by horizontal stirring has been evidenced by in situ experiments, such as NATRE (North Atlantic Tracer Release Experiment) [150] and SOIREE (Southern Ocean Iron RElcase Experiment) [151]. To rationalize this stirring, Abraham [152] proposes a mechanism based on the production of smaller and smaller tracer scales by mesoscale eddies, which is the classical direct tracer cascade. This mechanism is evidenced in a 2D numerical experiment where nutrients are injected at large scales within the euphotic layer and are subsequently affected by the direct cascade process (Fig. 10). In this experiment, the spatial variability of phytoplankton involves more energetic small scales than for nutrients (in other words, the phytoplankton spectrum slope is less steep than that of nutrients), and the variability of zooplankton involves even smaller scales than phytoplankton. This behavior is due to the fact that the e-folding time of the phytoplankton growth rate (1–2 days) is usually smaller than the cascade timescale (approximately 10 days in the ocean, [153]). Since nutrients are injected at large scales, smaller and smaller scales develop during the transformation of nutrients into phytoplankton, and ultimately into zooplankton. Thus the spatial variability of the successive biological populations reflect the different phases of the cascade process.

Abraham [152] also shows that the spectral slopes for zooplankton and phytoplankton vary as a function of model parameters (from −0.7 to −2.5 for zooplankton and from −1.5 to −2.5 for phytoplankton). Basically, when the parameter range favors faster zooplankton growth, lesser small scales appear in phytoplankton because the residence time of phytoplankton is reduced. Moreover, there is only a small range of parameter values where zooplankton exhibit a steeper spectral slope than phytoplankton. This atypic situation corresponds to that of rapid zooplankton growth (preventing the generation of small phytoplankton scales) combined with rapid zooplankton mortality (preventing the generation of small zooplankton scales).

Regarding the spectral slopes of phytoplankton and zooplankton, Lévy and Klein [144] illustrate a different situation, where they vary with time (between −1 and −2), over periods of a couple months (Fig. 9). These changes of slope are obtained for a unique set of parameters, and are such that the steepest slope is either that for zooplankton or for phytoplankton depending on the time in the simulation. Lévy and Klein [144] relate these changes of slope to the low-frequency variability of the eddy field (the equivalent of the atmospheric weather regimes).

One important difference between the experiment of Abraham [152] and that of Lévy and Klein [144] is that in the latter nutrients are injected at small scale by the vertical velocity field. These two contrasting experiments suggest that it is difficult to derive general conclusions on the processes that lead to the formation of small scales by comparing the relative spectra of phytoplankton and of zooplankton.

5.2.2 Impact of Stirring on the Rate of Vertical Advection

An underlying question is how much the horizontal stirring affects the vertical transport. This has been addressed by Martin et al. [154] and Pasquero et al. [155] in a model of two-dimensional turbulence in which vertical nutrient advection is externally imposed as a restoring flux. Pasquero et al. [155] show that PP is increased if upwelling is fragmented into many episodes of short duration and/or small size. The mechanism relies on the removal of nutrients from the active (i.e. upwelling) regions by horizontal advection on timescales shorter than the phytoplankton doubling time. This removal causes an increase of PP because it enables to maintain the strength of the restoring nitrate flux. The magnitude of the increase is shown to be diminished by the sheltering action of the eddies, which prevents horizontal dispersion in their core.

These two studies impose small-scale vertical advection in a somehow arbitrary manner, whereas submesoscale vertical advection is actually in phase with horizontal stirring [33, 145]. This arbitrary dimension highlights a double effect: increasing the model resolution increases the amplitude of w, but also causes w to be more fragmented. Both effects contribute to increase the vertical nutrient flux, and ultimately to increase PP.

6 Biophysical Interactions Through Stratification Changes

The impact of mesoscale eddies on primary production has also been reported when nutrients are plentiful in the euphotic layer : during the spring bloom [156] and in the HNLC (High Nutrient Low Chlorophyll) Antarctic Circumpolar Current [96]. In such a situation, light is the main limiting factor: PP is highly sensitive to the ML depth. ML shoaling can locally increase the mean exposure time of photosynthetic organisms and promote production. In this situation, the restratifying action of mesoscale eddies leads to beginning of bloom prior to seasonal stratification (as it has been reported in the North Atlantic by Townsend et al. [157]). Two illustrations of this mechanism, one at mesoscale, the other at submesoscale, are now presented.

6.1 Mesoscale Stratification

As demonstrated by Klein and Hua [158], mesoscale eddies generate heterogeneity of the mixed layer. It is now recognized that eddies have an important role in the restratification of the surface [159, 160]. When these eddies result from the baroclinic instability associated with the process of winter deep convection [161, 162, 163], they act against convection and tend to restratify the convective area, gradually cutting down its edges [146, 164]. This has been numerically evidenced in a regional model study of the spring bloom in a region of deep convection (northwestern Mediterranean, [30, 165, 166]). Figure 11b shows the surface density signature of eddies that are formed through baroclinic instability around a deep-mixing area in an idealized model of a convective patch. The formation of the eddies enable to release the available potential energy contained within the dense water patch. Basically, the eddies serve as vehicles for the transfer of water masses, by sinking the denser waters out of the convective zone and at the same time upwelling lighter peripheral waters toward the center. Consequently, these mesoscale instabilities are responsible for the collapse of the dense water patch. The axial symmetry of the problem allows to schematize the action of the eddies as a function of depth and distance from the center of the convective patch (Fig. 11a). The

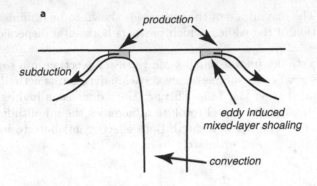

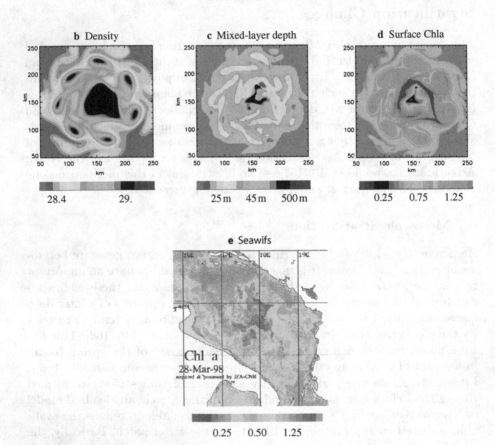

Fig. 11. (**a**) Schematic of eddy-induced stratification and eddy-induced subduction at the rim of a convective zone. Production is enhanced in the zone of eddy restratification. Model results [30] showing (**b**) the density, (**c**) the mixed-layer depth, and (**d**) the Chl concentration (in mgChl/m3) in an area of deep mixing (sea-surface view). (**e**) Satellite sea-color observations from Santoleri et al. [75] in the Adriatic, showing phytoplankton developing preferentially at the rim of the Adriatic convective area

transport of warmer waters from the stratified area toward the convective site across the frontal zone is responsible for the sloping of the isopycnals and hence for the shoaling of the mixed layer at the rim of the convective area (Fig. 11c). In winter, nutrients are plentiful within the euphotic layer and there is no obvious relationship between PP and nutrients, whereas PP increases when the mixed-layer depth decreases. Indeed, the mean exposure time of the phytoplankton cells to sunlight is inversely correlated with the mixed-layer depth. Consequently, the majority of phytoplankton production is obtained at the rim of the convective area, where the mixed layer is the shallowest (Fig. 11d). This "dynamical" stratification leads to a bloom which starts earlier that is induced by the more classical "seasonal" stratification. Satellite observations from Santoleri et al. [75] above the convective area of the Adriatic sea provide observational evidence of this eddy-fertilization process (Fig. 11e). Another process evidenced with the experiment of Lévy et al. [166] is the decorrelation in space between new and exported production. This decorrelation is induced by the eddies, which subduct phytoplankton rich waters (Fig. 11a). Hence, while NP is maximum at around 30 km from the center of the convective region (in the Mediterranean case), EP (through the subduction of phytoplankton) is maximum at 80 km from the center.

6.2 Submesoscale Stratification

In order to better assess the impact of submesoscale dynamics during the spring bloom in the north-east Atlantic, high-resolution numerical experiments were conducted in the frame of POMME (Programme d'Oceanographie Multidisciplinaire Meso Echelle) [167, 168, 169]. The domain of the experiment (16–22W, 38–45N) is covered by several eddies (Fig. 12a). Data collected during the first POMME survey were used for model initialization, and data from three other cruises for model validation. The model revealed much stronger space and time variability than could be seen with the resolution of the data (CTD stations were 50 km apart). Space variability during the onset of the bloom is illustrated by Fig. 12 which shows a snapshot of model outputs in March.

A very striking feature is the strong variability of the MLD on filamentary scales (Fig. 12b). The maximum MLD gradients are reached at the border of eddies (Fig. 12c); MLD changes from 200 to 50 m over 10 km. These fine scale structures in the MLD seem to result from the interplay between the mesoscale atmospheric forcing and the stirring induced by the eddies. The medium-scale picture is that MLD is shallower over the regions previously subjected to warming (in the southeast and northeast, Fig. 12b) and that it is deeper over the regions subjected to cooling (in the northwest, in the center, and in the southwest). This medium-scale picture is perturbed by the small-scale advection, which induces a direct cascade of the MLD toward smaller scales. Incidentally, the stratification of the upper ocean is also very sensitive

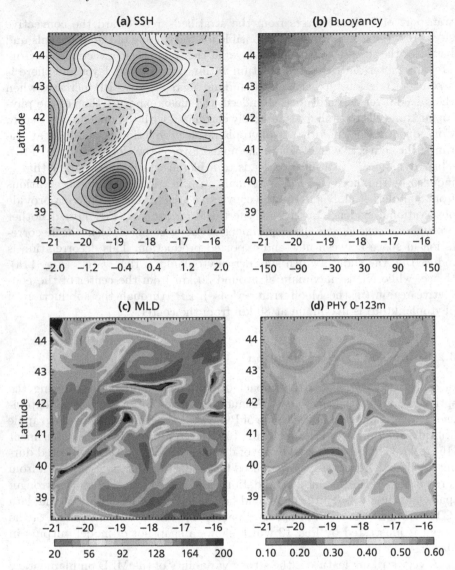

Fig. 12. Simulation of the spring bloom onset in the northeast Atlantic (adapted from [167]). (**a**) Sea-surface height (cm), with cyclones identified with *dashed lines*, and anticyclones with *plain lines*, (**b**) buoyancy (w/m^2), (**c**) mixed-layer depth (m), (**d**) Phytoplankton integrated over the euphotic layer (mmoleN/m^2)

to the ageostrophic submesoscale dynamics induced by baroclinic eddies in the absence of atmospheric forcing [170]. At medium scale, phytoplankton variability during the bloom is driven by the variability of the buoyancy flux and of the solar radiation. At small scale, phytoplankton patterns follow the MLD distribution (Fig. 12d). NP is maximum over the filamentary structures of

MLD minima, and conversely NP is minimum over the filaments of maximum MLD. The mesoscale dynamics therefore contribute to structuring the initial conditions for the onset of the spring bloom in specific submesoscale features. There is no nutrient limitation yet. The structuring is mainly the consequence of photosynthesis limitation by (lack of) light, which is greater when the MLD is deeper. It is worth noting that the phytoplankton distribution is not directly correlated with the eddies, rather, the deformation field associated with the presence of eddy induced submesoscale stirring (around and in between eddies) which structures the ML and the growth of phytoplankton.

7 Conclusions

This chapter attempts to review our current knowledge on the role of physical phenomena on primary production and export, on the horizontal scales of oceanic turbulence, i.e. from a few kilometers to a few hundred kilometers. It is focused on the modeling aspect of this problem and on the processes that control PP in the open ocean.

In Sect. 2, we gave an overview of the problem. More precisely, we have shown how biotic rates are modulated by transport and we have presented some features of the vertical and horizontal transport associated with oceanic turbulence. Some of the increasing observations that have led the research community to pay a particular attention at the (sub)-mesoscales were also presented. Section 3 presented how biogeochemistry is introduced in OGCMs. More precisely, we argue that the impact of (sub)-mesoscale dynamics on biogeochemistry appears in the form of advective Reynolds terms when the transport equation is solved on a grid coarser than the (sub)-mesoscale. In Sect. 4–6, the processes have been described in terms of how they perturb the system on the short term. Section 4 summarized the mesoscale transport process that provide nutrients to the euphotic layer. In it we distinguish the vertical transport from the horizontal transport because the ocean is strongly anisotropic in the scale range in consideration. In particular, we suggest that it is difficult to recognize eddy pumping from horizontal displacement on the basis of synoptic observations, and that knowledge on the time evolution of the eddy is required. Section 5 reviewed the impact of submesoscale transport. The submesoscale dynamics is associated with intense vertical velocities, in phase with the vorticity gradients. The consequence is the development of phytoplankton localized in filaments of negative vorticity, balanced by an export of phytoplankton in filaments of positive vorticity. Our results suggest that the net balance is toward a significant increase of PP when submesoscale vertical velocities are energetic, and a much more moderate increase associated with the QG mesoscale vertical velocities. Section 6 presented how (sub)-mesoscale features can result in a shoaling of the ML depth, and therefore provoke the bloom to begin prior to seasonal stratification.

This review is restricted to mid-latitude open ocean turbulence, where the mean productivity is driven by the cycling of the mixed layer. In particular, the equatorial and coastal regions, where productivity is driven by large-scale up-welling, have their own small-scale dynamics, and possibly different impacts on productivity [171, 172]. Also, the processes were described qualitatively. Dedicated studies combining observations and models can help to estimate more quantitatively the relative contribution of the small-scale physical pro-cesses and are more and more numerous [126, 164, 167, 173, 174, 175]. Finally, the modulation of air–sea CO_2 fluxes by oceanic submesoscale turbulence is not straightforward. As discussed by Mahadevan et al. [176], it results from a balance between the impact of mesoscale turbulence on biology, temperature and dissolved organic carbon.

Two major issues still need further investigations. The first issue is the role of the oceanic turbulence in the competition between different species. This aspect is crucial to the biogeochemical cycles since the efficiency of the bio-logical pump is very closely related with the phytoplankton species [147, 177]: large cells are more likely grazed by large grazers who produce fast sinking fe-cal pellets and an efficient carbon export, while small cells are involved in the regeneration network. A number of observation now provide evidence that the phytoplankton community is structured by (sub)-mesoscale turbulence [178, 179, 180, 181]. Modeling studies by Bracco and Provenzale [182], Mar-tin et al. [183], Lima et al. [184] and Pasquero et al. [185] also suggest that (sub)-mesoscale turbulence plays a role in the structuration of the ecosystem. The responses of those models depend very much on the choice of the parame-ters. A complete understanding of how (sub)-mesoscale turbulence structures the ecosystem now requires more systematic studies covering a large range of the parameter space. These studies should bring some new insight on the importance and variability of the biological Reynolds term.

The second issue is the role of the oceanic turbulence in the long-term and large-scale equilibrium of the nutrient distribution, i.e. in the subsurface nutrient reservoir. This far field effect of eddies has been shown to be impor-tant in idealized studies [186, 187]. The 1D view, appealing in simplicity, is that the upward flux of nutrients in the euphotic layer is balanced locally by a downward flux of organic material. This view relies on the premise that the remineralization of organic matter at depth occurs faster than the physical processes that advect the nutrients upward. This view neglects the lateral processes that deliver nutrient to the subsurface. These lateral processes can be due to the large scale Ekman transport [188] or associated with the for-mation and advection of mode waters [189, 190]. They can also be attributed to the eddy transport [4, 118], and possibly to the transport by submesoscale structures. Model studies suggest that eddies could modify the subduction rates [191, 192]; this also could have an impact on the subsurface nutrient pool.

Although our understanding of the impact of mesoscale turbulence on bio-logical production has been progressing very rapidly in the last decade, efforts

are still needed before we can get a complete picture, i.e. a precise quantification of this impact on the short-term, local impact and on the large-scale
maintenance of production. Today, the most powerful supercomputers (Los
Alamos, Earth Simulator, Oakridge) allow the first global simulations of the
ocean circulation at $1/10°$ [193]. The resolution at submesoscale, with a complete description of the biogeochemical cycles requires an increase in computer
power of approximately two orders of magnitude. Idealized basin scale studies
are affordable at very high resolution and are an alternative to reconcile and
quantify the various pathways through which mesoscale turbulence impacts
biogeochemical cycles and ultimately to derive and to test parametrizations
of these impacts for climate models.

In view of the different pathways through which mesoscale turbulence modifies marine productivity, it is very unlikely that a unique parametrization of
these impacts for climate models will be able to encompass all the processes.
Rather the identification and characterization of the different processes, as has
been attempted here, should be seen as a first step toward parametrization.
The second step is the parametrization of the physical transport alone. For
instance, regarding lateral transport at the mesoscale, Gent and McWilliams
[114] have proposed a parametrization of eddy-induced water mass exchanges
in the form of an additional eddy-induced advection flux which depends on
the large-scale slope of the isopycnals. This parametrization improved the
distribution of water masses simulated with an OGCM [194]. Treguier et al.
[195] warn however that this representation does not account for the transport within individual eddies traveling over long distances. Regarding lateral transport occurring at the submesoscale, Dubos [196] proposed a spatially selective parametrization which led promising results in 2D turbulence.
This parametrization remains to be tested in OGCMs. Regarding vertical
transport, I am not aware of parametrization of the strongly ageostrophic
vertical transport associated with mesoscale turbulence. The third step is
the imbrication of the physical parametrization for biogeochemical purposes.
For instance, Lévy et al. [197] have shown that the use of the Gent and
McWilliams [114] scheme could very nicely represent the eddy-induced restratification around a convective area and greatly improve the prediction
of the pre-bloom phytoplankton growth in the NW Mediterranean Sea. Pasquero [198] experiments suggest that when eddy diffusion is used to mimic
lateral turbulent transport, a smaller diffusion coefficient has to be used for
non-conservative tracers such as phytoplankton or nutrient. She convincingly
argues that turbulent transport can be significantly overestimated if the reaction timescale of the transported tracers is not accounted for. Finally, the
importance of the biological Reynolds terms has to be assessed and eventually
parametrized. Parametrization of small-scale transport and biology is a big
challenge for climate studies, and to take up this challenge will require the
strengthening of interdisciplinary approaches.

References

1. P. M. Holligan: Do marine phytoplankton influence global climate? In Primary productivity and biogeochemical cycles in the sea, pp. 487–501. Plenum press, New York (1992)
2. A. R. Longhurst: Role of the marine biosphere in the global carbon cycle, Limnol. Oceanogr. **36**, 1507–1526 (1991)
3. G. Shaffer: Effects of the marine biota on global carbon cycling. In The global carbon cycle. NATO ASI Series, Vol. I 15. Springer, Berlin (1993)
4. R. G. Williams and M. J. Follows: Physical transport of nutrients and the maintenance of biological production. In Ocean biogeochemistry: a JGOFS synthesis. Springer, Berlin (2003)
5. C. de Boyer Montégut, A. S. Fischer, G. Madec, A. Lazar and D. Iudicone: Mixed layer depth over the global ocean: an examination of profile data and a profile-based climatology, J. Geophys. Res. **109**, C12003 (2004)
6. A. P. Martin and K. J. Richards: Mechanisms for vertical nutrient transport within a North Atlantic mesoscale eddy. Deep Sea Res. Part II **48**, 757 (2001)
7. R. T. Pollard: Mesoscale (50–100 km) circulations revealed by inverse and classical analysis of the JASIN hydrographic data. J. Phys. Oceanogr. **13**, 377–394 (1983)
8. R. K. Shearman, J. A. Barth, J. S. Allen and R. L. Haney: Diagnosis of the three-dimensional circulation in mesoscale features with large Rossby number. J. Phys. Oceanogr. **30**, 2687–2709 (2000)
9. R. Davies-Jones: The frontogenetical forcing of secondary circulations. Part I: the duality and generalization of the Q vector. J. Atmos. Sci. **48**, 497–509 (1991)
10. R. T. Pollard and L. A. Regier: Large variations in potential vorticity at small spatial scales in the upper ocean. Nature, **348**, 227–229 (1990)
11. R. T. Pollard and L. A. Regier: Vorticity and vertical circulation at an ocean front, J. Phys. Oceanogr., **22**, 609–625 (1992)
12. M. A. Spall: Baroclinic jets in confluent flow. Journal of Phys Oceanogr. **27**, 1054–1071 (1997)
13. D.-P. Wang: Model of frontogenesis: subduction and upwelling. J. Mar. Res. **51**, 497–513 (1993)
14. R. G. Williams and M. J. Follows: Oceanography: eddies make ocean deserts bloom. Nature, **394**, 228 (1998)
15. D. J. McGillicuddy, L. A. Anderson, S. C. Doney and M. E. Maltrud: Eddy-driven sources and sinks of nutrients in the upper ocean: results from a 0.1 resolution model of the North Atlantic. Global Biogeochem. Cycles **17**(2), 1035 (2003)
16. A. Oschlies: Model-derived estimates of new production: new results point towards lower values. Deep Sea Res. Part II, **48**, 2173 (2001)
17. A. Oschlies: Can eddies make ocean deserts bloom? Global Biogeochem. Cycles **16**, 1106–1117 (2002)
18. A. Oschlies: Nutrient supply to the surface waters of the North Atlantic: a model study. J. Geophys. Res. **107**(C5), 3046 (2002)
19. Y. Dandonneau and A. Le Bouteiller: A simple and rapid device for measuring planktonic primary production in situ sampling, andJ 14 CJBJ injection and incubation, Deep Sea Res. **39**, 795–803 (1992)

20. G. L. Hitchcock: Methodological aspects of time-course measurements of 14C fixation in marine phytoplankton. J. Exp. Mar. Biol. Ecol. **95**, 233 (1986)
21. T. Platt and S. Sathyendranath: Oceanic primary production: Estimation by remote sensing at local and regional scales, Science, **241**, 1613–1620 (1988)
22. T. Dickey, D. Frye, H. Jannasch, E. Boyle, D. Manov, D. Sigurdson, J. McNeil, M. Stramska, A. Michaels and N. Nelson: Initial results from the Bermuda Testbed Mooring program. Deep Sea Res. Part I **45**, 771 (1998)
23. J. F. R. Gower, K. L. Denman and R. J. Holyer: Phytoplankton patchiness indicates the fluctuation spectrum of mesoscale oceanic structure. Nature **288**, 157 (1980)
24. E. M. Hood, L. Merlivat, and T. Johannessen: Variations of CO_2 and air–sea flux of CO_2 in the Greenland Sea gyre using high-frequency time-series data from CARIOCA drift buoys. J. Geophys. Res. **104**, 20571–20583 (1999)
25. A. J. Watson, C. Robinson, J. E. Robinson, P. J. L. B. Williams and M. J. R. Fasham: Spatial variability in the sink for atmospheric carbon dioxide in the North Atlantic. Nature **350**, 50–53 (1991)
26. J. A. Yoder, J. Aiken, R. N. Swift, F. E. Hoge and P. M. Stegmann: Spatial variability in near-surface chlorophyll a fluorescence measured by the Airbone Oceanographic Lidar (AOL). Deep Sea Res. **40**, 33–53 (1993)
27. W. J. Jenkins: Nitrate flux into the euphotic zone near Bermuda. Nature **331**, 521 (1988)
28. W. K. Nuttle, J. S. Wroblewski and J. L. Sarmiento: Advances in modeling ocean primary production and its role in the global carbon cycle. Adv. Space Res. **11**, 67 (1991)
29. J. L. Sarmiento, R. D. Slater, M. J. R. Fasham, J. R. Ducklow, J. R. Toggweiler and G. T. Evans: A seasonal three-dimensional ecosystem model of nitrogen cycling in the north Atlantic euphotic zone, Global Biogeochem. Cycles, **7**, 417–450 (1993)
30. M. Lévy, L. Mémery and G. Madec: The onset of a bloom after deep winter convection in the North Western Mediterranean sea: mesoscale process study with a primitive equation model. J. Mar. Syst. **16**, 7–21 (1998)
31. A. Mahadevan and D. Archer: Modeling the impact of fronts and mesoscale circulation on the nutrient supply and biogeochemistry of the upper ocean. J. Geophys. Res. **105**, 1209–1225 (2000)
32. A. Oschlies and V. Garcon: Eddy-induced enhancement of primary production in a model of the North Atlantic Ocean. Nature **394**, 266 (1998)
33. M. Lévy, P. Klein and A. M. Tréguier: Impacts of submesoscale physics on phytoplankton production and subduction. J. Mar. Res. **59**(4), 535–565 (2001)
34. M. R. Lewis: Variability of plankton and plankton processes on the mesoscale. In Phytoplankton productivity: carbon assimilation in marine and freshwater ecosystems, pp. 141–156. Blackwell, London (2002)
35. G. Flierl and D. J. McGillicuddy: Mesoscale and submesoscale physical–biological interactions. In The sea. Wiley, New York (2002)
36. A. P. Martin: Phytoplankton patchiness: the role of lateral stirring and mixing. Prog. Oceanogr. **57**, 125–174 (2003)
37. J. Marra: Phytoplankton photosynthetic response to vertical movement in the mixed layer. Mar. Biol. **46**, 203–208 (1978)
38. R. Dugdale and F. Wilkerson: Nutrient limitation of new production in the sea. In Primary productivity and biogeochemical cycles in the sea, Plenum Press, New York, pp. 107–122 (1992)

39. G. A. Riley: The relationship of vertical turbulence and spring diatom flowerings. J. Mar. Res. **5**, 67–87 (1942)

40. R. W. Eppley and B. J. Peterson: Particulate organic matter flux and planktonic new production in the deep ocean. Nature **282**, 57–70 (1979)

41. A. R. Longhurst and W. G. Harrison: The biological pump: profiles of plankton production and consumption in the upper ocean. Prog. Oceanogr. **22**, 47–123 (1989)

42. T. Kiørboe and B. R. MacKenzie: Turbulence-enhanced prey encounter rates in larval fish: effects of spatial scale, larval behaviour and size, J. Plankton Res. **16**, 2319–2331 (1995)

43. R. Murtugudde, J. Beauchamp, C. McClain, M. Lewis, A. Busalacchi: Effects of penetrative radiation on the upper tropical ocean circulation. J. Clim. **15**, 470–486 (2002)

44. A. Oschlies: Feedbacks of biotically induced radiative heating on upper-ocean heat budget, circulation, and biological production in a coupled ecosystem-circulation model. J. Geophys. Res. **109**, C12031 (2004)

45. D. B. Chelton, R. A. deSzoeke, M. G. Schlax, K. El Naggar and N. Siwertz: Geographical variability of the first-baroclinic Rossby radius of deformation. J. Phys. Oceanogr. **28**, 433–460 (1998)

46. D. Stammer: On eddy characteristics, eddy transports, and mean flow properties. J. Phys. Oceanogr. **28**, 727–739 (1998)

47. C. Wunsch: The vertical partition of oceanic horizontal kinetic energy. J. Phys. Oceanogr. **27**, 1770–1794 (1997)

48. D. Stammer: Global characteristics of ocean variability estimated from regional TOPEX/POSEIDON altimeter measurements. J. Phys. Oceanogr., **27**, 1743–1769 (1997)

49. J. Aristegui, P. Sangra, S. Hernandez-Leon, M. Canton, A. Hernandez-Guerra and J. L. Kerling: Island-induced eddies in the Canary islands, Deep Sea Res. Part I **41**, 1509 (1994)

50. E. Di Lorenzo, M. G. G. Foreman and W. R. Crawford: Modelling the generation of Haida Eddies. Deep Sea Res. Part II **52**, 853 (2005)

51. S. Herbette, Y. Morel and M. Arhan: Erosion of a surface vortex by a seamount. J. Phys. Oceanogr. **33**, 1664–1679 (2003)

52. T. Pichevin and D. Nof: The eddy cannon. Deep Sea Res. Part I, **43**, 1475 (1996)

53. T. Dubos and J. P. A. Babiano and P. Tabeling: Intermittency and coherent structures in the two-dimensional inverse energy cascade: comparing numerical and laboratory experiments. Phys. Rev. E **64**, 36302 (2001)

54. T. Dubos and A. Babiano: Cascades in two-dimensional mixing: a physical space approach. J. Fluid Mech. **467**, 81–100 (2002)

55. C. Pasquero, A. Babiano, A. Provenzale: Parameterization of dispersion in two-dimensional turbulence. J. Fluid Mech., **439**, 279–303 (2001)

56. A. Mariotti, B. Legras and D. G. Dritschel: Vortex stripping and the erosion of coherent structures in two-dimensional flows. Phys. Fluids **6**, 3954–3962 (1994)

57. C. Pasquero, A. Bracco, A. Provenzale and J. B. Weiss: Particle motion in a sea of eddies. In: A. Griffa, A. D. Kirwan, Jr., Arthur J. Mariano, Tamay M. Özgökmen and Thomas Rossby. (eds.) *Lagrangian analysis and prediction of coastal and ocean dynamics*. Cambridge University Press, (2007).

58. B. L. Hua, J.C. McWilliams and P. Klein: Lagrangian acceleration in geostrophic turbulence. J. Fluid Mech **35**, 1–22 (1998)

59. G. Lapeyre, P. Klein and B.L. Hua: Does the tracer gradient vector align with the strain eigenvectors in 2D turbulence? Phys. Fluids **11**, 3729–3737 (1999)
60. G. Lapeyre, B. L. Hua and P. Klein: Dynamics of the orientation of active and passive scalars in two-dimensional turbulence. Physics of Fluids **13**, 251–264 (2001)
61. J. Woods: Mesoscale upwelling and primary production. In *Toward a theory on biological–physical interactions in the world ocean*. B. Rothschild, Dordrecht (1988)
62. A. P. Martin, K. J. Richards, C. S. Law and M. Liddicoat: Horizontal dispersion within an anticyclonic mesoscale eddy. Deep Sea Res. Part II **48**, 739 (2001)
63. G. R. Halliwell, Jr and P. Cornillon: Large-scale SST anomalies associates with subtropical fronts in the Western North Atlantic during FASINEX. J. Mar. Res. **47**, 757–775 (1989)
64. M. A. Spall: Frontogenesis, subduction, and cross-front exchange at upper ocean fronts. J. Geophys. Res. **100**, 2543–2557 (1995)
65. B. J. Hoskins and F. P. Bretherton: Atmospheric frontogenesis models: mathematical formulation and solution. J. Atmos. Sci. **29**, 11–37 (1972)
66. J. A. Yoder, C. R. McClain, G. C. Feldman and W. E. Esaias: Annual cycles of phytoplankton chlorophyll concentrations in the global ocean: a satellite view. Global Biogeochem. Cycles **7**, 181–193 (1993)
67. M. Lévy, Y. Lehahn, J.-M. André, L. Mémery, H. Loisel, and E. Heifetz: Production regimes in the northeast Atlantic: a study based on sea-viewing wide field-of-view sensor chlorophyll and ocean general circulation model mixed layer depth. J. Geophys. Res. **110**, (2005)
68. A. R. Longhurst: Ecological geography of the sea. pp. 398 Academic Press, New York (1998)
69. S. Dutkiewicz, M. Follows, J. Marshall and W. W. Gregg: Interannual variability of phytoplankton abundances in the North Atlantic. Deep Sea Res. Part II **48**, 2323 (2001)
70. M. Follows and S. Dutkiewicz: Meteorological modulation of the North Atlantic spring bloom. Deep Sea Res. Part II **49**, 321 (2002)
71. H. U. Sverdrup: On conditions for the vernal blooming of phytoplankton. J. Cons. Int. Expor. Mer. **18**, 287–295 (1953)
72. D. W. Menzel and J. H. Ryther: Annual variations in primary production in the Sargasso Sea off Bermuda. Deap Sea Res. **7**, 282–288 (1961)
73. A. R. Longhurst: A major seasonal phytoplankton bloom in the Madagascar Basin. Deep Sea Res. Part I: Oceanogr. Res. Papers **48**, 2413 (2001)
74. D. J. McGillicuddy, V. K. Kosnyrev, J. P. Ryan and J. A. Yoder: Covariation of mesoscale ocean color and sea-surface temperature patterns in the Sargasso Sea. Deep Sea Res. Part II **48**, 1823 (2001)
75. R. Santoleri, V. Banzon, S. Marullo, E. Napolitano, F. D'Ortenzio, and R. Evans: Year-to-year variability of the phytoplankton bloom in the southern Adriatic Sea (1998P2000): Sea-viewing wide field-of-view sensor observations and modeling study. J. Geophys. Res. **108**(C9), 8122 (2003)
76. V. Lehahn, F. d'Oridio, M. Lévy and E. Heitzel: Stirring of the Northeast Atlantic spring bloom: a lagrangian analysis based on multi-satellite data. J. Geophys. Res. **112**, CO8005 (2007), doi: 10. 1029/2006JC003927
77. G. L. Hitchcock, C. Langdon and T. J. Smayda: Seasonal variations in the phytoplankton biomass and productivity of a warm-core Gulf Stream ring. Deep Sea Res. **32**, 1287–1300 (1985)

78. G. L. Hitchcock, C. Langdon and T. J. Smayda: Short term changes in the biology of a Gulf Stream warm-core ring: phytoplankton biomass and productivity. Limnol. Oceanogr. **32**, 919–928 (1987)

79. T. M. Joyce: Gulf Stream warm-core rings. J. Geophys. Res. **90**(C5), 8801–8951 (1985)

80. P. H. Wiebe and T. J. McDougall: Introduction to a collection of papers on warm-core rings. Deep Sea Res. Part A, **33**, 1455 (1986)

81. P. G. Falkowski, D. Ziemann, Z. Kolber and P. K. Bienfang: Role of eddy pumping in enhancing primary production in the ocean. Nature **352**, 55 (1991)

82. C. B. Allen, J. Kanda and E. A. Laws: New production and photosynthetic rates within and outside a cyclonic mesoscale eddy in the North Pacific subtropical gyre. Deep Sea Res. Part I **43**, 917 (1996)

83. D. J. McGillicuddy, Jr, A. R. Robinson, D. A. Siegel, H. W. Jannasch, R. Johnson, T. D. Dickey, J. McNeil, A. F. Michaels and A. H. Knap: Influence of mesoscale eddies on new production in the Sargasso Sea. Nature **394**, 263 (1998)

84. J. D. McNeil, H. W. Jannasch, T. Dickey, D. McGillicuddy, M. Brzezinski and C. M. Sakamoto: New chemical, bio-optical and physical observations of upper ocean response to the passage of a mesoscale eddy off Bermuda. J. Geophys. Res. **104** (C7), 15537–15548 (1999)

85. A. R. Robinson, D. J. McGillicuddy, J. Calman, H. W. Ducklow, M. J. R. Fasham, F. E. Hoge, W. G. Leslie, J. J. McCarthy, S. Podewski, D. L. Porter, G. Saure and J. A. Yoder: Mesoscale and upper ocean variabilities during the 1989 JGOFS bloom study. Deep Sea Res. **40**, 9–35 (1993)

86. J. Aristegui, P. Tett, A. Hernandez-Guerra, G. Basterretxea, M. F. Montero, K. Wild, P. Sangra, S. Hernandez-Leon, M. Canton and J. A. Garcia-Braun: The influence of island-generated eddies on chlorophyll distribution: a study of mesoscale variation around Gran Canaria. Deep Sea Res. Part I **44**, 71 (1997)

87. R. M. Letelier, D. M. Karl, M. R. Abbott, P. Flament, M. Freilich, R. Lukas, T. Strub: Role of late winter mesoscale events in the biogeochemical variability of the upper water column of the North Pacific Subtropical Gyre. J. Geophys. Res. **105**, (C12), 28723–28739 (2000)

88. X. A. G. Moran, I. Taupier-Letage, E. Vazquez-Dominguez, S. Ruiz, L. Arin, P. Raimbault and M. Estrada: Physical–biological coupling in the Algerian Basin (SW Mediterranean): influence of mesoscale instabilities on the biomass and production of phytoplankton and bacterioplankton. Deep Sea Res. Part I, **48**, 405 (2001)

89. J. A. Barth, T. J. Cowles and S. D. Pierce: Mesoscale physical and bio-optical structure of the Antartic Polar Front near 170 degrees W during austral spring. J. Geophys. Res. **106** (C7), 13879–13902 (2001)

90. C. A. E. Garcia, Y. V. B. Sarma, M. M. Mata and V. M. T. Garcia: Chlorophyll variability and eddies in the Brazil–Malvinas Confluence region. Deep Sea Res. Part II **51**, 159 (2004)

91. G. L. Hitchcock, A. J. Mariano and T. Rossby: Mesoscale pigment fields in the Gulf Stream: observations in a meander crest and trough. J. Geophys. Res. **98**, 8425–8445 (1993)

92. V. H. Strass: Chlorophyll patchiness caused by mesoscale upwelling at fronts. Deep Sea Res. **39**, 75–96 (1992)

93. F. F. Pérez, M. Gilcoto and A. F. Ríos: Large and mesoscale variability of the water masses and the deep chlorophyll maximum in the Azores Front. J. Geophys. Res. **108**(C7), 3215 (2003)

94. S. E. Lohrenz, D. A. Phinney, C. S. Yentch and D. B. Olson: Pigment and primary production distributions in a Gulf Stream meander. J. Geophys. Res. **98**, 14545–14560 (1993)

95. L. Prieur and A. Sournia: "Almofront-1" (April–May 1991): an interdisciplinary study of the Almeria-Oran geostrophic front, SW Mediterranean Sea. J. Mar. Syst. **5**, 187–203 (1994)

96. V. H. Strass, A. C. Naveira Garabato, R. T. Pollard, H. I. Fischer, I. Hense, J. T. Allen, J. F. Read, H. Leach and V. Smetacek: Mesoscale frontal dynamics: shaping the environment of primary production in the Antarctic Circumpolar Current. Deep Sea Res. Part II: Topical Studies Oceanogr. **49**, 3735 (2002)

97. C. J. Ashjian, S. L. Smith, C. N. Flagg, A. J. Mariano, W. J. Behrens and P. V. Z. Lane: The influence of a Gulf Stream meander on the distribution of zooplankton biomass in the Slope Water, the Gulf Stream, and the Sargasso Sea, described using a shipboard acoustic Doppler current profiler. Deep Sea Res. **41**, 23–50 (1994)

98. M. E. Huntley, M. Zhou and W. Nordhausen: Mesoscale distribution of zooplankton in the California current in late spring, observed by optical plankton counter, J. Mar. Res. **53**, 647–674 (1995)

99. B. Karrasch, H. G. Hoppe, S. Ullrich and S. Podewski: The role of mesoscale hydrography on microbial dynamics in the northeast Atlantic: results of a spring bloom experiment. J. Mar. Res. **54**, 99–122 (1996)

100. J. P. Labat, P. Mayzaud, S. Dallot, A. Errhif, S. Razouls and S. Sabini: Mesoscale distribution of zooplankton in the Sub-Antarctic Frontal system in the Indian part of the Southern Ocean: a comparison between optical plankton counter and net sampling. Deep Sea Res. Part I **49**, 735 (2002)

101. D. L. Mackas, M. Tsurumi, M. D. Galbraith and D. R. Yelland: Zooplankton distribution and dynamics in a North Pacific Eddy of coastal origin: II. Mechanisms of eddy colonization by and retention of offshore species. Deep Sea Res. Part II **52**, 1011 (2005)

102. P. Velez-Belchi, J. T. Allen and V. H. Strass: A new way to look at mesoscale zooplankton distributions: an application at the Antarctic Polar Front. Deep Sea Res. Part II, **49**, 3917 (2002)

103. P. P. Newton, R. S. Lampitt, T. D. Jickells, P. King and C. Boutle: Temporal and spatial variability of biogenic particles fluxes during the JGOFS northeast Atlantic process studies at 47!N, 20!W. Deep Sea Res. Part I: Oceanographic Res. Papers, **41**, 1617 (1994)

104. M. Lévy, A. Estubier and G. Madec: Choice of an advection schemeJ for biogeochemical models. Geophys. Res. Lett. **28**, 3725–3728 (2001)

105. A. Oschlies: Equatorial nutrient trapping in biogeochemical ocean models: the role of advection numerics. Global Biogeochem. Cycle, **14**, 655–667 (2000)

106. O. Aumont, E. Maier-Reimer, S. Blain and P. Monfray: An ecosystem model of the global ocean including Fe, Si, P colimitations. Global Biogeochem. Cycles **17**, 1060 (2003)

107. M. Lévy, A.-S. Krémeur and L. Mémery: Description of the LOBSTER biogeochemical model implemented in the OPA system, p. 13, Institut Pierre Simon Laplace, (2004)

108. L. Michaelis and M. L. Menten: Die Kinetik der Invertinwirkung. Biochemistry, Z (49), 333–369 (1913)
109. M. J. R. Fasham: Variations in the seasonal cycle of biological production in subarctic oceans: a model sensitivity analysis. Deep Sea Res. 42, 1111–1149 (1995)
110. B. Faugeras, M. Lévy, L. Memery, J. Verron, J. Blum and I. Charpentier: Can biogeochemical fluxes beJrecovered from nitrate and chlorophyll data? A case study assimilating data in the Northwestern Mediterranean sea at the JGOFS-DYFAMED station, J. Mar. Syst. 40–41, 90–125 (2003)
111. Y. H. Spitz, J. R. Moisan and M. R. Abbott: Configuring an ecosystem model using data from the Bermuda Atlantic Time Series (BATS) Deep Sea Res. Part II, 48, 1733 (2001)
112. B. Blanke and P. Delecluse: Variability of the tropical Atlantic Ocean simulated by a general circulation model with two different mixed-layer physics. J. Phys. Oceanogr. 23, 1363–1388 (1993)
113. P. Gaspar, Y. Gregories and J. M. Lefevre: A simple eddy kinetic energy model for simulations of the oceanic vertical mixing: tests at station papa and long term upper ocean study site. J. Geophys. Res. 95, 16179–16193 (1990)
114. P. Gent and J. McWilliams: Isopycnal mixing in ocean circulation models. J. Phys. Oceanogr. 20, 150–155 (1990)
115. P. R. Gent, J. Willebrand, T. J. McDougall and J. C. McWilliams: Parameterizing eddy-induced tracer transports in ocean circulation models. J. Phys. Oceanogr. 25, 463–474 (1995)
116. W. G. Large, J. C. McJWilliams and S. C. Doney: Oceanic vertical mixing: a review and a model with a non-local boundary layer parameterization. Rev. Geophy. 32, 363–403 (1994)
117. A. M. Tréguier, I. M. Held and V. D. Larichev: Parameterization of quasigeostrophic eddies in primitive equation ocean models. Journal Phys. Oceanogr. 27, 567–580 (1997)
118. M. Visbeck, J. Marshall, T. Haine and M. Spall: Specification of eddy transfer coefficients in coarse-resolution ocean circulation models. J. Phys. Oceanogr. 27, 381–402 (1997)
119. S. Edouard, B. Legras, F. Lefevre and R. Eymard: The effect of small-scale inhomogeneities on ozone depletion in the Arctic. Nature 384, 444 (1996)
120. J. F. Vinuesa and J. Vila-Guerau de Arellano: Introducing effective reaction rates to account for the inefficient mixing of the convective boundary layer. Atmos. Environ. 39, 445 (2005)
121. D. J. McGillicuddy and A. R. Robinson: Eddy-induced nutrient supply and new production in the Sargasso Sea. Deep Sea Res. Part I 44, 1427 (1997)
122. C. S. Yentsch and D. A. Phinney: Rotary motions and convection as a means of regulationg primary production in warm core rings. J. Geophys. Res. 90, 3237–3248 (1985)
123. D. J. McGillicuddy, R. Johnson, D. A. Siegel, A. F. Michaels, N. R. Bates and A. H. Knap: Mesoscale variations of biogeochemical properties in the Sargasso Sea. J. Geophys. Res., 104, C6, 13381–13394 (1999)
124. D. A. Siegel, D. J. McGillicuddy and E. A. Fields: Mesoscale eddies, satellite altimetry, and new production in the Sargasso Sea. J. Geophys. Res. 104(C6), 13359 (1999)
125. A. P. Martin and P. Pondaven: On estimates for the vertical nitrate flux due to eddy pumping. J. Geophys. Res. 108(C11), 3359 (2003)

126. D. J. McGillicuddy, A. R. Robinson and J. J. McCarthy: Coupled physical and biological modelling of the spring bloom in the North Atlantic (II): three dimensional bloom and post-bloom processes. Deep Sea Res. Part I **42**, 1359 (1995)

127. M. Kahru: Phytoplankton patchiness generated by long internal waves: a model. Mar. Ecol. **10**, 111–117 (1983)

128. P. Cipollini, D. Cromwell, P. G. Challenor and S. Raffaglio: Rossby waves detected in global ocean color data, Geophys. Res. Lett. **28**, 323–326 (2001)

129. B. M. Uz, J. A. Yoder and V. Osychny: Pumping of nutrients to ocean surface waters by the action of propagating planetary waves. Nature **409**, 597 (2001)

130. D. A. Siegel: Oceanography: the Rossby rototiller. Nature, **409**, 576 (2001)

131. Y. Dandonneau, A. Vega, H. Loisel, Y. du Penhoat and C. Menkes: Oceanic Rossby Waves acting as a "hay rake" for ecosystem floating by-products. Science **302**, 1548–1551 (2003)

132. P. D. Killworth: Comment on Oceanic Rossby Waves acting as a "hay rake" for ecosystem floating by-products. Science **304** (2004)

133. Y. Dandonneau, C. Menkes, T. Gorgues and G. Madec: Response to comment on oceanic Rossby waves acting as a "hay rake" for ecosystem floating by-products, Science **304**, 390 (2004)

134. P. D. Killworth, P. Cipollini, B. M. Uz, J. R. Blundell, Physical and biological mechanisms for planetary waves observed in sea-surface chlorophyll. J. Geophys. Res. **109**, C07002 (2004)

135. E. L. McDonagh and K. J. Heywood: The origin of an anomalous ring in the Southeast Atlantic. J. Phys. Oceanogr., **29**, 2050–2064 (1999)

136. P. L. Richardson: Gulf Stream Rings. In Eddies in marine science, pp. 19–45 Chapter 2, Springler, Berlin (1993)

137. A. Provenzale: Transport by coherent barotropic vortices, Annu. Rev. Fluid Mech. **31**, 55–93 (1999)

138. M. Lévy: Mesoscale variability of phytoplankton and of new production: JB-Jimpact of the large scale nutrient distribution. J. Geophys. Res. **108**(C11), 3358 (2003)

139. W. R. Crawford, P. J. Brickley, T. D. Peterson and A. C. Thomas: Impact of Haida Eddies on chlorophyll distribution in the Eastern Gulf of Alaska. Deep Sea Res. Part II **52**, 975 (2005)

140. G. Flierl and C. S. Davis: Biological effects of Gulf Stream meandering. J. Mar. Res. **51**, 529–560 (1993)

141. C. L. Smith, K. J. Richards and M. J. R. Fasham: The impact of mesoscale eddies on plankton dynamics in the upper ocean. Deep Sea Res. Part I **43**, 1807–1832 (1996)

142. A. Yoshimori and M. J. Kishi: Effects of interaction between two warm-core rings on phytoplankton distribution. Deep Sea Res. **41**, 1039–1052 (1994)

143. S. A. Spall and K. J. Richards: A numerical model of mesoscale frontal instabilities and plankton dynamics – I. Model formulation and initial experiments. Deep Sea Res. I, **47**, 1261 (2000)

144. M. Lévy and P. Klein: Does the low frequency variability of mesoscale dynamicsJ explain a part of the phytoplankton and zooplankton spectral variability? Proc. R. Soc. Lond. **460**, 1673–1683 (2004)

145. P. Klein, A.-M. Tréguier and B. L. Hua: Three-dimensional stirring of thermohaline fronts. J. Mar. Res. **56**, 589–612 (1998)

146. A. J. G. Nurser and J. W. Zhang: Eddy-induced mixed layer shallowing and mixed layer/thermocline exchange, J. Geophys. Res., **105** (C9) 851–868 (2000)
147. P. W. Boyd and P. P. Newton: Does planktonic community structure determine downward particulate organic carbon flux in different oceanic provinces? Deep Sea Res. Part I **46**, 63 (1999)
148. W. Koeve, F. Pollehne, A. Oschlies and B. Zeitzschel: Storm-induced convective export of organic matter during spring in the northeast Atlantic Ocean. Deep Sea Res. Part I **49**, 1431 (2002)
149. A. Mahadevan and J. W. Campbell: Biogeochemical patchiness at the sea surface. Geophys. Res. Lett. **29**, 19 (2002)
150. J. R. Ledwell, A. J. Watson and C. S. Law: Evidence for slow mixing across the pycnocline from an open ocean tracer release experiment. Nature **364**, 701–703 (1993)
151. E. R. Abraham, C. S. Law, P. W. Boyd, S. J. Lavender, M. T. Maldonado and A. R. Bowie: Importance of stirring in the development of an iron-fertilized phytoplankton bloom. Nature **407**, 727 (2000)
152. E. R. Abraham: The generation of plankton patchiness by turbulent stirring. Nature **391**, 577 (1998)
153. P. Klein and B. L. Hua : The mesoscale variability of the sea surface temperature : an analytical and numerical model. J. Mar. Res. **48**, 729–763 (1990)
154. A. P. Martin, K. J. Richards, A. Bracco and A. Provenzale: Patchy productivity in the open ocean. Global Biogeochem. Cycles **16**(2), 1025 (2002)
155. C. Pasquero, A. Bracco and A. Provenzale: Impact of the spatio-temporal variability of the nutrient flux on primary productivity in the ocean. J. Geophys. Res. **110**, C07005 (2005)
156. D. Antoine, A. Morel and J.-M. André: Algal pigment distribution and primary production in the Eastern Mediterranean as derived from coastal zone color scanner observations. J. Geophys. Res. **100**, 16193–16209 (1995)
157. D. W. Townsend, L. M. Cammen, P. M. Holligan, D. E. Campbell and N. R. Pettigrew: Causes and consequences of variability in the timing of spring phytoplankton blooms. Deep Sea Res. Part I **41**, 747 (1994)
158. P. Klein and B. L. Hua: Mesoscale heterogeneity of the wind-driven mixed layer: influence of a quasigeostrophic flow. J. Mar. Res. **46** 495–525 (1988)
159. C. C. Henning and G. K. Vallis: The effects of mesoscale eddies on the stratification and transport of an ocean with a circumpolar channel. J. Phys. Oceanogr. **35**, 880–896 (2005)
160. C. A. Katsman, M. A. Spall and R. S. Pickart: Boundary current eddies and their role in the restratification of the Labrador Sea. J. Phys. Oceanogr. **34**, 1967–1983 (2004)
161. A. J. Hermann and W. B. Owens: Energetics of gravitational adustment for mesoscale chimneys. J. Phys. Oceanogr. **23**, 346–371 (1992)
162. G. Madec, M. Chartier and M. Crépon: The effect of thermohaline forcing variability on deep water formation in the western Mediterranean Sea: a high-resolution three-dimensional numerical study. Dyn. Atmosp. Oceans **15**, 301–332 (1991)
163. G. Madec, F. Lott, P. Delecluse and M. Crépon: Large scale pre-conditioning of deep water formation in the north-western Mediterranean sea. J. Phys. Oceanogr. **26**, 1393–1408 (1996)

164. V. H. Strass, A. C. N. Garabato, A. U. Bracher, R. T. Pollard and M. I. Lucas: A 3-D mesoscale map of primary production at the Antarctic Polar Front: results of a diagnostic model. Deep Sea Res. Part II **49**, 3813 (2002)

165. M. Lévy, L. Memery and G. Madec: Combined effects of mesoscale processes and atmospheric high-frequency variability on the spring bloom in the MEDOC area. Deep Sea Res. Part I **47**, 27 (2000)

166. M. Lévy, L. Mémery and G. Madec: The onset of the Spring Bloom in the MEDOC area: mesoscale spatial variability. Deep Sea Res. Part I: Oceanographic Res. Papers **46**, 1137 (1999)

167. M. Lévy, M. Gavart, L. Mémery, G. Caniaux and A. Paci: A four-dimensional mesoscale map of the spring bloom in the northeast Atlantic (POMME experiment): results of a prognostic model. J. Geophys. Res. **110**, C07S21, (2005)

168. A. Paci, G. Caniaux, M. Gavart, H. Giordani, M. Lévy, L. Prieur, and G. Reverdin: A high-resolution simulation of the ocean during the POMME experiment: simulation results and comparison with observations. J. Geophys. Res. **110**, (2005), doi:10.1029/2004JC002712

169. A. Paci, G. Caniaux, H. Giordani, M. Lévy, L. Prieur and G. Reverdin: A high-resolution simulation of the ocean during the POMME experiment: mesoscale variabllity and near surface processes. J. Geophys Res. **112**, C04007 (2007), doi: 10.1029/2005JC003389

170. G. Lapeyre and P. Klein and B. L. Hua: Oceanic restratification forced by surface frontogenesis. J. Phys. Oceanogr **36**, 1577–1590 (2006)

171. N. Gruber, H. Frenzel, S. C. Doney, P. Marchesiello, J. C. McWilliams, J. R. Moisan, J. Oram, G. K. Plattner and K. D. Stolzenbach: Simulation of plankton ecosystem dynamics and upper ocean biogeochemistry in the California current system: Part I: Model description, evaluation, and ecosystem structure, Deep-Sea Research **53**(9), 1483–1516 (2006)

172. C. E. Menkes, S. C. Kennan, P. Flament, Y. Dandonneau, S. Masson, B. Biessy, E. Marchal and A. Herbland: A whirling ecosystem in the equatorial Atlantic, Geophys. Res. Lett. **29**, 48 (2002)

173. L. A. Anderson, A. R. Robinson and C. J. Lozano: Physical and biological modeling in the Gulf Stream region: I. Data assimilation methodology. Deep Sea Res. Part I **47**, 1787 (2000)

174. L. A. Anderson and A. R. Robinson: Physical and biological modeling in the Gulf Stream region: Part II. Physical and biological processes. Deep Sea Res. Part I **48**, 1139 (2001)

175. E. E. Popova, C. J. Lozano, M. A. Srokosz, M. J. R. Fasham, P. J. Haley and A. R. Robinson: Coupled 3D physical and biological modelling of the mesoscale variability observed in North-East Atlantic in spring 1997: biological processes, Deep Sea Res. Part I, **49**, 1741 (2002)

176. A. Mahadevan, M. Lévy and L. Mémery: Mesoscale variability of sea surface PCO2: JBJWhat does it respond to? Global Biogeochem. Cycles **18**, 1017 (2004)

177. S. J. Bury, P. W. Boyd, T. Preston, G. Savidge and N. J. P. Owens: Size-fractionated primary production and nitrogen uptake during a North Atlantic phytoplankton bloom: implications for carbon export estimates. Deep Sea Res. Part I **48**, 689 (2001)

178. S. D. Batten and W. R. Crawford: The influence of coastal origin eddies on oceanic plankton distributions in the eastern Gulf of Alaska. Deep Sea Res. Part II **52**, 991 (2005)

179. H. Claustre, M. Babin, D. Merien, J. Ras, L. Prieur, S. Dallot, O. Prasil, H. Dousova and T. Moutin: Toward a taxon-specific parameterization of bio-optical models of primary production: a case study in the North Atlantic. J. Geophys. Res. **110**, C07S12, (2005) doi:10.1029/2004JC002634

180. E. N. Sweeney, D. J. McGillicuddy and K. O. Buesseler: Biogeochemical impacts due to mesoscale eddy activity in the Sargasso Sea as measured at the Bermuda Atlantic Time-series Study (BATS). Deep Sea Res. Part II **50**, 3017 (2003)

181. R. D. Vaillancourt, J. Marra, M. P. Seki, M. L. Parsons and R. R. Bidigare: Impact of a cyclonic eddy on phytoplankton community structure and photosynthetic competency in the subtropical North Pacific Ocean. Deep Sea Res. Part I: Oceanogr. Res. Papers **50**, 829 (2003)

182. A. Bracco and A. Provenzale: Mesoscale vortices and the paradox of the plankton. Proc. R. Soc. B **267**, 1795–1800 (2000)

183. A. P. Martin, K. J. Richards and M. J. R. Fasham: Phytoplankton production and community structure in an unstable frontal region. J. Mar. Syst. **28**, 65–89 (2001)

184. I. D. Lima, D. B. Olson and S. C. Doney: Biological response to frontal dynamics and mesoscale variability in oligotrophic environments: a numerical modeling study. J. Geophys. Res. **107**(C8) (2002)

185. C. Pasquero, A. Bracco and A. Provenzale: Coherent vortices, Lagrangian particles and the marine ecosystem: vortical shelter. In Shallow flows. pp. 399–412 Balkema Publishers, Leiden, NL (2004)

186. M. M. Lee, D. P. Marshall and R. G. Williams: On the eddy transfer of nutrients: Advective or diffusive? J. Mar. Res. **55**, 483–505 (1997)

187. M. M. Lee and R. G. Williams: The role of eddies in the isopycnic transfer of nutrients and their impact on biological production. J. Mar. Res. **58**, 895–917 (2000)

188. R. G. Williams and M. J. Follows: The Ekman transfer of nutrients and maintenance of new production over the North Atlantic. Deep Sea Res. Part I **45**, 461 (1998)

189. M. Lévy: Oceanography: nutrients in remote mode. Nature **437**, 628–629 (2005)

190. J. B. Palter, M. S. Lozier and R. T. Barber: The impact of the nutrient reservoir in the North Atlantic subtropical gyre. Nature **437**, 687–692 (2005)

191. W. Hazeleger and S. S. Drijfhout: Eddy subduction in a model of the subtropical gyre. J. Phys. Oceanogr. **30**, 677–695 (2000)

192. M. Valdivieso Da Costa, H. Mercier and A. M. Tréguier: Effects of the mixed layer time variability on kinematic subduction rate diagnostics. J. Phys. Oceanogr. **35**, 427–443 (2005)

193. M. Nakamura and T. Kagimoto: Potential vorticity and eddy potential enstrophy in the North Atnatic Ocean simulated by a global eddy-resolving model. Dyn. Atmos. Oceans, **41**, 28–59 (2006)

194. G. Danabasoglu, J. McWilliams and P. Gent: The role of mesoscale tracer transports in the global ocean circulation, Science **264**, 1123–1126 (1994)

195. A. M. Tréguier, O. Boebel, B. Barnier and G. Madec: Agulhas eddy fluxes in a 1/6! Atlantic model. Deep Sea Res. Part II **50**, 251 (2003)

196. T. Dubos: A spatially selective parameterization for the transport of a passive or active tracer by a large scale flow. C. R. Ac. Sci. (Paris) **329**, 509–516 (2001)

197. M. Lévy, M. Visbeck and N. Naik: Sensitivity of primary production toJ different eddy parameterizations: a case study of the spring bloom development in the northwestern Mediterranean Sea. J. Mar. Res. **57**, (1999)
198. C. Pasquero: Differential eddy diffusion of biogeochemical tracers. Geophys. Res. Lett. **32** (2005)